Video and Camcorder Se
and Technology

Video and Camcorder Servicing and Technology

Steve Beeching
I.Eng A.M.I.E.E

Newnes

OXFORD   AUCKLAND   BOSTON
JOHANNESBURG   MELBOURNE   NEW DELHI

Newnes
An imprint of Butterworth-Heinemann
Linace House, Jordan Hill, Oxford OX2 8DP
225 Wildwood Avenue, Woburn, MA 01801-2041
A division of Reed Educational and Professional Publishing Ltd.

A member of the Reed Elsevier plc group

First published 2001

© S. Beeching 2001

All rights reserved. No part of this publication
may be reproduced in any material form (including
photocopying or storing in any medium by electronic
means and whether or not transiently or incidentally
to some other use of this publication) without the
written permission of the copyright holder except
in accordance with the provisions of the Copyright,
Designs and Patents Act 1988 or under the terms of a
licence issued by the Copyright Licensing Agency Ltd,
90 Tottenham Court Road, london, England W1P 9HE.
Applications for the copyright holder's written permission
to reproduce any part of this publication should be addressed
to the publishers.

British Library Cataloguing in Publication Data
A catalogue record for this book is available from the British library.

ISBN 0 7506 5039 7

Library of Congress Cataloguing in publication Data
A catalogue record for this book is available from the Library of Congress.

Printed and bound in great Britain

Durham County Council Arts, Libraries & Museums	
CO 1 59 03O56 23	
Askews	
621.38932	

CONTENTS

	Preface	vii
1.	Magnetism and magnetic recording	1
2.	Azimuth tilt technique	18
3.	Frequecy modulated recording and playback	20
4	Motor control and servomechanisms	60
5.	Colour recording and playback	106
6.	Systems control	135
7.	Long play	150
8.	Hi Fi audio	154
9.	Still pictures and slow motion	166
10.	Camcorders	178
11.	Digital camcorders	230
12.	MPEG and D-VHS	271
13.	Editing	278
14.	PDC and IEEE1394	288
15.	Soldering and Desoldering	297
	Index	314

Preface

My original book *Servicing Video Cassette Recorders* was last published as the fourth edition in 1993. This book was going to be an updated version and the fifth edition, but so much has changed over the last few years and those changes will progress at an increasing rate.

The aim is the same as with the previous book, that is, to provide that basic technological information that is not given in manufacturers' servicing data, to the extent that the reader may gain an understanding as to what is going on inside the ICs that make up a block diagram now called a service manual.

Production of video recorders utilising magnetic tape continues for the time being although the end may be just over the horizon. Recording of video as an MPEG 2 signal onto hard drives may eventually replace VHS as a record/replay system unless DVD-RW reduces in price to be a more commercially viable option. I can confidently predict that 'Holiday' recordings will be edited by consumers on their PC and written onto DVD as the optical storage medium. Possibly those early Sci-Fi programmes where the character drops a small optical disc into a slot and views the message may not be so far fetched now.

I have not listed any specific test equipment: it may be obvious that a test meter is inadequate and that a dual beam digital storage oscilloscope with a minimum 100MHz bandwidth is the least requirement for measurement purposes. I have included some examples of soldering and desoldering techniques and equipment: this can and will be updated as this equipment evolves.

A PC will be required for setting up purposes along with expensive interface test jigs and connection cables for digital products. These will be outside the scope and capabilities of amateur repairs and standard equipment for manufacturers' service agencies.

A PC and a 19" monitor is needed to view service manuals and accompanying data that will no longer be supplied in paper format. Service manuals are issued on CDs and possibly made available for download via the Internet.

I have re-written the whole book and added chapters on current digital technology up to D-VHS recording of MPEG-2 signals. Some 300 drawings have been re-drawn in a manner that illustrates the points that I have made in the accompanying text.

A big thank you to JVC, Panasonic and Sony who generously supplied me with detailed information, some of which was not generally made available.

Any mistakes in this book are down to me and I must thank the numerous friends that have patiently read through the text and pointed out the major grammatical errors 'wot I have wrote' that the spell checker did not pick up.

Steve Beeching I.Eng. AMIEE
Grove Farm, Newark. Jan 2001

Steve@grovefarm.force9.co.uk
Steve@newarkvideoservices.co.uk
www.newarkvideoservices.co.uk

1
Magnetism and magnetic recording

Video and audio recording utilises the properties of magnetism to transfer electrical signals onto a magnetic medium that can retain the information. The medium is usually tape but it can be a floppy or hard disc drive. In order to understand the limitations imposed on video and audio recording signals by the magnetic medium it is necessary to understand the principles involved.

Certain materials, once subjected to a magnetising force, can retain some magnetism. An example is steel. A bar of steel can retain a high value of magnetism and is considered to be a 'hard' magnetic material as shown in Figure 1.1. A 'soft' material such as iron does not retain very much magnetism. Another 'soft' material is copper: it can be magnetised, but does not retain any magnetism once the driving force has been removed. Other materials, for example glass, cannot retain any magnetism nor can it be magnetised. It can be used as a magnetic 'insulator'.

A magnetic field is created by passing a current through a length of wire, generating a magnetising force. The amount of magnetic force is determined by the value of electric current flowing through

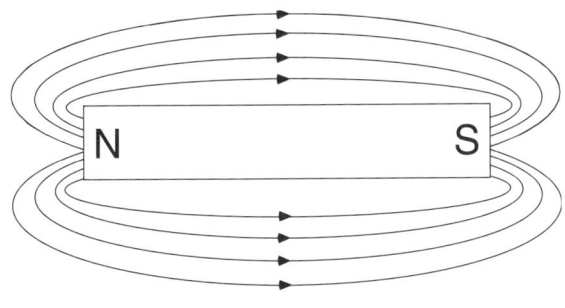

Figure 1.1 *Bar magnet*

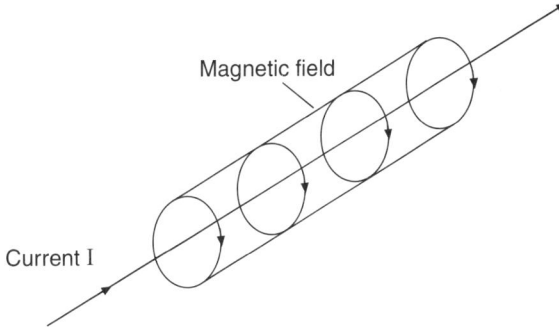

Figure 1.2 *Magnetic field generated by current flow*

the length of wire. Limiting factors are the resistance of the wire and how much current it can carry without overheating.

We can consider that the internal structure of a bar of steel is similar to a box of nails, all mixed up and lying in all directions. When a magnetising force is applied all of the 'nails' align in the same direction and become uniform small bar magnets all contributing to being one big one. The structure of the iron bar becomes magnetically uniform and it retains some of the magnetism applied to it. In Figure 1.1 the steel bar magnet has a low density of flux and is not a powerful magnet. If the number of lines of flux are increased and have a higher density then the bar could be considered a powerful magnet.

Magnetic force

The Magneto Motive Force (MMF) is the force required to produce a magnetic field. Its symbol is the Ampere (F). This must not to be confused with the electric current Ampere (I). The magneto motive force (F) generates a magnetic field around a straight piece of wire, shown in Figure 1.2.

The force (F) is proportional to the current flowing in a single length of wire, therefore:

$$F \propto I$$

If the wire is wound into a coil then the length can be increased and (F) becomes cumulative. The force (F) is determined by the number of turns and the value of the current. F is therefore proportional to the current (I) and the number of turns (N), therefore:

$$F \propto NI$$

Magnetising force

The Magnetising Force (H) is measured in amperes per metre and is dependent upon the physical length of a wound coil (l). If the coils are long and spread apart like a stretched spring, (l_1) in Figure 1.3a, then (H) is low in value. If the coils are closely wound as a compact short coil (l_2) in Figure 1.3b, then value of H is high and so is the flux density. (H) is proportional to the magnetising force F divided by the length of the coil.

$$H \propto F/l$$

For a given winding of coils that is fixed in the number of turns and in the length of the winding; (H) becomes variable only in proportion to the current flowing through the windings.

Then (H) is proportional to (I).

$$H \propto I$$

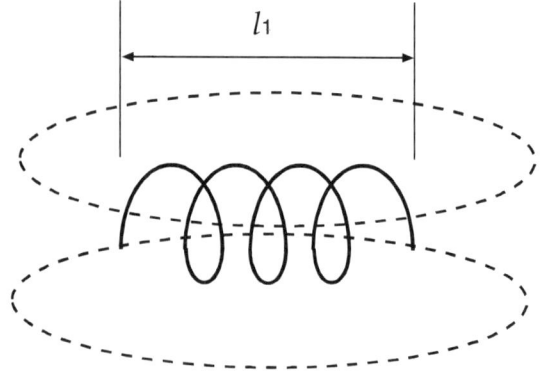

Figure 1.3 *(a) Low level magnetising force (H)*

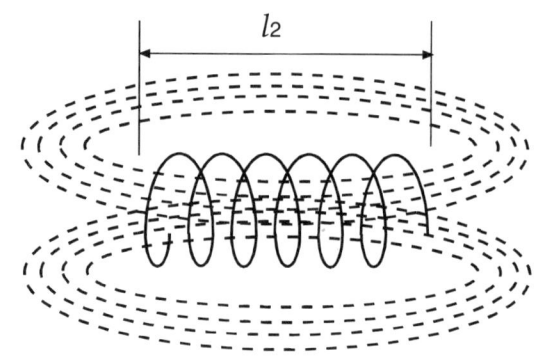

Figure 1.3 *(b). High level magnetising force (H)*

Magnetic field

A magnetic field is therefore generated by an electric current passing through a coil, wound around a steel bar, or an equivalent 'hard' material. This material is then magnetised. Once the current is removed a certain amount of magnetism remains in the bar it is permanently magnetised and is refered to as a **permanent magnet.**

The retained magnetism in a material is measured by the density of the lines of flux surrounding the magnetic material, (see Figure 1.1). This flux density is much higher nearer the bar than further away. The measurement is in Gauss (B) or lines per square centimetre.

A graph can be produced to see the relationship between the magnetising force (H) applied to a material and the flux density (B) remaining once the magnetising force is removed.

Figure 1.4 is the graph of the Magnetising force (H) along the horizontal axis, and the Flux density (B) on the vertical axis. Starting from point 0, as the magnetising force (H) is increased so does the flux density (B). However, it is not linear. The flux density increases slowly at first for an increase in magnetising force. Then as (H) is further increased there is a steep rise in the graph as the flux density rapidly increases, up to a limiting point Sp. This is where the material saturates and cannot absorb any more flux. Increasing the magnetising force beyond this point does not significantly increase the flux density, as the material is 'full' or saturated. When the magnetising force (H) is reduced back to 0 the flux density in the material

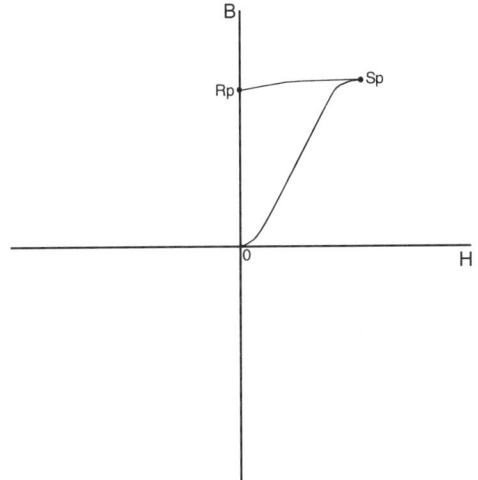

Figure 1.4 *B/H curve*

does not reduce to zero but remains at a value Rp. This is the retained flux density of a permanent magnet.

The B/H loop

Once the point of retained flux is reached the current that provides the magnetising force can now be reversed. The polarity of the magnetising force also reverses. Flux density now gradually reduces until it passes through the zero axis of (B) at (-H). Then the flux is reversed until it saturates in the negative direction at point Sn. When the magnetising current is reduced back to zero the reversed polarity flux remains at point Rn. Increasing the magnetising current again in the positive direction reduces the reversed polarity flux density. It passes through (H) axis and the material saturates in the positive direction once again. This gives rise to the important B/H loop for magnetising a given 'soft' material such as iron as shown in Figure 1.5.

A B/H loop for a 'hard' material is shown in Figure 1.6 and it is much squarer. Flux retention is much higher than for a softer material as required by a magnetic tape to store information. This is called the hysteresis loop and is the magnetising curve followed by the flux for an alternating current.

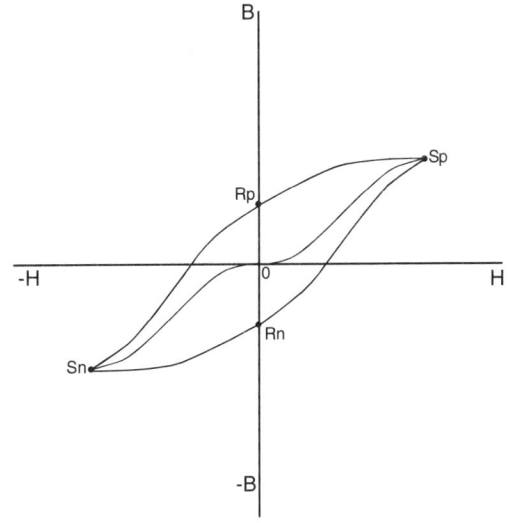

Figure 1.5 *'Soft' B/H loop*

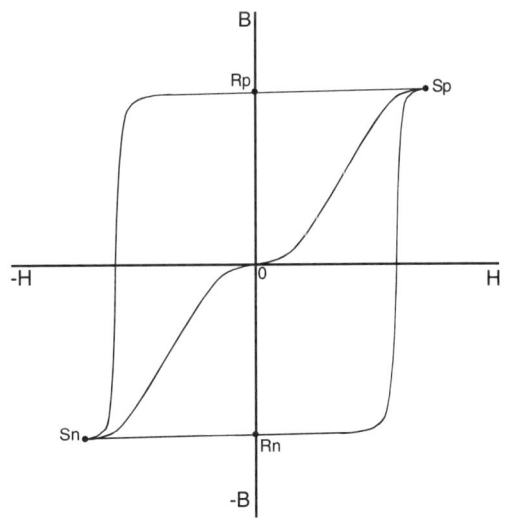

Figure 1.6 *'Hard' B/H loop*

Demagnetising

In order to remove the stored flux and demagnetise the material it would be necessary to apply an AC current and then gradually reduce its amplitude to zero. In this way the B/H loop would get smaller and smaller until it was just a dot at zero with no residual magnetism. This is the way in which TV tubes are degaussed.

Transfer characteristic

Now we have established the basic parameters of a B/H loop, we have to consider the problems of magnetising tape to get a recording onto it. A tape will not be magnetised by a single B/H hysteresis loop, as already shown, but by varying levels of signal current and therefore varying levels of magnetising force (H).

The information that is required is to determine what the residual values of flux are for the different values of (H).

This is obtained by applying incremental values of (H) starting at zero and plotting the respective values of residual flux as shown in Figure 1.7. The value of residual flux is drawn on the vertical axis of a graph as values of Rp against the drive force (H) on the B/H characteristic. Each value of (H) is applied to the material and then removed. The remaining flux (Rp) for that value of (H) is then plotted against the original value of (H) as shown by the short vertical lines. The resulting plot is called the Transfer Characteristic; it is non linear at the bottom and top, but very linear in the middle. A typical transfer characteristic is shown in Figure 1.8. For small values of (H) there is no transfer curve at all. From B=0 to saturation the transfer curve is very linear. A transfer curve is the bridge between the electrical signal, audio or video, and the magnetic equivalent upon the tape. To avoid distortion of the signal the transfer curve of a tape medium must be very straight between B=0 and saturation. A problem exists in the lower signal levels, as there is no magnetism on the tape due to the absence of the transfer curve. The recording signal must be kept within the linear part of the transfer curve; it must be biased into this area.

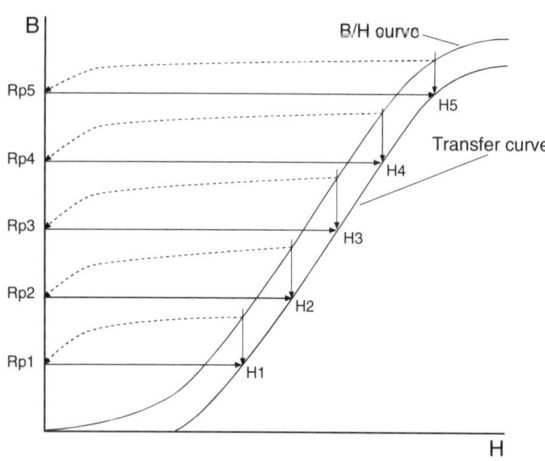

Figure 1.7 *Transfer characteristic curve*

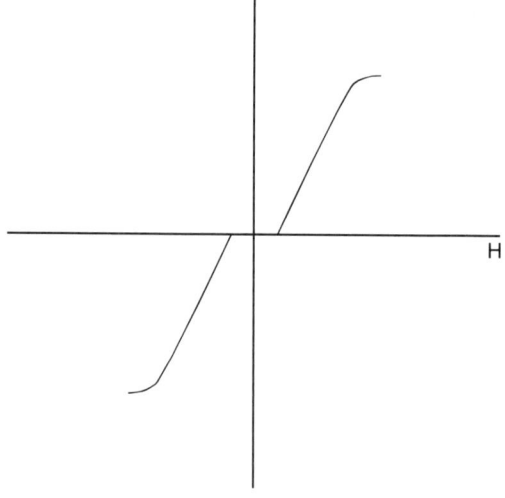

Figure 1.8 *Transfer characteristic*

Magnetism and magnetic recording

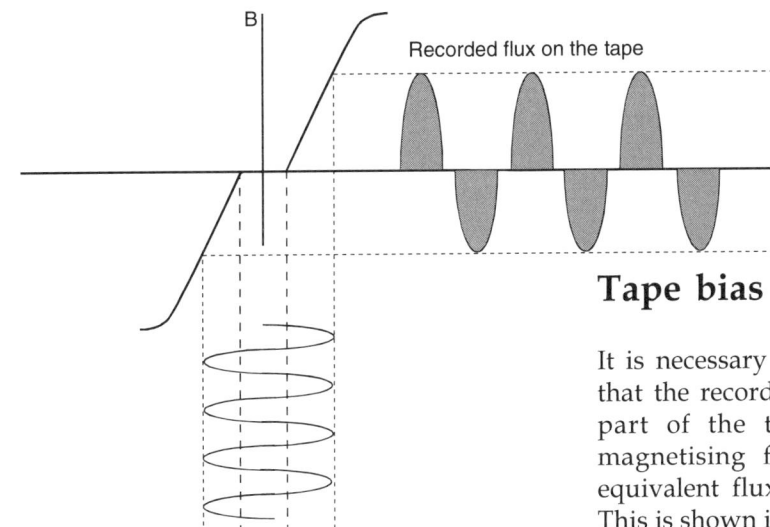

Figure 1.9 *Crossover distortion*

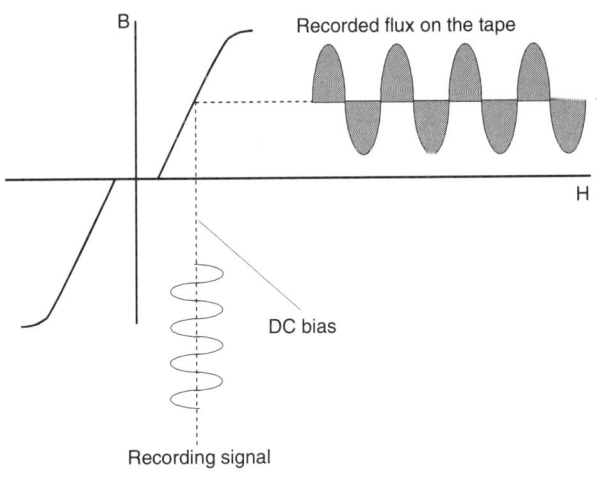

Figure 1.10 *Recorded magnetism with DC bias*

Tape bias (audio)

It is necessary in magnetic recording to ensure that the recording signal is kept to the straight part of the transfer curve. Otherwise, the magnetising force (H) will not produce an equivalent flux with the same signal linearity. This is shown in Figure 1.9. If an AC input signal were applied as a single drive waveform there will be parts of the (H) axis at low levels, around 0, that will not produce any residual flux at all.

The equivalent magnetic flux version will have no residual flux until (H) reaches a level to produce some. Steps are formed in the magnetic flux waveform where it crosses the zero axis and this is called crossover distortion.

The first method of bias is to pre-magnetise the tape with a magnet before it passes over the recording head. This method is not suitable for high quality recordings as this DC magnetisation causes high tape noise. Audio signal-to-noise ratio (s/n) is poor.

A second method is to provide a DC bias for the AC waveform that centers the signal on the transfer curve. The advantage is that the AC signal is kept in the linear section of the transfer curve as illustrated in Figure 1.10. There are some disadvantages: a standing magnetic bias is left on the tape, this also creates noise in replay. Neither of the methods described takes full advantage of the magnetising capabilities of the tape, as the reverse polarity area is not used. The effect of not using this area is a reduced playback level; this contributes to a reduced signal-to-noise ratio. If the level of the AC signal is high compared to the DC level another problem occurs. Signal peaks will be crushed or flattened by the non-linear upper and lower portion of the transfer curve.

The best way to bias the tape is by adding or

Video and Camcorder Servicing and Technology

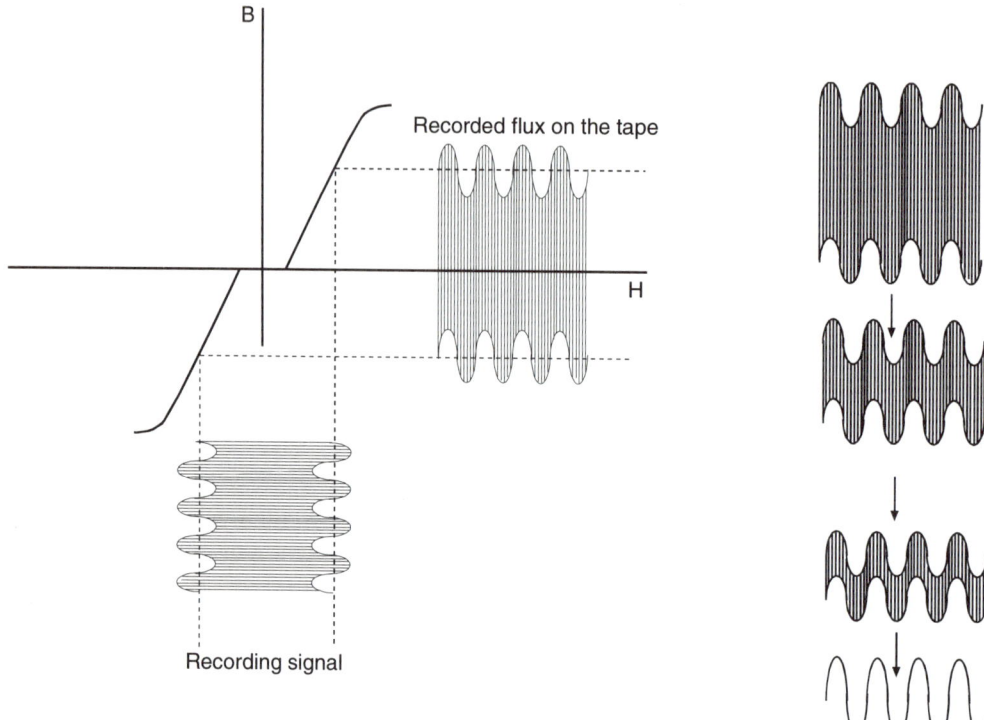

Figure 1.11 *Recorded magnetism with AC bias*

Figure 1.12 *Reduction in bias level due to high frequency losses in the record/playback process*

mixing the recording signal with a high frequency AC bias signal. Audio is a much lower frequency recording signal compared to the bias signal; up to 15 kHz with a 60 kHz bias oscillator. It is kept on the linear parts of the transfer curve in both positive and negative quadrants, as shown in Figure 1.11. The audio signal, by virtue of the mixing process, sits on the top and bottom of the bias waveform. This gives a larger replay output that has a good signal-to-noise characteristic. If the AC bias is large enough to maintain the signal on the linear part of the transfer curve the distortion is very low.

Any crossover distortion in the lower parts of the transfer curve effects only the AC bias waveform. As the recording signal is added to the AC bias and does not modulate it, then the recording signal is unaffected by this distortion of the AC bias waveform.

Care is taken in the design of the tape head to ensure that it is not saturated by the bias signal; the lower frequency signal component will be otherwise distorted. This also applies to the tape that it is being recorded onto. High frequency bias is not reproduced and is lost by record and replay bandwidth limitations, as illustrated in Figure 1.12. When the audio signal, at approximately 2 V peak to peak, is mixed with the bias signal at about 50 V peak to peak, the audio component is very small in comparison to the Bias. Figure 1.12 has been drawn to clarify, rather than be accurate. As the high frequency losses involved in the recording and replay process take effect, the bias signal, about 60 kHz, is gradually diminished. The audio signal merges into a single component as the bias reduces to zero in replay off the tape via a low pass filter.

In the recording process losses, occur in the transfer from electrical to magnetic signal. There are four main reasons for the reduction in electric/magnetic efficiency: magnetisation loss, eddy current loss, impedance and penetration loss.

Magnetisation loss

Magnetisation loss refers to the initial amount of energy required to establish the magnetic field in the first instance, to a level high enough such that transfer to the tape can take place.

Eddy current loss

Electrical currents are generated in the core of the recording head from the magnetic flux that is produced by the signal current passing through the windings. These electrical currents themselves generate a magnetic flux opposing that which is produced by the head windings. An energy loss results from the subtraction of the eddy current flux from the signal flux, this loss is minimised by reducing the eddy currents in the head design. This is achieved by making the electrical resistance of the core a very high value. Ferrite has a very high electrical resistance and it is commonly used in the manufacture of recording heads.

Recording head impedance loss

Impedance (Z) is the combination of electrical resistance, inductance (inductive) and capacitance (capacitive reactance) components of the recording head, and is frequency dependant.

Electrical resistance is (R), inductive reactance is (XL), capacitive reactance is (XC).

$$Z = R + XC + XL$$

Where $XL = 2\pi f L$
and $XC = \dfrac{1}{2\pi f C}$

The components R and XC are very small and so the equation can be simplified to:

$$Z = 2\pi f L$$

This shows that the impedance loss is mostly the result of the inductance of the head and the signal frequency.

For a constant voltage signal drive to the recording head, the recording current will reduce as the frequency of the signal increases, due to rising impedance of the head with frequency. Consequently, the magnetic recording will also reduce in level with increased signal frequency.

To reduce the impedance losses the inductance of the recording head is kept low. Recording amplifiers are designed to provide constant current drive to the recording head, this minimises the effect of the impedance losses with rising signal frequency.

Penetration loss

Low frequency signals magnetise the coating of the tape much more deeply than higher frequencies. A physical property of magnetic domains causes the flux to penetrate the tape by a ripple effect. Each domain magnetises the next and the field gradually penetrates the tape. It takes a finite time for the magnetic flux across the head gap to penetrate the tape. Thus for higher frequencies the time that the flux is present is less than for lower frequencies, consequently penetration is less. The difference in remnant magnetism is therefore less for higher frequencies than for lower frequencies. This represents a high frequency loss as shown in Figure 1.13.

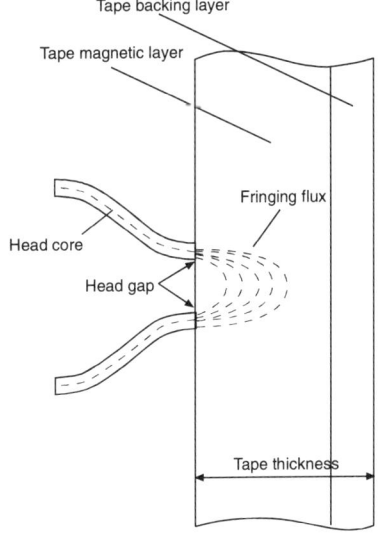

Figure 1.13 *Penetration loss*

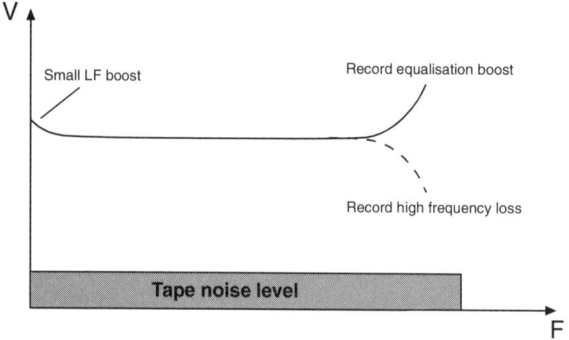

Figure 1.14 *Record pre-emphasis*

Note: This effect is utilised for hi-fi deep layer recordings in video recorders (see p158).

Compensation for high frequency losses in recording cannot be applied in replay by increasing the gain of the pre-amplifiers. This will result in more gain for the high frequency tape noise (hiss) and degrade the signal-to-noise ratio. Compensation for higher signal frequency losses must be applied in the recording equalisation process to maintain the highest signal-to-noise ratio.

Recording equalisation

Pre-emphasis is applied in recording by boosting the high frequencies (Figure 1.14) with some lift at the low frequency end. Low frequency boost is to compensate for replay losses in this area due to the relatively long wavelengths of low frequency signals (more on this subject later).

Recording frequency response

The recording frequency response of a magnetic tape recorder, audio or video, is more a function of tape speed rather that of head gap. The rate of change of flux across the head gap is not only dependent upon the length of the gap but high frequencies can be achieved as long as the tape is moving fast enough. That is, fast enough to move the small magnetised portion out of the way before the signal changes to the next cycle.

Each complete cycle of the recording signal consists of a positive and a negative section each representing 180°. On the tape the cycle is represented by a small bar magnet. This bar magnet element is formed by two sections; N-S and S-N of opposite polarity, each being equivalent to the positive and negative half cycles as shown in Figure 1.15. Each bar magnet element has length representing the recording width, or the wavelength lambda (λ). The recording width is dependent on two factors; one is the speed of the tape (V) and the other is the period of the cycle (t).

Recording width of the bar element is equal to the wavelength (λ).

$$\lambda = V t$$

where: $t = 1/f$ then: $\lambda = V/f$

It can be proved then, that the tape speed and not the head gap limits the frequency response. Tape speed has to be high enough to move the tape past the head by a distance equal to the recording width in time t, where t is the period of the highest frequency.

As the frequency rises, a point is reached when the positive cycle has not moved far enough along when the negative cycle starts. Partial erasure of the positive cycle occurs as the negative cycle records over it, limiting the amplitude of the flux on the tape and reducing recording efficiency.

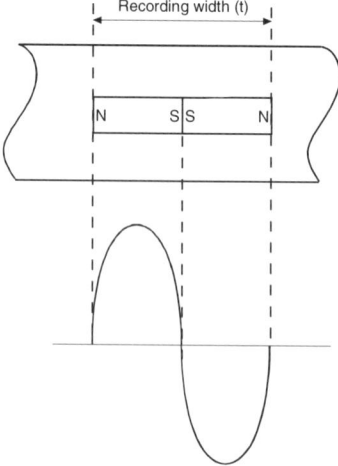

Figure 1.15 *Recording wavelength width*

Magnetism and magnetic recording

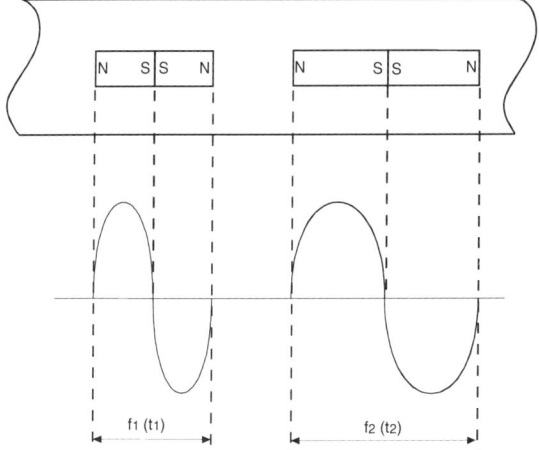

Figure 1.16 *Replay flux changes increase the output (V) with frequency*

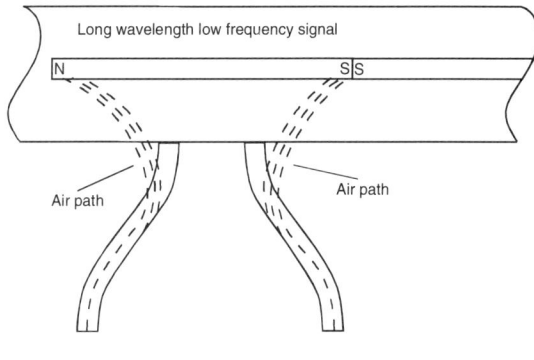

Figure 1.17 *Low frequency loss*

Replay signal level

In replay the tape passes over the head at a constant speed. The output from the head, however, increases with signal frequency. It has been shown that the recorded signal is a row of bar magnets on the tape. At low frequencies the bar magnets are very long and they are obviously much shorter at high frequencies.

Two frequencies, f_1 and f_2, are compared in Figure 1.16, where $f_2 = 2f_1$; this is equivalent to one octave. In replay f_2 is replayed by the equivalent of a bar magnet on the moving tape and it takes time 't2' to pass by the replay head. Frequency f_1 has an equivalent bar magnet that is shorter and takes half the time, 't1', to pass by the head.

The output level (V) from a coil within a moving magnetic field is proportional to the rate of change of flux (F).

$$V \propto \frac{d\Phi}{dt}$$

The flux change for the bar magnets of f_1 is twice that of f_2 in the given time 't2'. This is because there are two bar magnets from f_1 as opposed to one bar magnet for f_2 in the period 't2'. Consequently, the output level is doubled. This means that over the available bandwidth the output from the replay head doubles for each octave, which is an increase in level of 6 dB/octave.

At very low frequencies the bar magnets are very long in comparison to the replay head gap and this causes a reduction of replay level. This is due to the flux following a path external to the replay head, as illustrated in Figure 1.17. Due to the very long bar magnet, the flux takes a path through air rather than through the head core. A loss occurs through the higher reluctance of the air path. Compensation is applied by the small amount of boost in the low frequency end of the record pre-emphasis curve as already mentioned.

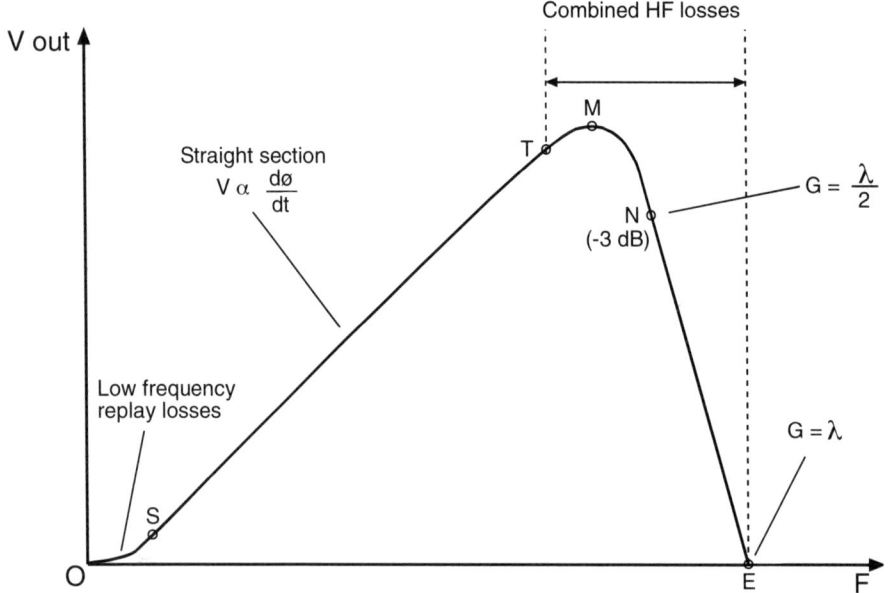

Figure 1.18 *Replay response curve*

Replay characteristic

A replay response graph can be drawn to show the frequency response over the complete bandwidth. This is illustrated in Figure 1.18. From point O the curve of the replay head output rises slowly up to point S; this is due to those losses caused by very long wavelengths at low frequencies. Between S and T the rise in output is steady and linear at the rate of 6 dB/octave.

After point T, as the higher frequencies are reached, the peak reaches a maximum, at point M. M is the practical maximum. N is the theoretical maximum as it is the real point of maximum output, but due to increasing high frequency losses the result is down by 3 dB. Between the peak of the signal and -3 dB the high frequency losses are a combination of head-to-tape contact, replay head inductance, eddy current losses and head alignment tolerances.

Magnetism and magnetic recording

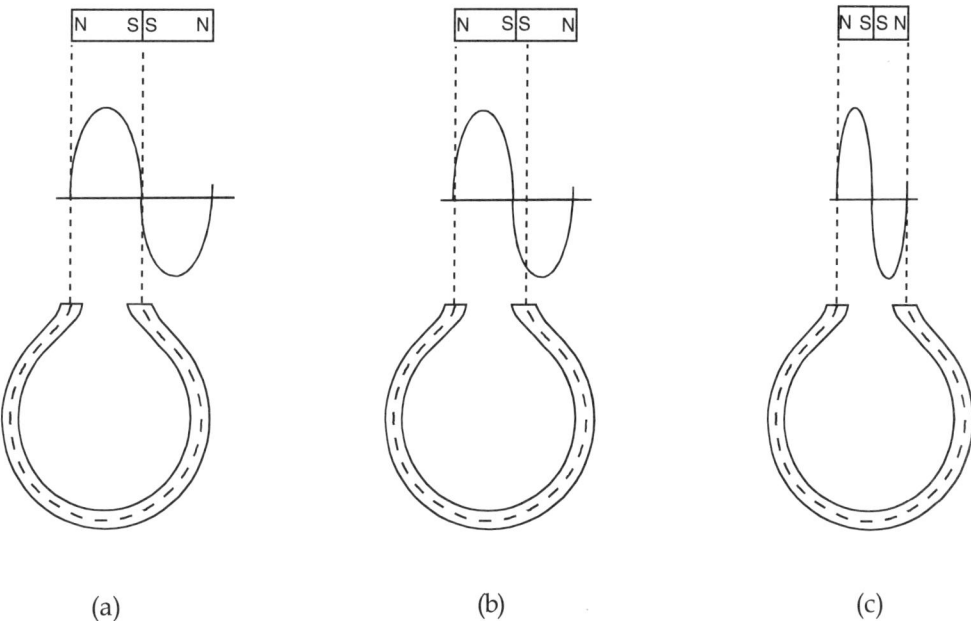

(a)   (b)   (c)

Figure 1.19 *Wavelength compared to head gap*

Extinction frequency

The curve now falls rapidly to zero at the extinction frequency, point (E).

At the -3 dB point the high frequencies are very small bar magnets, the length of which is twice that of the replay head gap (G) That is:

$$G = \frac{\lambda}{2}$$

This means that the wavelength of the frequency is twice the gap (G), as shown in Figure 1.19(a). As the frequency increases further then the wavelength gets smaller and some of the negative half cycle occurs in the same domain as the positive one, as shown in Figure 1.19(b). Some subtraction takes place and the overall signal output is decreased. The extinction frequency is reached when the width of the head gap is the same as the wavelength of the replayed signal, as shown in Figure 1.19(c). This happens when the bar magnet is so short that the length of it, or the wavelength, is equal to the width of the head gap. Both magnetic polarities of the bar magnet exist across the gap and thus cancel each other. There is no output from the head at this point, hence the term extinction frequency.

Dynamic range

In order to obtain the maximum output and the best signal-to-noise ratio, the replay signal is optimised between two constraints. Saturation of the tape limits the maximum level of signal that can be recorded and replayed without distortion. Residual tape noise will limit the signal-to-noise ratio at low levels. By using a high quality tape, a signal-to-noise ratio exceeding 60 dB is theoretically possible without overloading, although a typical dynamic range of audio magnetic recording is not much more than 50–55 dB.

For video recording a practical signal to noise ratio is more like 45–50 dB, as the wider bandwidth introduces more noise as more frequencies are added into the spectrum. If an average limitation is 60 dB for the dynamic range and the output level varies by 6 dB/octave, it follows that the maximum usable frequency range is about 10 or 11 octaves. This is illustrated in Figure 1.20. Between the maximum output and the residual noise level the dynamic range is 60 dB. With a slope of 6 dB/octave in the replay of a tape, where the output V is proportional to the rate of change of the flux, then a 60 dB dynamic range equates to a frequency range of 10 octaves.

Ten octaves is sufficient for audio signals with a frequency response from 25 Hz to 25 kHz but not enough for a video signal. A video signal has a frequency response extending from 25 Hz to over 5 MHz; this is a dynamic range that is more than 18 octaves. It is impossible to record and replay the bandwidth of a video signal by using the same techniques as that for an audio signal, as the frequency range of the video signal is 8 octaves over the theoretical and practical limitations. Some kind of bandwidth compression technique is required to reduce the 18 octave range down to a value that is more manageable; this is where frequency modulation comes in to

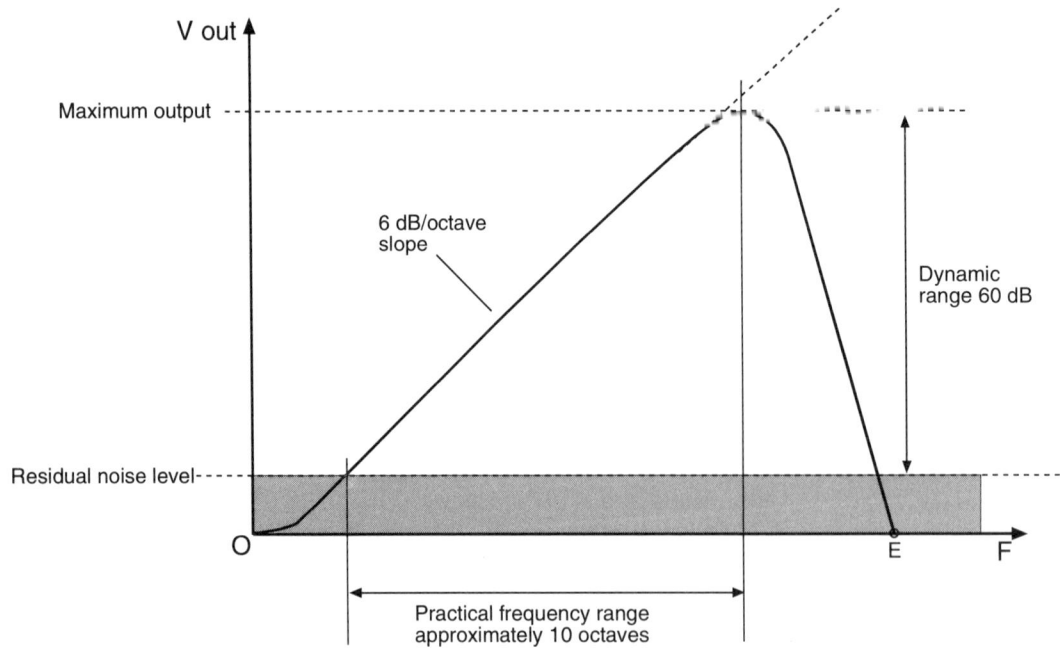

Figure 1.20 *Dynamic range*

Magnetism and magnetic recording

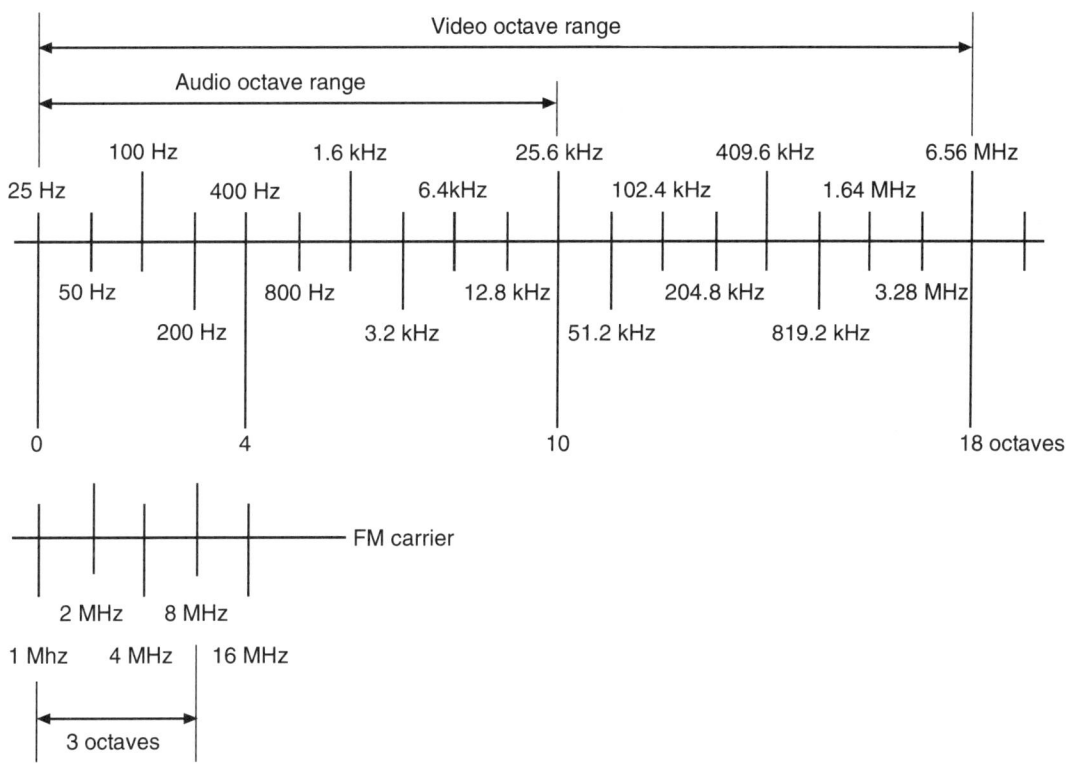

Figure 1.21 *Frequency/octave range*

play. A frequency/octave range is shown in Figure 1.21. From a starting point of near DC, taken as 25 Hz, a range of 10 octaves covers up to 25 kHz, which is sufficient for audio recording and playback. It can be seen that a video signal with an upper frequency limit of 6.5 MHz is equivalent to a range of 18 octaves; this is too much for an analogue recording.

By taking a carrier signal with a centre frequency of 4.2 MHz and modulating it with a video signal which is restricted to a bandwidth of 3.2 MHz, then the frequency spectrum of the resultant FM carrier, including upper and lower sidebands, is between 1 and 8 MHz.

It can be seen from Figure 1.21 that this frequency spectrum of 3 octaves will be within normal record and replay parameters of magnetic tape.

Frequency modulation

A very common example of frequency modulation is that of FM radio broadcasts for high quality reception. An oscillator that can vary its centre frequency in response to an audio input generates the high frequency carrier signal. A conversion from voltage to frequency takes place in the modulation process. The higher the audio signal level, then the greater the oscillator shifts from its centre frequency. A shift in frequency is called deviation. The audio signal causes the oscillator to deviate from its centre frequency, and the amount of deviation will depend upon the amplitude of the audio signal.

In terms of frequency response of the audio signal the higher frequencies will cause the oscillator to deviate at a faster rate. This creates

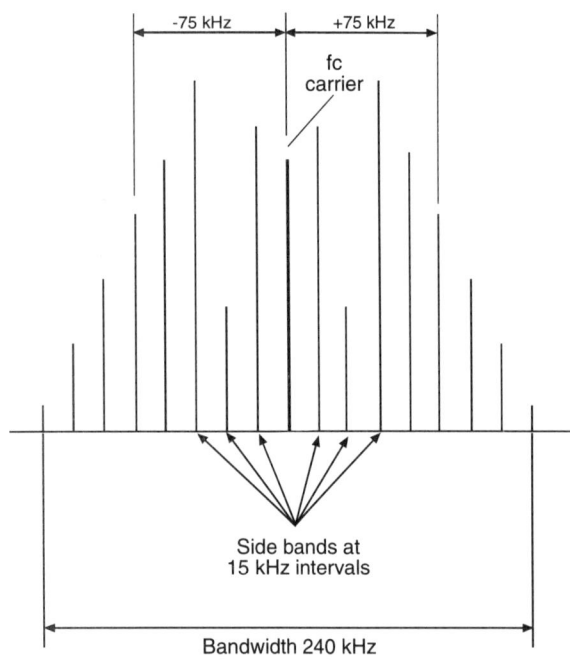

Figure 1.22 *FM sidebands*

sidebands that spread out from the centre frequency. This means that the highest audio signal level deviates the carrier frequency to the highest value. The highest frequency of the signal then adds to the carrier frequency spectrum to create the extended sidebands. An example of the audio sidebands is shown in Figure 1.22, fc is the carrier frequency somewhere in the region of 89 to 108 MHz, in the FM radio band. The maximum amplitude of the audio signal deviates the carrier frequency by 75 kHz, so for positive and negative swings the deviation is ±75 kHz or a total deviation swing of 150 kHz. Given that the audio frequency response is limited to 15 kHz then the sidebands occur at 15 kHz intervals.

Modulation index

When a carrier signal (fc) is frequency modulated by a single frequency (fs) the sidebands occur at multiples of the frequency both above and below the carrier frequency, i.e., (fc + fs), (fc+2fs), (fc+3fs), (fc+4fs) etc. The energy level of these sidebands diminishes as the multiple increases and so there is a point where the energy levels of the sidebands are too low to be significant. It is possible to determine the number of significant sidebands required to ensure that the FM signal can be recovered by demodulation without distortion or noise. This is done by the mathematical calculation of complex 'Bessel functions' and confirmed by measurement to obtain the modulation index table.

The modulation index is calculated from the formula:

$$\text{modulation index} = \frac{\text{maximum deviation}}{\text{highest modulation frequency}}$$

For the FM radio transmission the maximum deviation is 75 kHz, and the highest modulation frequency is 15 kHz. Therefore the modulation index is 5.

From the table below it can be seen that 16 sidebands are needed to carry all of the signal information without distortion. The 16 sidebands are made up of eight upper sidebands and eight lower sidebands.

With channel spacing, a full channel bandwidth of 240 kHz is specified for FM radio broadcasting. However, the very end sidebands have low energy levels and are not significant, so that a bandwidth of 200 kHz is adequate.

Modulation Index	Number of significant sidebands
<0.5	2
0.5	4
1	6
2	8
5	16
10	28

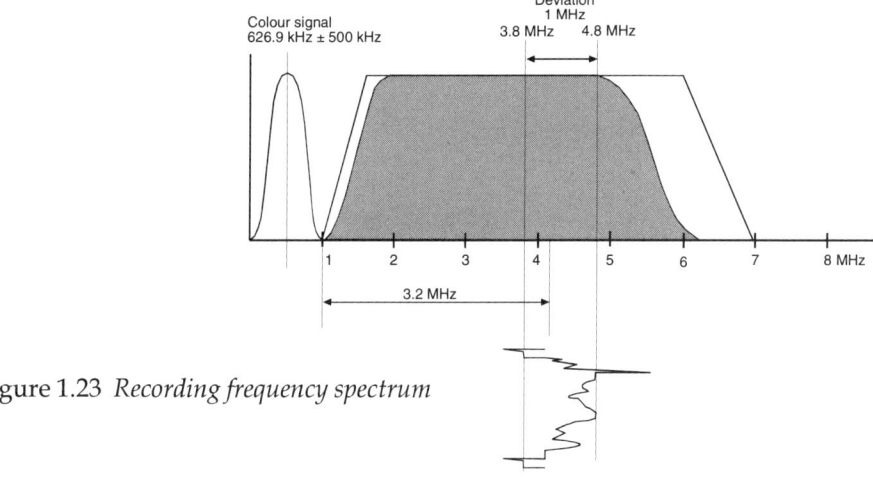

Figure 1.23 *Recording frequency spectrum*

Video frequency modulation

For the VHS video system the frequency modulation of the carrier signal is from sync tip at 3.8 MHz to peak white at 4.8 Mhz. That is a deviation of 1 MHz. Bandwidth of the luminance signal is limited to 3.2 MHz. It can be calculated then that the modulation index is about 0.3. As this value is below 0.5 then it can be seen from the table that only two sidebands are required to successfully carry the video information in its FM form: one upper sideband and one lower sideband. The lower sideband carries most of the video information. The upper sideband carries mainly the high frequency components, as such it requires less energy and can be reduced by a low pass filter to 5/6 MHz.

Calculations given on p.8 show that there is a relationship between the wavelength (λ) recorded onto the tape, the tape speed (V) and head gap;

$$G = \lambda/2.$$

By rearranging this formula an expression for (λ) can be given as $\lambda = 2G$, i.e. twice the head gap.

We now have two formulae for λ one that relates it to frequency and tape speed, and another that relates it to head gap.

$$\lambda = V/f$$
$$\lambda = 2G$$

Now by combining the two formulae we can determine one single formula that relates tape speed and frequency to head gap:

$$2G = V/f$$

Rearranging this gives: $V = 2Gf$.

In a videorecorder the head drum rotates fast, whereas the linear tape speed is quite slow, this means that V is not the tape speed but the 'head to tape' speed.

If the drum diameter is 62 mm, the circumference is ($\pi \times 0.062$) m = 0.1948 m. Rotation speed is 25 revolutions per second, so the head to tape speed (V) is 25 × 0.1948 = 4.87 m/s.

We now have to take into account that the linear tape speed is 0.02 m/s (23.39 mm/s) and as this is in the same direction as the head it must be subtracted to give about 4.85 m/s

For VHS the head/tape speed is calculated as 4.85 m/s and the video head gap is 0.3 µm.

$$f = V/2G = 4.85/0.6 \times 10^{-6} = 8 \times 10^6$$
$$= 8 \text{ MHz}$$

However, due to the normal HF losses the practical range is 6 MHz on most video recorders

In the case of S-VHS the head gap G = 0.2 µm, the bandwidth available can be calculated to 12 MHz, in which the practical range is about 10 MHz.

FM modulator

A video recorder frequency modulator converts the amplitude modulated luminance signal into a frequency modulated carrier signal. By using high frequency video heads and a fast head-to-tape speed the high frequency FM carrier can be successfully recorded and replayed on magnetic tape.

A video frequency modulator is a voltage to frequency converter; Figure 1.24 shows the basic circuit and Figure 1.25 the characteristic of the voltage to frequency conversion. The curve is non linear the bottom end and it is necessary to ensure that the video signal does not come within this area. This is achieved by the use of a DC bias level corresponding to the minimum sync tip frequency of 3.8 MHz and clamping the video sync tips to this bias. This can then be preset to calibrate the voltage to frequency converter free running frequency to 3.8 MHz. Operation and calibration of the video AGC section sets the video signal amplitude level. The AGC level is set to maximise the peak white part of the signal to 4.8 MHz, there is an allowance for a small amount of peak white overshoot caused by the pre-emphasis section. If the white clip is not set up correctly then the peak white overshoot can overdrive the modulator and the FM signal is either absent, or very small, for a short time period. In replay this causes inversion of the signal or black spots to the right of the peak white, most easily seen on white captions. A similar effect can be seen if the video heads are worn and attenuating the higher frequencies of the carrier.

A basic voltage controlled astable oscillator produces the carrier signal, a preset control sets

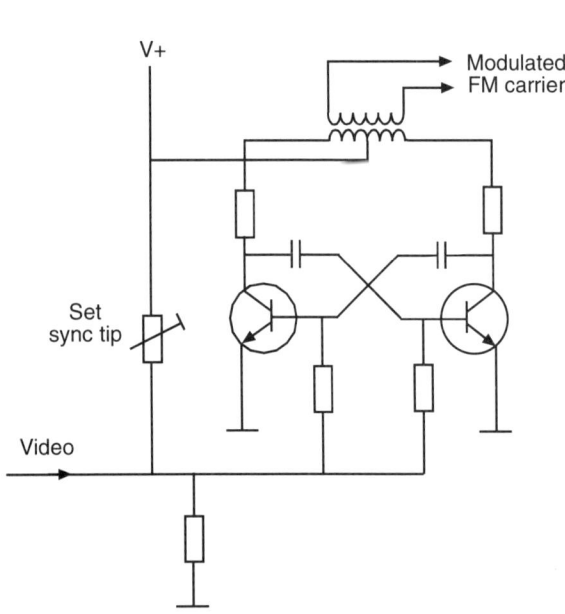

Figure 1.24 *Frequency modulator circuit*

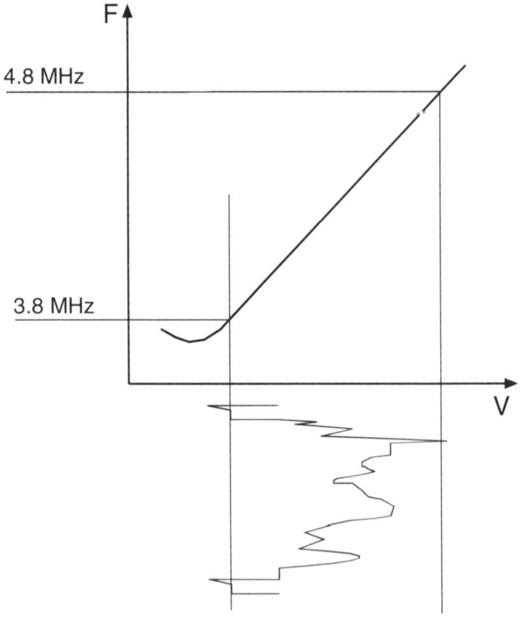

Figure 1.25 *Video FM modulator characteristic*

Magnetism and magnetic recording

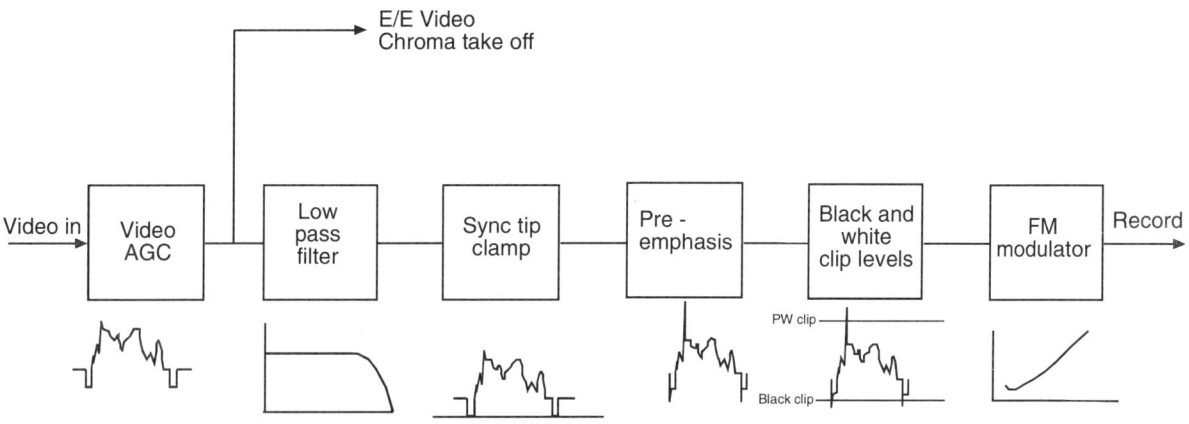

Figure 1.26 *Record processing path*

the oscillator to 3.8 MHz, and there may also be a symmetry control for equalising the mark/space ratio, although usually this is not present. An input video signal raises the oscillator frequency with increasing amplitude; this is limited by controlling the maximum amplitude of the video signal in the preceding AGC stage, see Figure 1.26.

After the AGC stage is a low pass filter to limit the maximum frequency of the video signal to around 3 MHz. If any higher frequency signals were to get through to the modulator then the lower sideband would extend down too far and interfere with the downconverted colour signal. Any shift downward of the preset 3.8 MHz would cause similar problems, any shift upward would increase the peak white to a frequency higher than the record/replay can handle causing white inversion interference. To overcome HF losses in the record/replay process the video signal is pre-emphasised; this creates overshoots of transient parts of the signal. Sync tips have a negative-going overshoot and the peak whites a positive one; these have to be limited to avoid over driving the modulator. Sync tip overshoot is restricted to 40% of the signal amplitude and peak white overshoot is limited to 60%–80% depending on the make and manufacture.

2
Azimuth tilt technique

Originally video recorders, mostly monochrome reel to reel machines, had single video tracks with a gap between them; that gap was called a guard band.

The video tracks and guard bands can be seen in Figure 2.1. One video head is shown properly aligned on track and replaying maximum FM signal, the other is shown off the track with improper tracking. Although the second head is not scanning the full width of the FM magnetic track there will still be a reasonable quality replay for two reasons. First, there will be sufficient signal picked up from the magnetic track that is covered to enable the replay amplifiers and limiters to operate satisfactorily and demodulate the signal. Second, the part of the video head that is off the magnetic track is scanning blank tape in the guard band and not the adjacent field track.

If the guard band were not present and the video head wandered off its track onto the next one there would be a replay of both tracks together, i.e. a replay of two fields at the same time, and considerable crosstalk and patterning would occur. This is the reason for the guard band, at least originally. Philips used this tape format in the late 1970s for a cartridge colour video recorder called the N1500. It had a 1-hour tape recording capability and was marketed for industrial and educational use.

At that time the cost was too high for domestic use at around £500, although there were a few enthusiasts who did purchase one for domestic recording as the source was only the tuner, no external inputs.

As the VHS and Betamax systems were being developed in Japan for the American and Japanese markets, Philips were developing their N1700 format. The aim of these parallel developments was to produce formats with longer playing times. In order to achieve longer times, bearing in mind that the N1500 was limited to 1 hour, the tape would have to be slowed down and the magnetic track 'packing density' increased, that is to put more information on the same area of tape. The most obvious development was to use the blank guard band between each track. Whilst the guard band served a useful purpose for protection against crosstalk it was in fact wasted tape space that could be better utilised to carry a signal. A technique was therefore developed which allowed the guard band to be eliminated, i.e. the azimuth tilt technique, shown in Figure 2.2. To understand how it works you must appreciate that the FM carrier, onto which the luminance signal is modulated, is a high frequency signal.

Turning for a moment to audio tape recorders, for the best signal on replay the audio head must

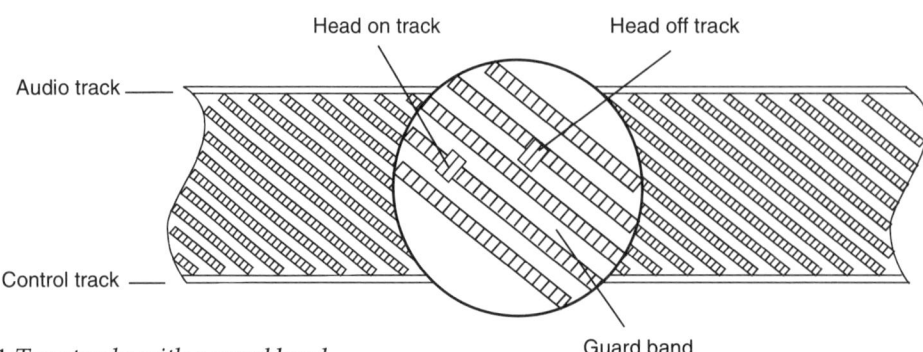

Figure 2.1 *Tape tracks with a guard band*

Video and Camcorder Servicing ansd Technology

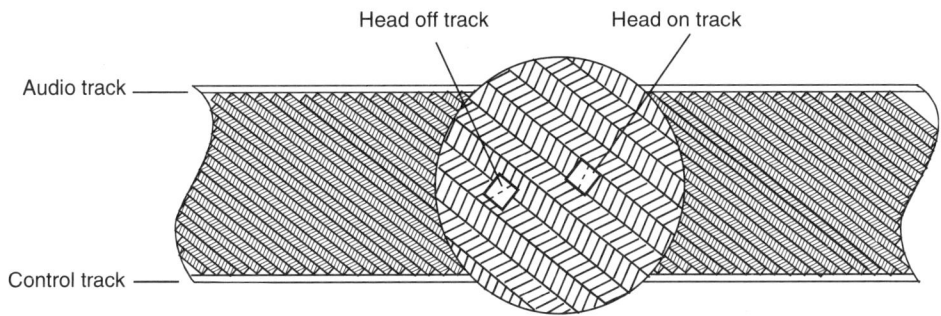

Figure 2.2 *Adjacent tracks have a different azimuth*

be vertical, that is at an angle of 90° to the tape. The adjustment to obtain maximum reproduction of the high frequencies, or the treble, from the tape is called the azimuth adjustment. If the audio head is not vertical, that is if it is tilted one way or the other, then the efficiency of reproduction is impaired and the high frequencies are reduced, resulting in a loss of treble content.

The high frequencies are not reproduced by a head that is tilted at an angle different to the angle of azimuth used in the recording mode. The technique applied to video recorders is to cut the gaps in the video heads at an angle, one way for one head and the opposite way for the other head as illustrated in Figure 2.3, shown for the VHS system. Video head A records its video track with the azimuth of its air gap at +6° and video head B records its magnetic track with the azimuth slanted at -6°, resulting in a total difference of 12° between the two video tracks. The result of this difference in azimuth is that each video head can only record and replay its own magnetic track of FM luminance carrier. For the high frequency FM luminance carrier, each head can reproduce only its own track, the one with the same azimuth angle as the playback head.

It can be seen from Figure 2.2 that the appearance of the magnetic tracks is like a herringbone pattern due to the azimuth difference between the video heads. Obviously this pattern is diagrammatic and cannot be seen in reality, except by special processes. The advantage here is that the tape can be slowed down, the guard bands filled with signal and advantage taken of longer playing times.

When a video head replays the track that it recorded the FM signal and the colour under carriers are reproduced normally. If that video head wanders off track slightly it will cover an adjacent track. The magnetic FM carrier of the adjacent track will have been recorded at a different azimuth angle by the other video head so that it will not be reproduced. However, some pick up does occur but it is about 30 dB down, too low to create problems, except for the higher quality replay requirements. However, there are noise reduction systems to reduce this FM carrier crosstalk still further (see the next chapter). The colour under-carrier is a much lower frequency and is not affected by the azimuth slant so colour crosstalk will occur during replay and so colour crosstalk cancellation techniques have to be used to eliminate it (see Chapter 5). In later video recorders there is an active luminance crosstalk comb filter to 'clean up' any luminance signal that is not sufficiently attenuated by the azimuth slant.

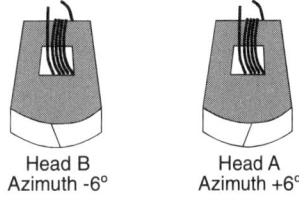

Figure 2.3 *Typical video heads with different azimuths*

3
FM recording and playback

The video signal is a wide bandwidth signal of varying amplitude, but for recording purposes it is converted to a frequency modulated high frequency carrier signal. There are two main advantages in this technique.

The bandwidth of the TV signal is from near DC to 3 MHz, which is greater than 12 octaves, and a magnetic recording head cannot practicably handle such a range. By frequency modulating the signal onto a high band carrier and by using a high frequency video recording head the recorded octave range can be greatly reduced.

The second advantage of using a frequency modulated carrier is that it is unaffected by amplitude variations that occur due to head-to-tape contact problems in both record and replay. The video heads are subject to bounce, and variations in the tape tension cause the tape to stretch and flex as it passes through the mechanics, both contributing to amplitude fluctuations of the FM carrier.

The bandwidth of the system is not flat over the recording frequency spectrum; high frequencies around the 4 – 5 MHz will also cause a reduction in FM carrier level. An FM modulated carrier can be amplified and then limited to eliminate amplitude variations leaving a clean and level FM carrier for demodulation.

Figure 3.1 illustrates the frequency modulation relationship. Sync tips correspond to the carrier free run frequency of 3.8 MHz and peak whites correspond to maximum modulation of 4.8 MHz. This is a result of converting an amplitude-modulated signal to a frequency-modulated signal. Frequency spectrums are illustrated in Figure 3.5. The frequency spectrum of the video signal, from DC to 3 MHz, creates sidebands extending above and below the modulated carrier frequency of 3.8 – 4.8 MHz. These sidebands are limited to 1.3 MHz on the lower sideband and 5.75 MHz on the upper sideband.

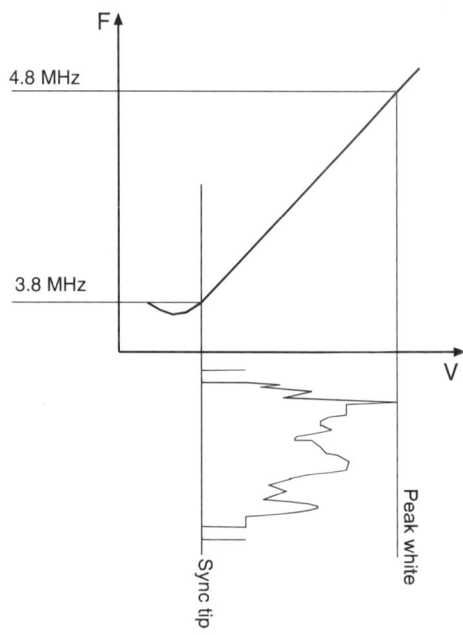

Figure 3.1 *Video signal voltage to frequency modulation*

The figures given are for the VHS system with a total recording spectrum of 1.3 –5.75 MHz, which is just over two octaves, and with the colour under carrier centred on 627 kHz (see Figure 3.5).

Frequency modulated recording

The three main advantages of FM recording are as follows.

1. It permits the recording and replay of a wide band signal on a relatively narrow band video head. This means that a video signal with a bandwidth from 0V (DC) to 3 MHz can be recorded using an FM carrier deviated from about 3.8 MHz to 4.8 MHz, these values depending on the format, much higher for S-VHS.

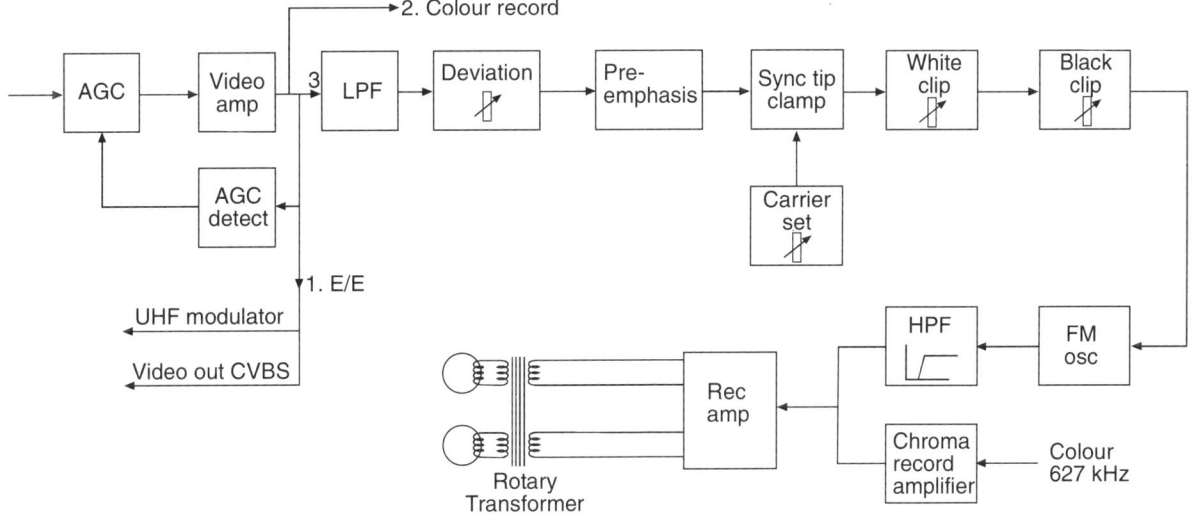

Figure 3.2 *Typical recording system*

2. The replay amplitude variations arising from irregularities of head-to-tape contact can easily be removed by using an FM carrier limiter.

3. The AC bias signal, as required for audio tape recorders, is not needed as the FM carrier forms its own bias and acts as the bias for the colour under carrier signal.

Prior to recording, the video signal has to undergo an amount of processing and the example we shall look at first is a typical VHS system. (Figure 3.2).

A video signal from either the tuner/IF or the external input is fed to an AGC (automatic gain control) network, which stabilises the mean or average level of the video signal. Correction is available against inputs from cameras, other recorders and variations in received signal level that is not at the standard 1 V p-p.

AGC output via a video buffer is split up into three paths as the signal contains full CVBS (Colour, Video, Blanking and Syncs). First it provides the E/E signal (electric to electric, bypassing record/replay) output to the UHF modulator and video output socket for monitoring purposes. Note that when a recording is monitored all the viewer sees is the AGC output. Secondly the video signal is sent to the colour processing circuits where chroma is filtered out for the colour recording signal. Third, the signal passes via a low pass filter, which cuts off at 3.38 MHz to filter out the 4.43 MHz colour signal that may cause beat patterning with the FM carrier. The filter is used in dual roles in both record and replay paths; its response curve is shown in Figure 3.3.

Out of the filter in record mode the signal is fed to a level control, which determines the amplitude of the signal to the FM modulator circuit. In the modulator this control sets the peak white FM

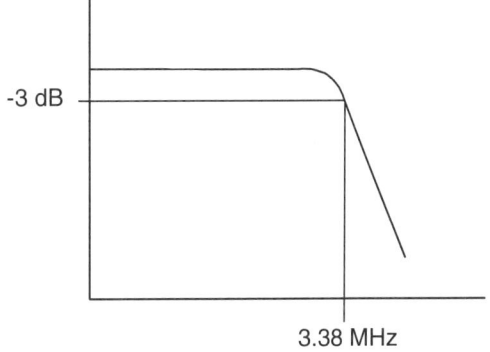

Figure 3.3 *Low pass filter response*

carrier deviation frequency of 4.48 MHz and is therefore called the FM deviation control. The signal passes next into a pre-emphasis circuit, which makes it rather spiky with overshoots at peak whites and sync tips (Figure 3.4).

After pre-emphasis the signal is clamped so that the line sync pulse tips are held at a clamped DC level into the FM modulator corresponding to 3.8 MHz. The precise frequency can be set by the 'set carrier' control to adjust the clamped level. Video signal amplitude thus extends in the positive direction up from this level, stabilising the signal to the FM oscillator against DC drift. Before the video signal is fed to the modulator the overshoots created by the pre-emphasis stage have to be removed to protect the FM oscillator from over modulation. Figure 3.4 also illustrates the problem associated with over modulation. Peak white overshoots drive the modulator to a frequency much higher than that which the record/replay circuits and the video heads can handle. The amplitude of this high frequency is too low for the limiter to amplify, because is it is below the clipping level. In replay, then, there will be part of the FM carrier missing and the resultant output is 'black'. As we shall see, in replay mode there is a circuit to alleviate this problem under normal working conditions. In abnormal conditions, such as incorrect video record signal levels or worn video heads, the correction circuits cannot cope. The visual effect in replay is black speckling on white edges, most noticeable on mid-afternoon transmissions of horse racing when white titles of high amplitude appear. A high pass filter takes the modulated signal from the modulator to the record amplifier where it is joined by the chroma colour under carrier record signal. For details see Chapter 6.

Figure 3.5 shows the spectrum of the composite video signal converted to the spectrum of the FM and colour under carriers for recording, the colour signal being centred around 627 kHz and the FM carrier between 1.3 and 5.75 MHz.

Record amplifier

The record amplifier and its response curves are shown in Figure 3.6. It can be seen from the graph of playback output versus record current that there is an optimum current that passes through the record/playback curves at all frequencies. This current is set by the record FM carrier level control; the colour record current is also optimised by a colour record level control.

The most difficult setting up procedure in the record amplifier circuits is that of the FM carrier deviation and carrier levels; however, with a little thought and the correct equipment it can be achieved. Equipment required is an RF generator to cover 3 – 5 MHz, a video pattern generator with a white pattern (grey scale or a chess board pattern could also be used) two 1K resistors and a dual beam oscilloscope. As this may need to be carried out on any format video recorder, the procedure is generalised.

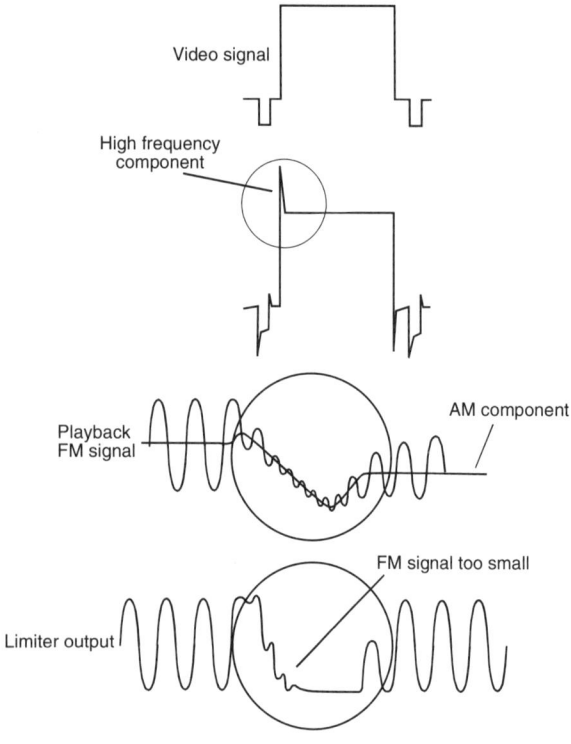

Figure 3.4 *Signal loss due to over modulation*

FM recording and playback

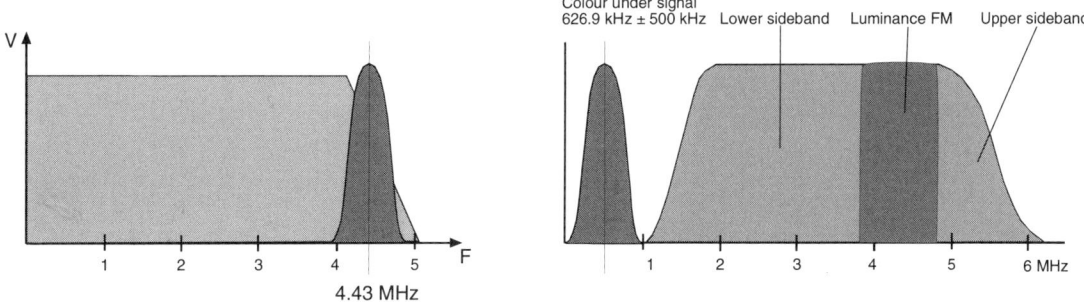

Figure 3.5 *Frequency spectrums: left, PAL spectrum; right, FM modulation with colour under carrier*

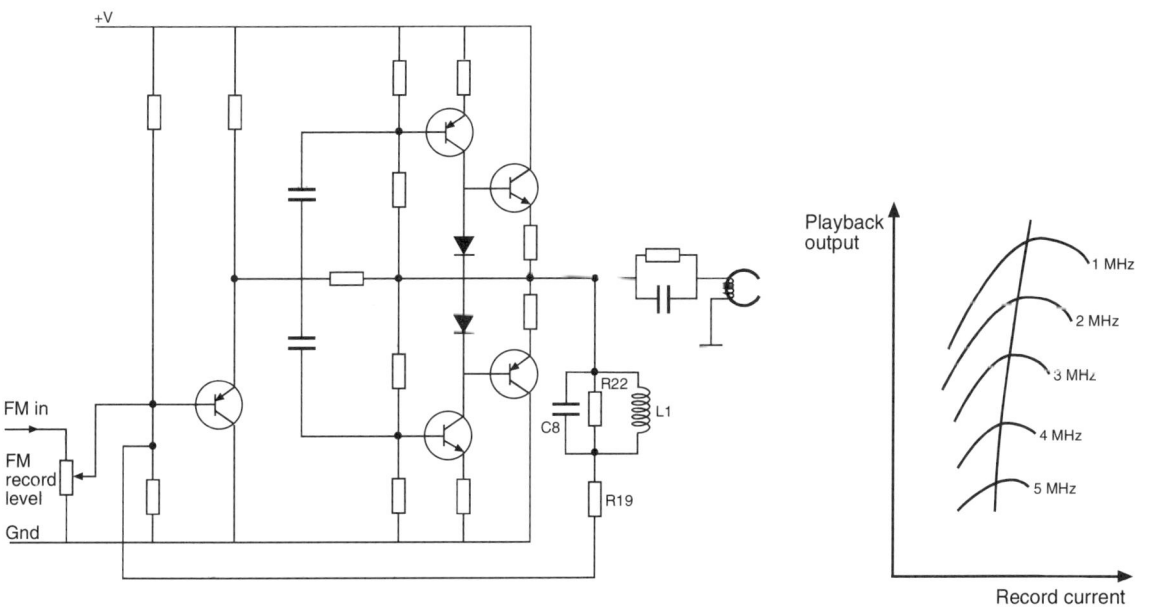

Figure 3.6 *Typical FM carrier record amplifier with HF boost feedback and its response curves*

(a)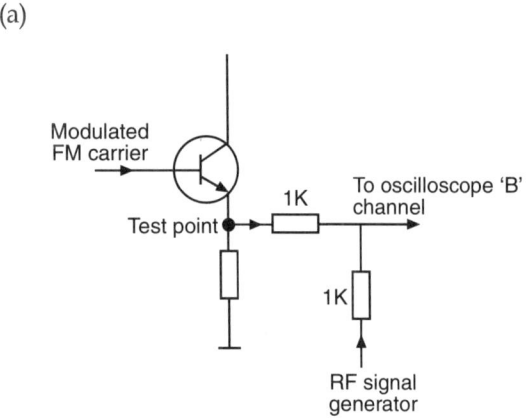

(b)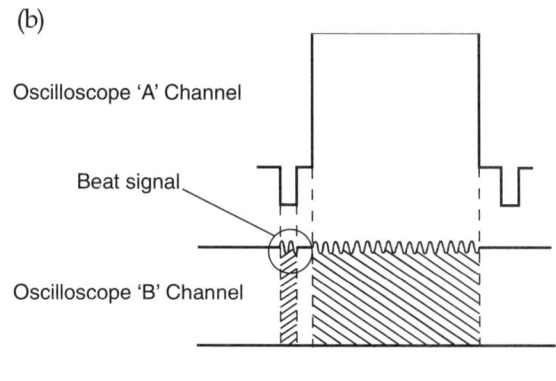

Figure 3.7(a) *Test signal mixer* (b) *Example beat waveforms at sync tip*

First locate a point within the video recorder which corresponds to the output of the FM modulator through a buffer stage, this is usually the output of the luminance record panel or the input side of the video head drive amplifier before the colour under carrier is added.

Connect this point and the output of an RF generator via 1K resistors connected together so that the two signals are mixed. Then connect an oscilloscope to this junction by one of its inputs, say channel B, as shown in Figure 3.7(a). The other input to the oscilloscope is connected to the video output socket.

Connect a pattern generator to the video recorder's line input and put the machine into the 'record' mode.

Synchronise the oscilloscope to the video signal at line rate and compare the two beams. Set the RF generator frequency and level to display a beat signal on the edge of the FM carrier that corresponds to the line sync tip (Figure 3.7(b)). Once reasonable conditions for the display have been established set the RF generator to the sync tip frequency, which would be 3.8 MHz for a VHS video recorder, then adjust the clamp level or 'set carrier' control for minimum beat in this area of the sync tip/FM carrier.

Re-adjust the RF generator for peak white frequency, 4.8 MHz in the case of VHS, and examine the edge of the FM carrier in the area corresponding to the peak white signal. Adjust the video AGC level for zero beat in this area.

Check that the video level in from the pattern generator is 100% peak white and that it corresponds to peak white off air; if not re-adjust the level control accordingly.

The frequency modulator

The discrete circuit frequency modulator is shown in Figure 3.8. A tuned transformer forms the output stage with a secondary winding to feed the FM carrier to the record drive amplifiers. Tuning is done by capacitor Ct, and as it can influence the frequency of oscillation it will operate in conjunction with the 'Set carrier' sync tip clamp level potentiometer. Ct should not normally be adjusted as twiddling can severely alter the FM modulator range and cause attenuation at the higher frequencies. Ct tunes the transformer for maximum bandwidth in the range of 3–5 MHz and is set for maximum output in this range. A sweep generator and detailed manufacturer's instructions are required to set Ct; fortunately it is very rarely neccessary.

The main adjustment in the FM oscillator is the potentiometer Rs, which ensures that there is symmetrical drive to the pair of transistors in the oscillator. It is set by viewing the output on an

FM recording and playback

oscilloscope with no modulation, to see clearly each cycle of the oscillations and then adjusting for the most symmetrical waveform; the positive half of the cycle is a mirror image of the negative half. Note that you will not see a pure sine wave as it will have small imperfections. On some machines, the waveform is best displayed as a carrier band using a lower oscilloscope sweep rate; in this way faint horizontal bands can be seen above and below the centre line of the band. These bands are caused by the imperfections in the waveform, and Rs is adjusted so that the faint bands are symmetrical about the centre of the FM carrier, above and below its centre line. The video signal applied to the modulator is also shown in Figure 3.8. The clamped DC level, the peak white clip and dark clip levels which trim off the overshoot spikes, created by pre-emphases, top and bottom of the waveform can also be seen.

Discrete record processing circuit

So much for FM recording theory, let us now consider a practical approach on circuits from an early video recorder using discrete components. Some simplifications have been made so that the techniques can be clearly seen. Modern video recorders have the signal processing carried out inside integrated circuits, so much so that the circuit diagram starts to look more like a block diagram and the real signal processing is hidden. Figure 3.9 is the basic circuit diagram of an early video FM modulator. Whilst being straightforward it does contain the necessary features for signal processing in FM recording.

Video comes in from the tuner/IF circuits via a level control. Note that there is no automatic gain circuit. This early video recorder did not have

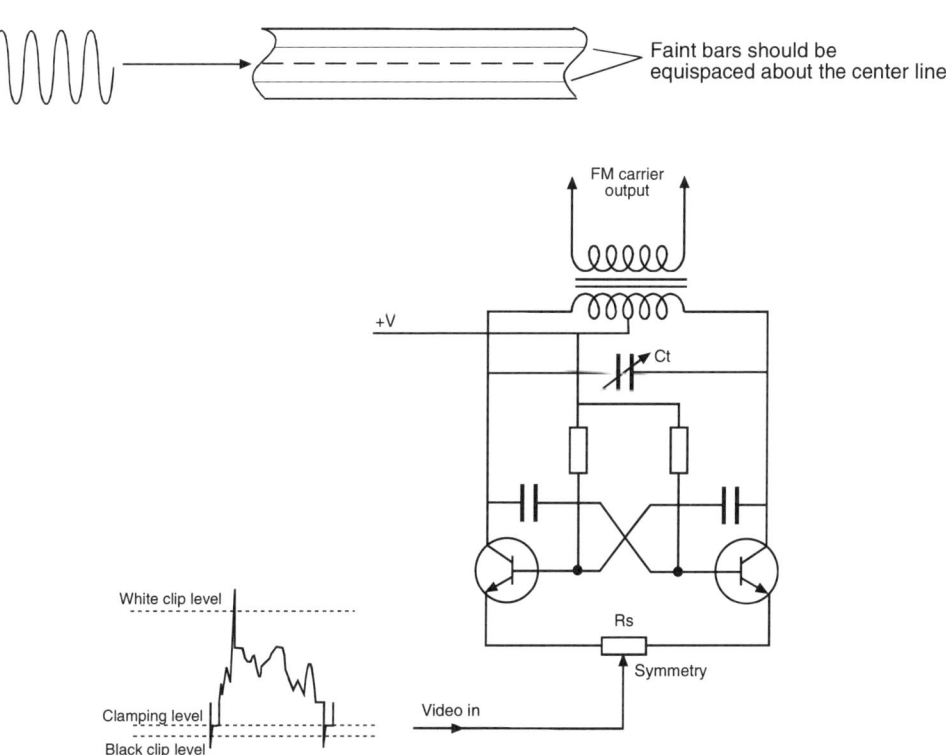

Figure 3.8 *FM modulator circuit with the video modulating signal and the FM carrier output*

Video and Camcorder Servicing and Technology

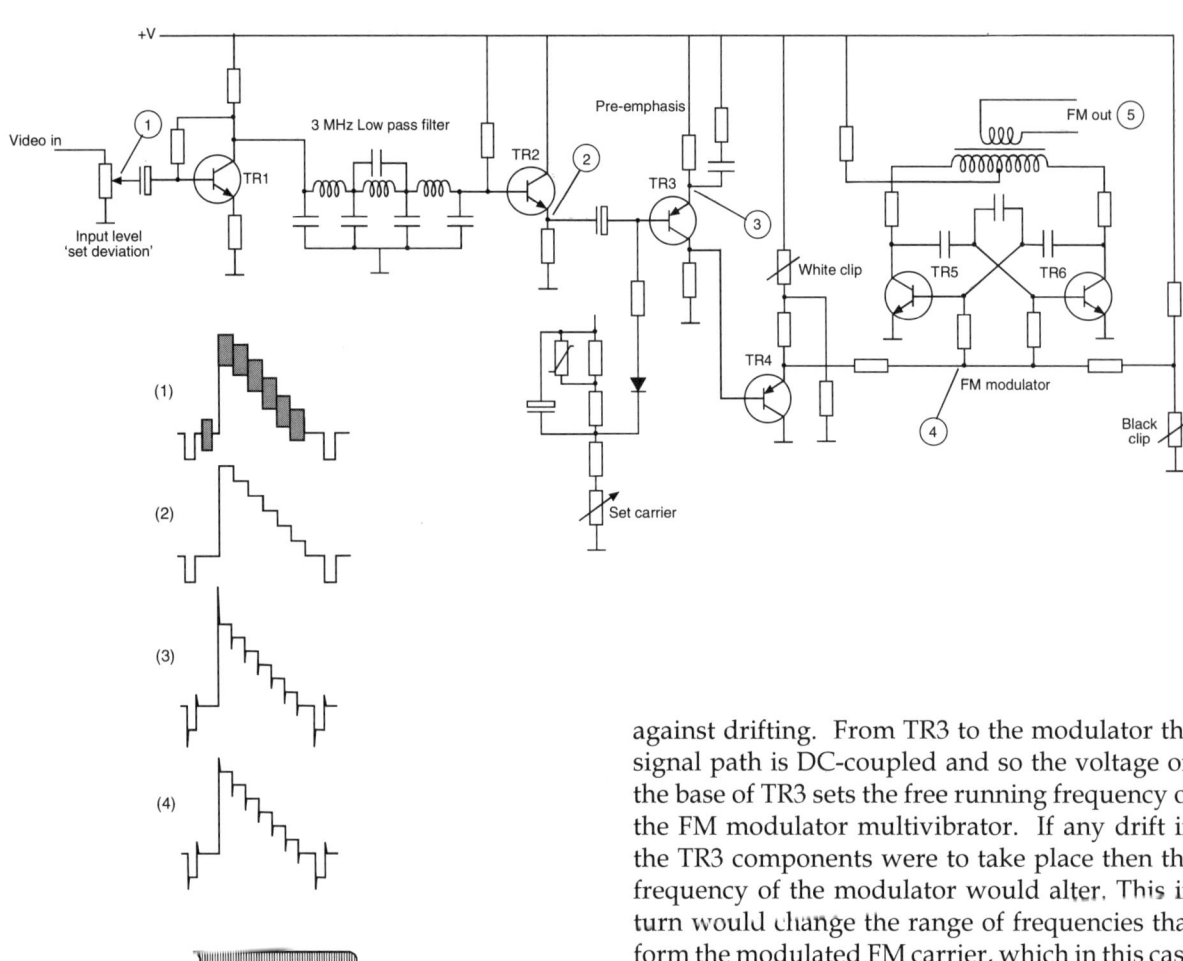

Figure 3.9 *A discrete example of video signal processing and typical waveforms*

any auxiliary or external inputs so it relied on the IF AGC to produce a fixed level video signal to the recording amplifiers. The input level control determines the amplitude of the signal, and at this point it is a full colour bar signal, but it passes through the 3 MHz low pass filter that limits the luminance bandwidth and removes the colour content. The filter section is buffered by an emitter follower, TR2, which provides low impedance drive to TR3. The base of TR3 is clamped to a set DC voltage, adjusted by the 'set carrier' control and temperature compensated against drifting. From TR3 to the modulator the signal path is DC-coupled and so the voltage on the base of TR3 sets the free running frequency of the FM modulator multivibrator. If any drift in the TR3 components were to take place then the frequency of the modulator would alter. This in turn would change the range of frequencies that form the modulated FM carrier, which in this case are 3 MHz (sync tip) to 4.4 MHz (peak white).

Pre-emphasis is applied by the use of the decoupling components in the TR3 emitter and creates the 'spiky' signal with lots of overshoots in it. The peak white overshoot is removed by TR4, whose emitter is held at a fixed DC voltage that is adjusted by the 'white clip' potentiometer. The emitter of TR4 cannot go any more positive than this set voltage, peak white overshoots applied to the TR4 base that are more positive than the fixed emitter voltage and will cause TR4 to cut off and so limit the value of peak white.

Black clip is a 'soft' control and it is set to limit the lower level transitions of the signal. If we assume the standing DC voltage is 2 V and that this corresponds to sync tip, then black overshoots may go down as low as 0.5 V. The black clip control is set to 1 V and prevents the signal voltage going lower than this, effectively clipping the sync tip black overshoots.

The varying level on TR4 emitter formed by the video signal modifies the frequency of oscillation of TR5 and TR6, and the output of the oscillator is taken from a secondary winding on the tuned transformer. This is the modulated FM carrier that now has to be mixed with the colour under carrier and fed to the video heads by the recording amplifier.

Discrete record amplifier

The FM recording amplifier has to add the luminance FM carrier to the colour under carrier at their correct optimum respective levels, and provide symmetrical low impedance current drive to the video heads, as well as give some HF lift to overcome head-to-tape bandwidth reductions. A typical discrete circuit is shown in Figure 3.10. The FM signal input will contain some amplitude variations due to impedance changes of the coupling transformer in the oscillator causing level variations with frequency changes, also there will be a length of screened cable from the modulator to the record drive amplifier which will be susceptible to pick up. Diodes D1 and D2 clip the top and bottom of the FM signal thus removing such variations.

Prior to mixing both the FM and the colour under carriers undergo HF lift, and resistive mixing is used at the base of TR3. A class B amplifier provides recording current drive that is measured at TP1, across a very low value resistor (typically 1 Ω). The measurement is in mV, and the colour is adjusted first by shorting the luminance to chassis across the test point at the level control to remove the FM carrier. This is to enable the colour signal which is very small to be seen clearly at around 2 mV. The FM record current is then set at around 25 mV. Great care has to be taken to terminate the coax leads to the meter at 75 Ω, and a wide band millivoltmeter is specified. The author uses the oscilloscope and sets the levels at 5 mV p/p for colour and 70 mV p/p for luminance (peak to peak values).

The FM recording drive on some machines could be optimised to a particular tape, especially as newer high energy tapes that are available. The method is as follows.

The recorder is set up in record mode with a microphone connected to the audio input and colour bars to the video input. An oscilloscope is connected to the FM drive amplifier output; the level will be found to be around the 2–3 V mark. The FM drive is then turned down to a suitable

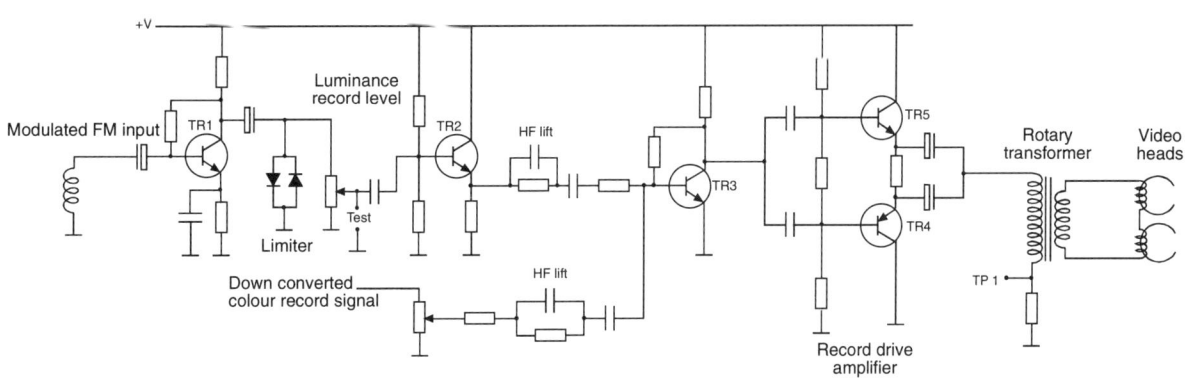

Figure 3.10 *An example of a discrete FM carrier recording amplifier*

minimum, for example 0.25 V, it is then advanced 0.25 V at a time whilst announcing the value into the microphone up to a maximum of 4 V. Having completed this, the tape is then replayed with the oscilloscope connected to the replay pre-amp output. The replayed FM carrier will then increase with each announced record level and after reaching a peak value it will fall again as the recording level is further increased. All you do then is set the recording level to that value which produced the maximum replay level. Most recorders will flatten off around certain values, others will not alter the replay level value between recording values of 2.5 V and 3.5 V. These record drive levels are set at 3 V p/p which is usually close enough.

The colour under carrier record level is best set out as follows. Replay the manufacturer's test tape at the colour bar section and measure the replay colour level on the output of the replay pre-amp – do not measure it on the colour PCB as it may have gone through an ACG level control amplifier.

Having noted this replay level, record some colour bars, preferably not on the test tape – don't forget to put a new one in the machine! It is then a matter of recording and replaying colour bars, each time adjusting the colour record level until such time as the replayed level meets the value measured off the test tape.

Keyed AGC

Input composite video is passed to a variable gain amplifier, the AGC amplifier, the output of which is measured against a constant and any variation against this constant is compensated. A typical circuit is shown in Figure 3.11 with its waveforms.

To measure the output the signal is fed back via a low pass filter to a sync tip clamp and to a mixer. After the low pass filter, to reduce the chrominance component, the horizontal sync pulses are separated (B) and delayed to form the

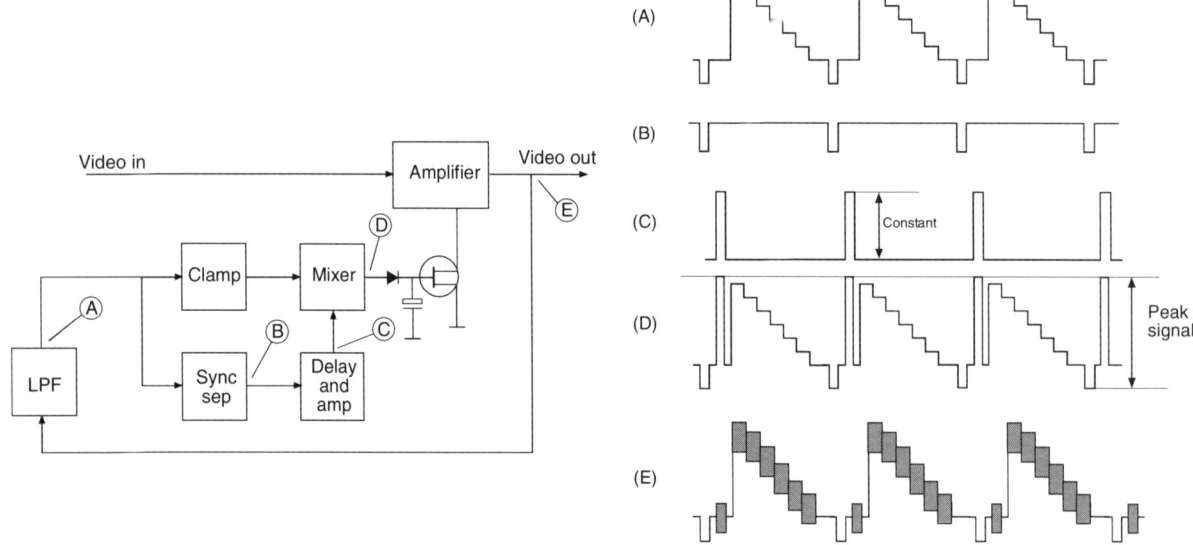

Figure 3.11 *Keyed AGC circuit and waveforms*

key pulse (C). In the mixer this key pulse is added to the sync back porch. There is no burst as it has been removed by the LPF and the output (D) is sent to a peak rectifier. It can be noted that the key pulse is of a higher level than peak white; therefore the peak rectifier output is dependent only upon the peak level of the key pulse. As the amplitude of the key pulse is constant then the output of peak level rectifier is varied only by any changes in the amplitude of the sync pulse. If the video signal level rises due to increased input then the sync pulse level will increase. This increases the peak level of the key pulse and Q1 conducts, reducing the gain of the AGC amplifier to compensate.

Some models of video recorder only measure the key pulse once every field and hold to AGC until the next field.

Macrovision spoils tape copying by adding a variable key pulse to the video signal. This adds and subtracts from the inserted AGC key pulse, seriously affecting the gain of the AGC amplifier.

A comparison of FM carrier frequency spectra

Within the FM carrier spectrum two main parameters are set: one is the bandwidth of the video signal and hence the resolution of the replayed picture; the other is the signal-to-noise ratio of the video signal replayed off tape. Bandwidth is determined by the range of the FM lower sideband between the lower limit around 1 MHz to the centre frequency of the modulation band. Signal to noise is determined by the bandwidth of the modulation frequency and the coercivity of the videotape. Resolution, i.e. the crispness of the picture, is influenced by the upper side band and the maximum FM carrier frequency achievable.

TV lines

Video signal bandwidth can be looked at in two ways. The first is equivalent to the audio specification of -3 dB points with respect to a sine wave. This is never quoted, however, as it would appear very low indeed and anyway enhancement techniques are typically used which can produce a picture far better than if only the signal's bandwidth is considered.

The other way is to consider bandwidth in terms of picture resolution. Resolution is quoted in lines per picture height that can be resolved in a picture. Although the topic of picture resolution is highly subjective, as a guide, 625-line TV tubes are generally considered to have a horizontal resolution of about 400 lines.

When referring to VCRs, because resolution is so subjective an approximation can be given (1 MHz » 80 lines) relating bandwidth to the number of lines that can be resolved. VHS, with its bandwidth of about 3.2 MHz, for example, gives a resolution of about 250 lines.

$$(3.2 \times 80 = 256)$$

S-VHS can resolve 400 lines and digital video 500 lines.

VHS

The VHS system (Figure 3.13) has a colour under carrier of about 40fh which includes a slight frequency shift; 40fh is 625 kHz, the shift being 1.952 kHz results in a carrier of 626.952 kHz. The FM modulation is between 3.8 and 4.8 MHz and

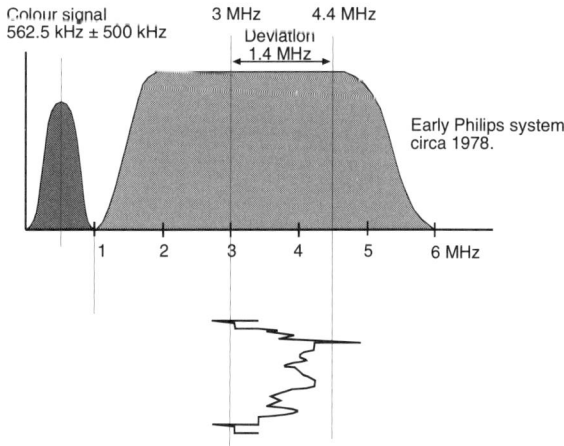

Figure 3.12 *Early VCR recording frequency spectrum*

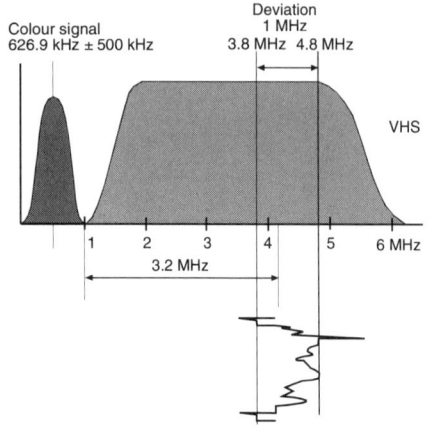
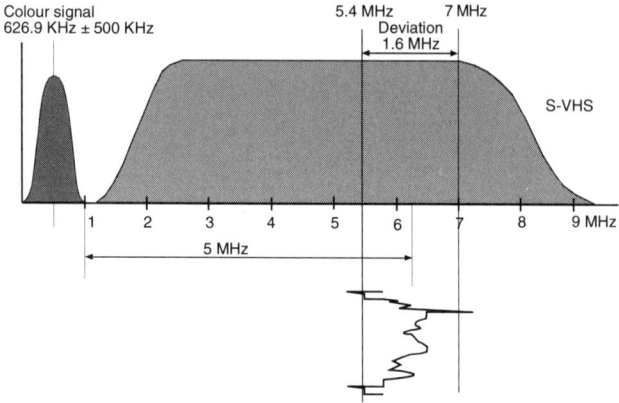

Figure 3.13 *VHS recording frequency spectrum*

Figure 3.14 *S-VHS recording frequency spectrum*

the modulator incorporates both dark and peak white clip adjustments. As the lower side band is from 1 to 4.2 MHz the bandwidth is 3.2 MHz giving a resolution of 250 lines.

S-VHS

For S-VHS the bandwidth has been increased while retaining the same colour under frequency as VHS. To improve the luma/chroma cross-talk, a greater margin has been obtained between the chroma upper side band and the FM carrier lower side band. To improve resolution and signal-to-noise ratio the deviation is increased to 1.6 MHz between 5.4 MHz at sync tip to 7 MHz at peak white. This allows the lower side band to increase to 5 MHz and a resolution of 400 lines.

Hi Fi audio is added as a deep layer recording of FM audio carriers centred on 1.4 MHz (L) and 1.8 MHz (R). These do not appear as part of the video recording spectrum as the deep layer recording essentially keeps the video and audio carriers apart. For further information see Chapter 8, Hi-fi audio.

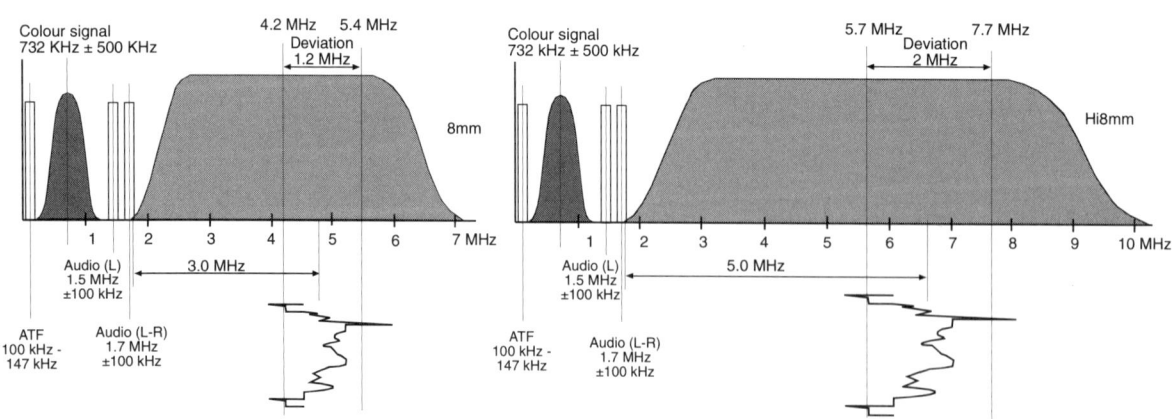

Figure 3.15 *8mm and Hi8 recording frequency spectra*

8mm video recording spectrum

The 8mm video system spectrum is shown in Figure 3.15. FM deviation is approximately 1 MHz due to the inclusion of a (L–R) audio FM carrier at 1.7 MHz. Originally the 8mm system was a mono camcorder format and stereo capability was added as an after thought. This is shown by the audio FM carrier at 1.5 MHz being the left channel and the carrier at 1.7 MHz being a difference channel to accommodate the audio right-hand channel.

Hi8 is an enhanced version of 8mm, as with S-VHS the carrier frequencies have been extended up to 10 MHz and provided an increased resoltuion of 400 lines.

FM replay and demodulation

In replay, each video head scans the tape in turn, replaying the magnetic tracks and reproducing the FM carrier signal one field at a time. It is important to note at this point that the videotape wrap around the video head drum is in excess of 180°. In replay this will provide for an overlap of replayed signal from each head, to avoid any discontinuity of the replayed FM carrier.

First, a look at the overall block diagram of a typical replay system taken from a VCR with discrete circuitry in Figure 3.16.

The replayed FM carrier is amplified in a tuned pre-amplifier, the response of which is tailored to peak the carrier in a 'Q' network. A dropout compensator is used in this part of the system. A drop out is a white spot or even a horizontal line caused by magnetic particles missing from the tape itself. A 1H delay line is used; in the event of a drop out occurring, the missing FM carrier is 'filled in' by switching to the carrier stored in the delay line from a previously replayed line. After the dropout compensator, considerable limiting is employed to eliminate any amplitude variations in the signal resulting from head-to-tape bounce and as the VCR ages, and video head wear. A balanced modulator is employed to demodulate the FM carrier, the resultant signal being a video signal full of carrier, which is then filtered out by a de-emphasis network and a low pass filter.

The cleaned-up video is sent via a play back level control to an aperture corrector, i.e. a crispener circuit to enhance the black to white, and the white to black transients. It is sometimes

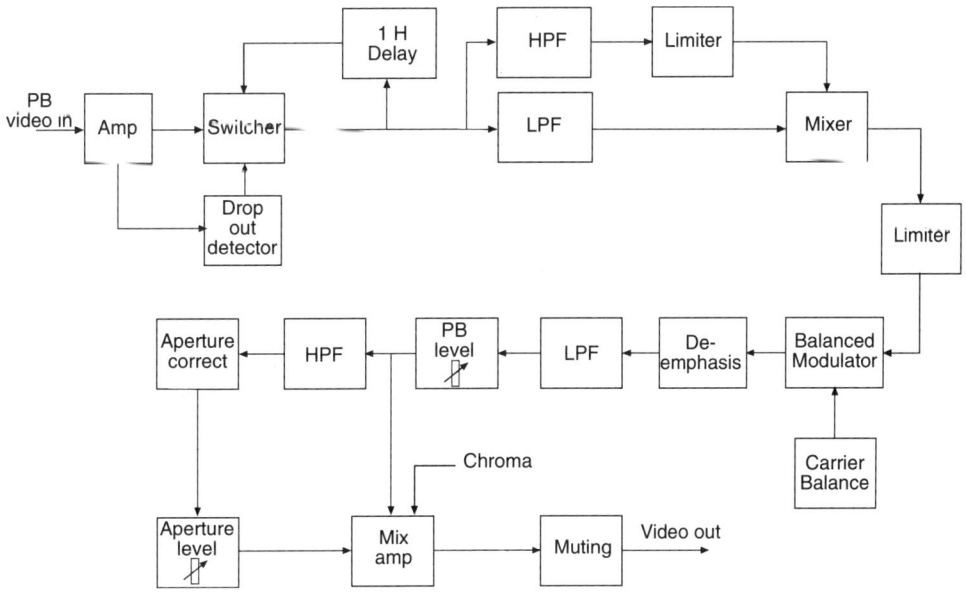

Figure 3.16 *Block diagram of a typical video replay system*

called edge enhancement, too much enhancement will create overshoots and ghosting, or reflections, and a control may be fitted to set the level of crispening. It may also be called a sharpness control.

After aperture correction the replayed chroma signal is added The composite signal is then buffered and fed to the modulator via a muting circuit, which blanks the video signal for a few seconds after the play button has been selected or if the servos are out of lock. This allows the servo circuits and replay picture to settle before the video is allowed to the output.

The pre-amplifier and video head switching

The pre-amplifier is split into two pre-amplifier stages, see Figure 3.17, usually low noise field effect transistors, or low noise IC, one for each video head. Although the video heads are coupled to the pre-amplifiers via a single rotary transformer, they are not interactive. This is because the rotary transformer is split into two separate sections each having two close coupled windings; an upper winding connected to the video head, which is rotating, and a lower winding connected to the pre-amplifier, which is static.

This is in effect two rotary transformers constructed as one, one rotary coupling for each video head. Across the secondary of the rotary transformer is the 'Q' network, to peak the response of the video head and rotary transformer to a frequency of 5.0 MHz (see Figure 3.18).

It is necessary to align each of the 'Q' networks, if fitted, each time a head assembly is replaced; they must be balanced for frequency and amplitude. If the 'Q' network for each video head is not balanced a selective flashing effect can be seen on the replayed picture, which under normal circumstances cannot be readily defined or explained. However, replay of frequency graticules, that can be obtained from the more expensive pattern generators that reproduce bands of vertical lines, from 0.5 to 3.0 MHz, will define the flashing. The effect, a 25 Hz flashing, will be confined to certain frequencies on replay or sets of frequency gratings on a test pattern, because of imbalance in the 'Q' networks.

It is important not to meddle with the 'Q' of each pre-amp unless you have a manufacturer's test tape that contains a sweep frequency section. A frequency sweep is the best way to peak 5Mhz or 4.5 MHz with the capacitor and then balance the amplitude with the pre-set damping resistors. It is normal to have at least one of the two variable damping resistors set at maximum resistance.

The outputs of the pre-amps are connected

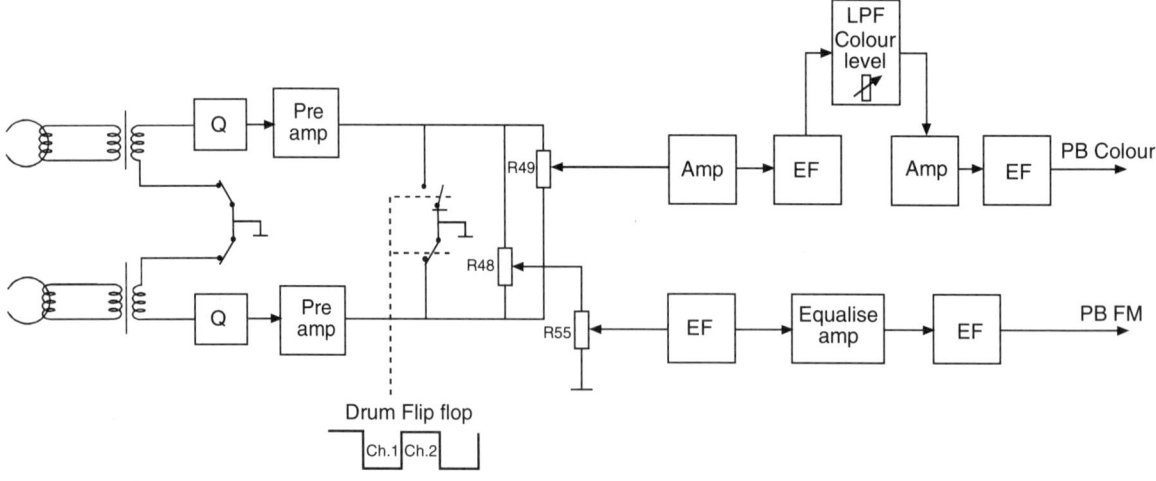

Figure 3.17 *Typical VHS FM carrier replay signal path*

FM recording and playback

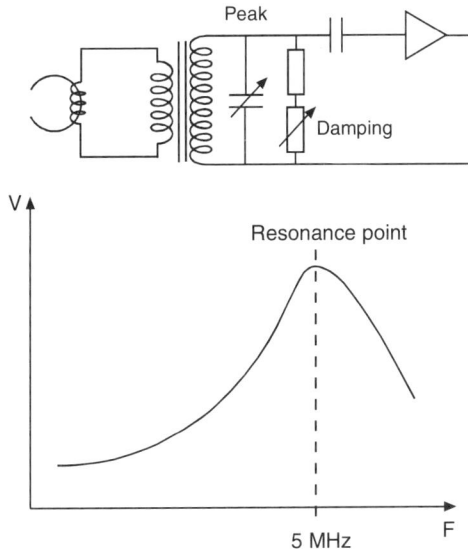

Figure 3.18 *'Q' network and frequency response*

occurs. The crossover point is positioned 6½ lines prior to field syncs, depending on the system. On the replayed picture the crossover point is at the bottom of the screen, out of sight in the over scan area.

Figure 3.20 illustrates the Channel 1 and Channel 2 video head replayed signals. Switching is provided by the flip-flop signal, which is derived from the rotating drum, in the servo circuits.

Transistors Q1 and Q2 are switched 'on' 180° out of phase. When the flip-flop is 'low' then Q2 is on and Ch.2 video head is clamped to ground. Ch.1 video head is replaying out via R1 (colour under) and R2 (FM carrier). At the changeover point both heads are reproducing the FM carrier and colour under carrier so there is no loss of signal. There is a small amount of tolerance across two variable resistors. One is used to tap off the replayed colour under carrier (R49) and balance its level to be equal from each head. Setting up must be done by replaying the manufacturer's standard test tape again.

The other variable (R48) taps off replayed FM carrier and balances each output to a second variable (R55), which adjusts the overall level of the modulated carrier. The replayed colour under signal is fed through an amplifier and a low pass filter to remove the residual FM carrier, level adjusted, and then buffered out to the colour replay processing circuits. Replayed FM carrier is passed through an equalising stage to peak up the HF upper sideband; this is additional to the 'Q' tuning which aids signal-to-noise ratio.

Each video head replays more FM carrier than is actually needed to provide an overlap period (see Chapter 4 for more information). In this overlap period the signal information is duplicated so it is only necessary to switch one head off and the other head on to maintain continuity of the replay signal. The demodulated video signal will not then have any gaps in it where parts of the picture would be missing. There is only the head crossover point that could be seen in the picture where the transition from one head to the other

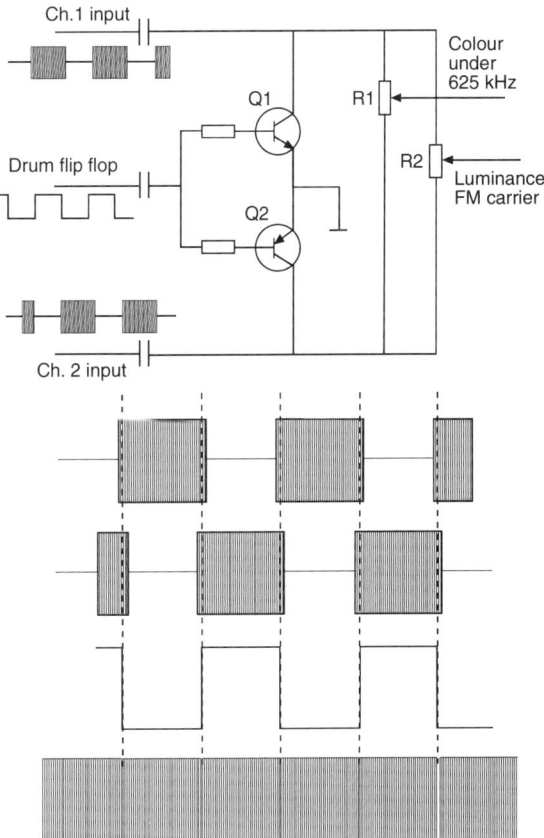

Figure 3.20 *Video head switching for contiguous carrier signal*

allowed of the switching point, ±1 line. More on head switching in Chapter 4, p.71.

The accuracy required for the switching point is why the switching edges of the flip-flop have to be adjusted after a head change, again using a manufacturer's test tape as a reference.

A comparison of replay head switching and amplifiers is shown in Figure 3.20. The original video recorders had only one rotary transformer winding because the video heads were in series. In consequence only a single pre-amplifier was used with the Q network fixed. There was no active video head switching and a sufficiently good signal-to-noise ratio was achieved in spite of the alternate head producing noise in addition to the scanning head.

In Figure 3.20(b) the VHS video recorder has dual rotary transformer windings, Q networks and pre-amps switched from a flip-flop, or RF switching signal as it is sometimes called. Early European video recorders produced the most complex switching. First, in the record mode, S2a and S2b are both closed to ground, record FM carrier is driven to the centre of the two windings so that both video heads are driven continuously in parallel. In replay S1 is closed. S2a and S2b are open and RF switching signal is applied in anti-phase to switches S3a and S3b. When head Ch.1 is replaying, S1 and S3a are closed and the signal is transferred to the single replay amplifier via the upper loop. When head Ch.2 is replaying, S1 and S3b switches are closed and the signal is transferred via the lower loop.

From this information it is possible to deduce from the fact that both heads are recorded in parallel that during replay Ch.2 head is inverted with respect to Ch.1 head. Whilst this 180° phase inversion does not affect the FM carrier or its demodulation it does have a bearing on the colour under carrier. It causes the replayed colour under carrier to be inverted when replayed by the Ch.2 video head. A compensation switching signal at 25 Hz is applied to the replay colour processing circuitry to compensate for this inversion.

In Figure 3.20(d) the video recorder uses two pre-amplifiers, one for each video head in which the Q networks are fixed. Replay FM carrier and colour under signals are balanced and level controlled in a single AGC amplifier.

In most video recorders both heads are driven in series for recording purposes by a single amplifier.

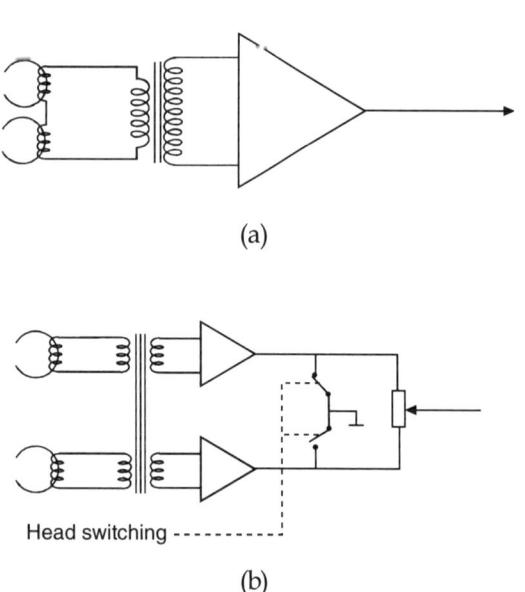
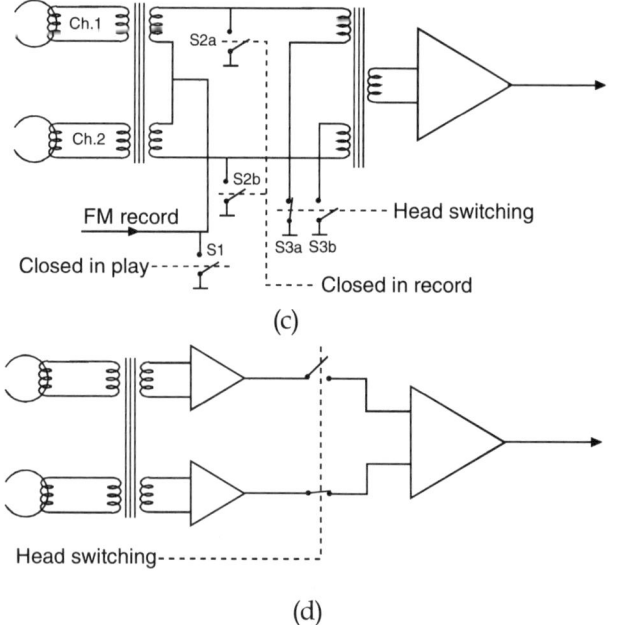

Figure 3.20 *Comparison of pre-amplifers*

FM recording and playback

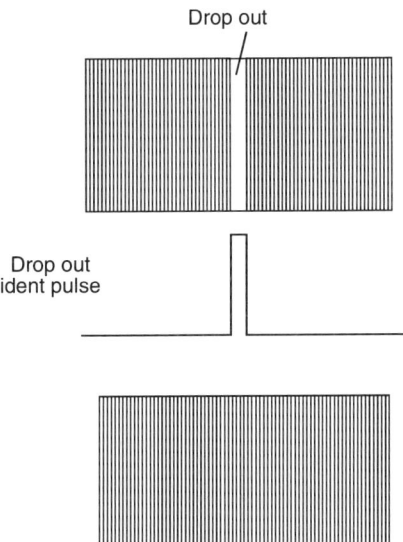

Figure 3.21 *Drop out ident pulse is used to 'fill in' a drop out with FM from a previous line*

The drop out compensator

A drop out compensator is used to reduce random white spots that would be visible on the TV screen during replay. The engineer has only to inhibit the drop out compensator to see the quantity of drop out spots, which will increase with the age of the tape or its environmental storage conditions. Dropouts occur due to the tape shedding oxide particles and even new tape contains a certain amount of drop out. The drop out compensator, DOC as it is usually called, relies on the fact that the drop outs are random in nature and that they will not occur at the same position visually along subsequent lines, which would produce an effect similar to vertical white lines.

Figure 3.22 is a standard type of drop out compensation system used on early VCRs. The direct path is through a limiter and demodulator 1, via switch S1, and out as a video signal. A parallel path is through a 64 ms, one-line (1H) glass delay line and demodulator 2. The FM carrier is also fed to a drop out detector circuit, which monitors the FM carrier. When a drop out occurs an output switching signal is developed which is of very short duration lasting not much longer than the drop out and S1 changes over for the duration of the drop out. When switch S1 changes over the output from demodulator 2 is fed out as a video signal and 'fills in' for the drop out duration.

This small piece of video signal that is switched in, in place of the drop out, was replayed one line earlier and so it is used twice and prevents an obvious white spot, or black spot. Demodulators 1 and 2 are either matched or have separate gain and black level adjustments. In the extremes of drop out, when it is excessive due to high oxide shedding, it is possible for the delayed signal, which is switched to the output, to also contain drop outs. If this is the case then demodulator 2

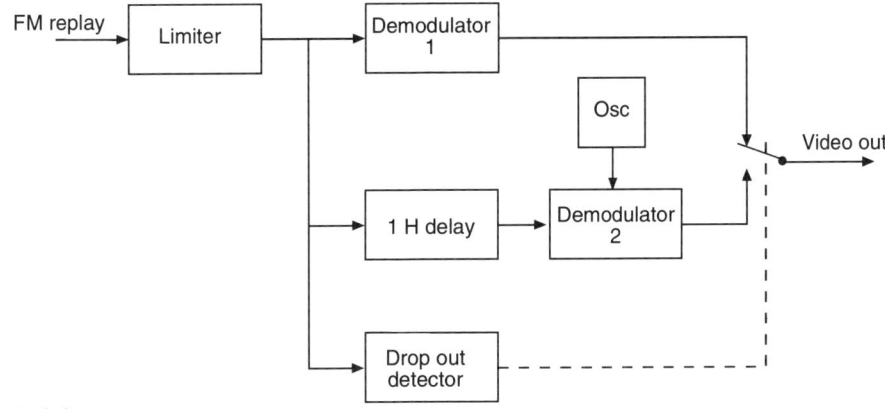

Figure 3.22 *Typical drop out compensator*

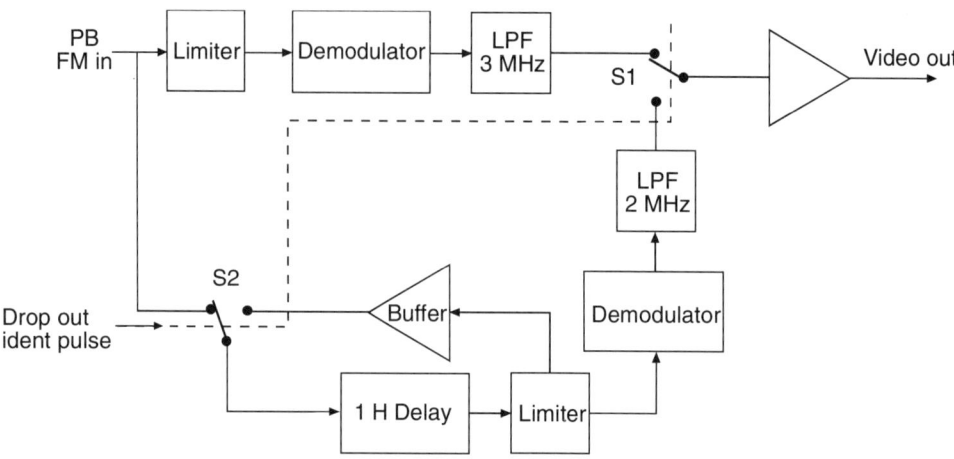

Figure 3.23 *Cyclic drop out compensator*

is supplied by an oscillator at about 3.5 MHz, to produce a mid-grey video level which is less obtrusive than pure white or black spots.

Cyclic drop out compensator

The example of the cyclic drop out compensator in Figure 3.23 is taken from a video recorder manufacturer in the mid 1980s.

The direct signal path for the FM carrier is through a limiter and demodulator to a low pass filter at 3 MHz to remove residual carrier from the video signal. Switch S1 connects the demodulated video signal to an output buffer amplifier. The parallel path is to the 1H delay line, the delayed output passing back into a limiting amplifier. The limiter has two outputs for the FM carrier: one to a demodulator, and the other to switch S2 via a buffer amplifier, to compensate for delay line and limiter attenuation. The demodulated and delayed video signal passes through another low pass filter at 2 MHz to switch S1. Both switches S1 and S2 change over, driven by a drop out detector ident pulse for the period of the drop out, and the detector is a Schmitt trigger circuit.

During a drop out, S1 switches between direct and delayed luminance video. This is similar to the basic drop out compensator, in so much as the drop out is filled with delayed video. Also during a drop out, S2 changes over and the output of the delay line is re-fed back to its input, thus preventing the delay line from being fed with a drop out by re-circulating its output.

The delay line is then always full of FM carrier without any drop out. However, in the limits of very poor tape both switches may stay over, the delay line re-circulates the FM carrier continuously and the same video line is continuously demodulated and repeated at the video output. This is confirmed by the fact that a blank section of tape following a recording is displayed as vertical grey and white stripes from the last video line being re-circulated.

Double limiter

If the level of the FM carrier is too low for a limiter to fully amplify and limit the output, then a severe loss of carrier results, the lost signal when 'demodulated' being seen as a black element on the television screen. This effect can be seen when the FM record/replay frequency response is insufficient or the peak white clip level is set too high. A peak white will over-modulate the carrier

FM recording and playback

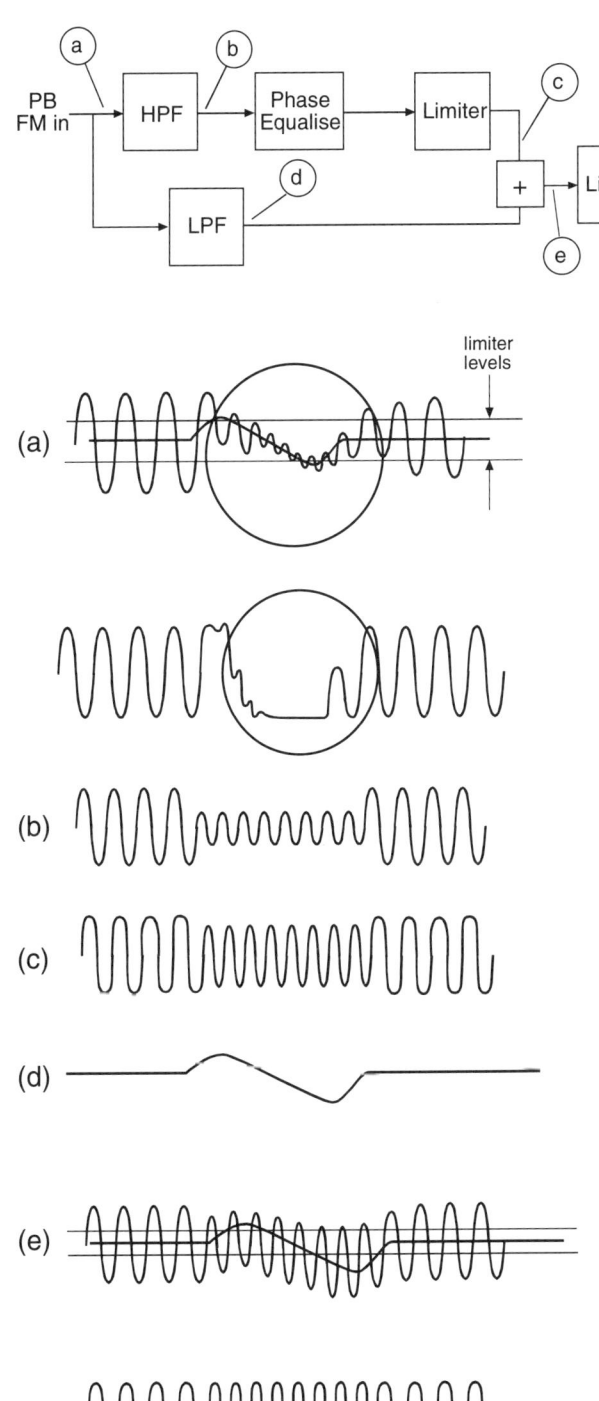

Figure 3.24 *Double limiter and waveforms*

and the modulator will produce very high frequencies. Bandwidth limitations will reduce the amplitude of the carrier and in replay the carrier will be too low for the limiter. As it is the peak white signals that create the highest frequency FM carriers, the visual effect on the screen during replay is black speckles after whites, most clearly seen on early VCRs when the video heads are wearing low. The speckling is less noticeable on VHS video recorders due to a double limiter technique, as shown in Figure 3.24.

The waveform into the double limiter is shown Figure 3.24(a). The high frequency component of the FM carrier, corresponding to the peak whites, is mixed onto a lower frequency component, forming a complex signal. Without double limiting only the low frequency component will be present at the limiter output, as shown below it; the high frequency component is lost and black is the result as no carrier is present.

The double limiter technique is seen in the small block diagram in Figure 3.4. The high frequency component is split off through a high pass filter and is shown in waveform (b). It is amplified and phase corrected for shifts introduced by the filter and then limited to give waveform (c) which is then mixed with the low frequency component (d) to produce the complex waveform (e). The complex signal is then further limited in a second limiter and the output is as shown in (f); the high frequency component has been recovered. Consequently (f) is similar to (a) as it should have been had there not been any loss of HF component.

The VHS system can cope with a wide range of high frequency carrier levels and it is not subject to black after white speckling, except in the extreme cases of video head wear.

The best all round test for this is recording of a live transmission where the white titling is often not kept within the broadcasting maximum limits.

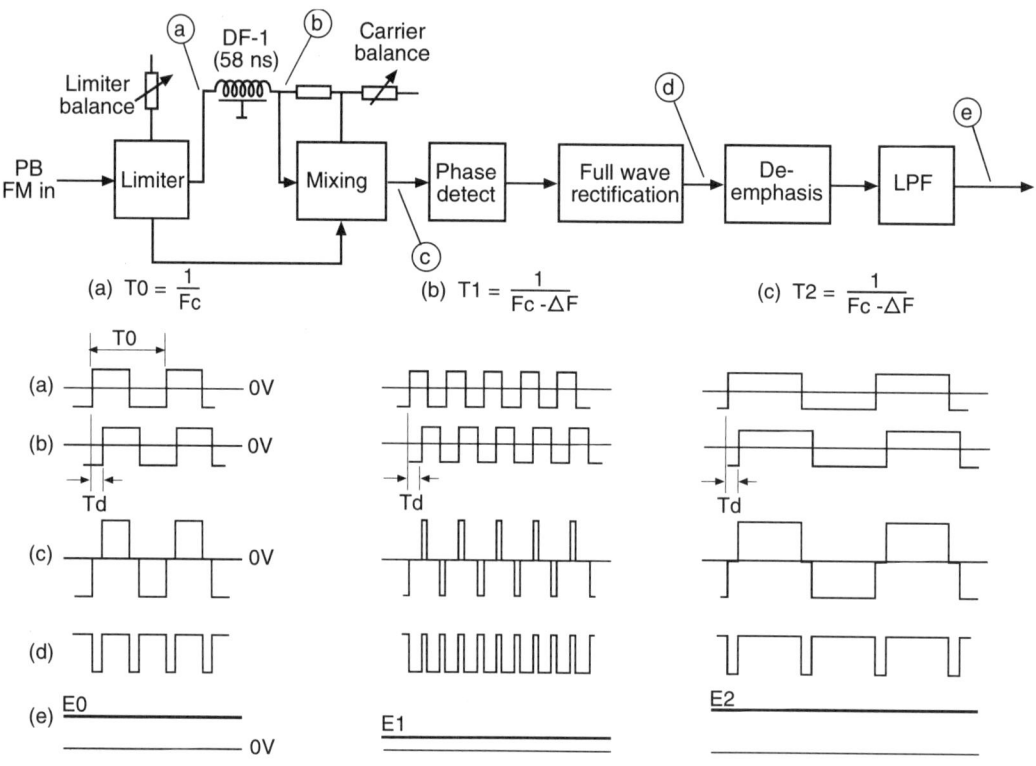

Figure 3.25 *Aperture correction circuit and waveforms*

Demodulation

The FM carrier in record mode is frequency modulated between 3.8 and 4.8 MHz, for VHS. From this carrier the modulation frequencies up to 3 MHz have to be recovered by demodulation and filtering. In order to simplify the design of the filters the FM carrier is effectively doubled in frequency to produce a frequency modulated band from 7.6 to 9.6 MHz. Filters are needed to remove residual carrier and are designed as low pass filters around 3 MHz to pass the demodulated video and reject the higher frequency carrier. For a 3 MHz bandwidth signal a better rejection is obtained from a higher frequency carrier at 8 MHz than one at 4 MHz.

A VHS demodulator is shown in Figure 3.25, and the basis of the demodulator is a fixed delay line, DF-1; with a very short delay time Td (where Td is 58 ns or 58×10^{-9} s). As a further note, a 58 ns delay is equivalent to the period of a 17 MHz signal. The FM carrier enters the demodulator circuit where it is limited. The carrier is then split into two paths to an additive mixer, one path direct and the other via the delay line DF-1. It can be seen that this delayed signal, when added back to the direct signal, doubles the frequency. Figure 3.25(a) shows the input carrier waveforms at three sample frequencies: T0 = 1/FC, which is the centre frequency, T1 = 1/(FC + ΔF) which is the highest frequency, and T2 = 1/(FC - ΔF) which is the lowest frequency. If we deal with the centre frequency T0 = 1/FC then the others can easily be followed.

Replayed FM carrier (a) is the input direct to the additive mixer and (b) is the output of the delay line and the second input to the mixer so; (b) is (a) delayed by the short time Td. In the mixer the two inputs are voltage added and the castellation waveform (c) results. This derived waveform is then passed to a full wave rectifier to give a higher frequency output (d). This is double the input frequency (a) but does not have

FM recording and playback

a symmetrical mark/space ratio; an alteration of the symmetry of the rectified output occurs. When passed through de-emphasis and filters it is integrated to a video level, E0. As the carrier is frequency modulated the output E0 will vary in level, according to the modulated frequency shift, and will be the recovered video signal and not a DC level (it is only shown as such for clarity). The highest modulated carrier frequency results in an output level of E1 and the lowest E2

It can be reasonably assumed that for a carrier that is amplitude/frequency modulated the output level E will vary with respect to the carrier modulation. Sync tips that are the lower carrier frequency will be demodulated to the highest voltage level; the demodulated video signal will therefore be inverted (upside down). Further filtering removes the residual carrier leaving a clean inverted video signal to be re-inverted, corrected and buffered out.

Aperture correction

The term aperture correction may seem a little strange, but its history lies in the earlier video camera equipment which suffered from defocusing of the picture when the iris or aperture was opened wide. The soft picture was corrected by crispening the edges to enhance it, and this was called aperture correction. It is used in the replay path of video recorders to make a picture with a bandwidth of 2 MHz (-3 dB) look as though it had a bandwidth of 3 MHz (-10 dB) by enhancing the transient edges of the signals and effectively crispening the picture.

One aperture corrector is shown in Figure 3.26. The replayed video signal is split into two paths; one direct to a mixing amplifier and the other to a high pass filter. The input signal (a) is shown as a trapezoid waveform. The transient edges have been softened by low bandwidth and they no longer have sharp rise times due to the high frequency losses. After passing through the high pass filter only the transient edges and noise are left as in waveform (b). These signals are then passed through an amplifier, the noise and transient edges are then limited, to limit the transient peaks only. The limiter is designed to produce pulse type outputs from the short duration of the input spikes. Following the limiter is a buffer driving a diode arrangement which removes the noise content as it is not of sufficient level to overcome the diode's forward voltage, leaving the transient edge pulses. An aperture level taps off The required level of the pulses to the mixer input. the transient pulses are then added back to the original waveform (a) to sharpen up the edges and to crispen up the signal as in (e). Note that (e) has very small overshoots and the original amount of noise; this is not a noise reduction circuit.

Adjustments of the aperture level control will increase, or decrease, the overshoots. This

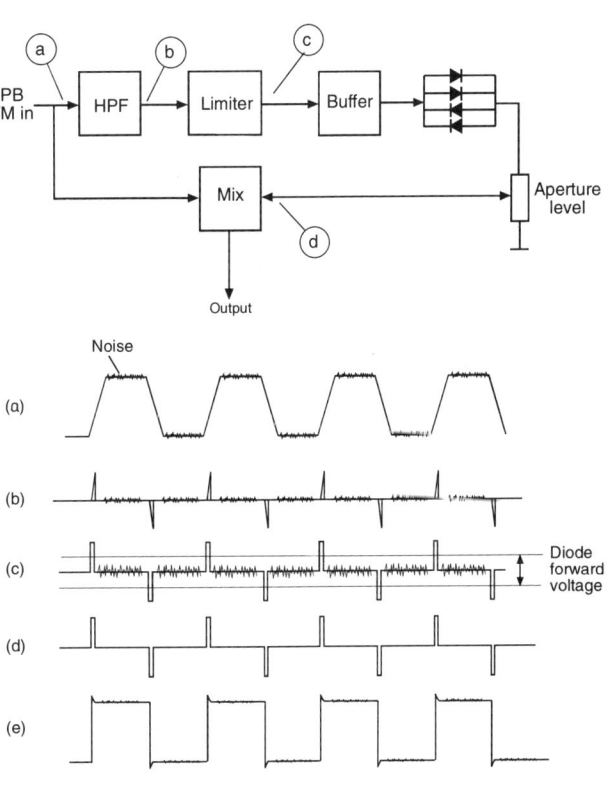

Figure 3.26 *Aperture correction circuit and waveforms*

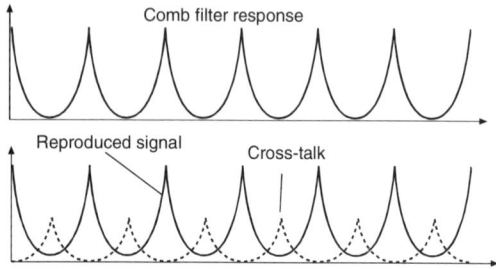

Figure 3.27 *Comb filter response*

correction is sometimes referred to as edge enhancement and it will be found in some form or other in all domestic video recorders. As mentioned earlier, some video recorders have an external sharpness control for the user to adjust to taste.

Comb filter cross-talk cancellation

In replay, as each video head scans its magnetic track it also picks up some FM carrier from an adjacent track. Chapter 2 discussed adjacent FM carrier rejection due to the azimuth differential in the video head gaps; this attenuation is not total and so some adjacent carrier is picked up. This carrier appears as crosstalk and is responsible for an effect that looks like symmetrically patterned noise, often referred to as an orange-peel effect, as it looks like the surface of an orange. This crosstalk of FM carrier will also contain other line frequency related noise components and all of this 'noise' can be reduced, if not eliminated, by a technique called 'Y comb filter'.

The basis of a comb filter is to shift one of the recorded FM carriers by 0.5 fh (7.8 kHz), half line frequency, in record, so that each magnetic track has a half line frequency shift of its FM carrier frequency spectrum with respect to its adjacent tracks.

Figure 3.27 shows the replayed signal and noise spectrum; the black line peaks are the replayed signal spectrum at line rate. The spectrum of the wanted signal is identical to that of the comb filter and passes through it. The dotted line is the adjacent track crosstalk; because of the 0.5fh shift in record, the noise spectrum is interleaved between the wanted signals and rejected by the comb filter.

Frequency shift of an FM carrier by 7.8 kHz causes its vector to rotate at a rate of 360°/2 lines, or 180°/line. By shifting the frequency of only one head FM track causes carriers on two adjacent tracks to rotate with respect to one another by 180°/line. This produces a line by line shift of 180° in the adjacent track carrier pick-up known as cross-talk, which is not fully cancelled by the head azimuth technique.

If we then used a filter which has a spectral response similar to the teeth of a comb, where the teeth are attenuators, then the filter could be laid on top of the replay spectrum and eliminate the crosstalk peaks. Hence the name 'comb filter', which interleaves within the replay spectrum and attenuates the dotted line crosstalk. The essential component of a line frequency comb filter is a one-line delay line.

Figure 3.28 illustrates the use of a delay line.

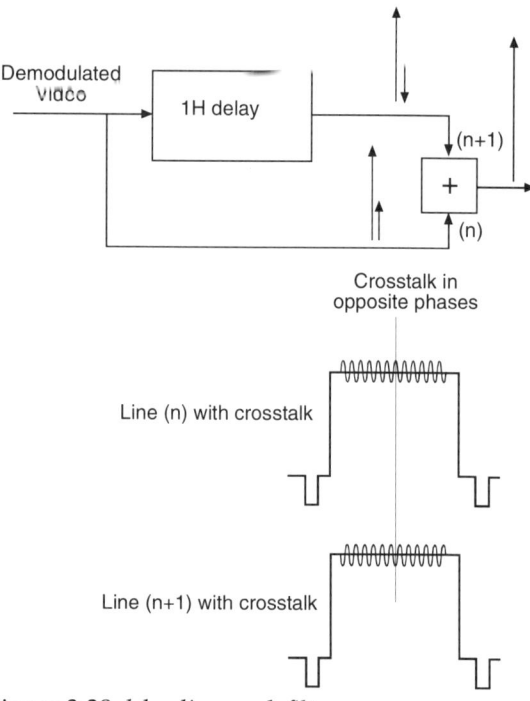

Figure 3.28 *delay line comb filter*

FM recording and playback

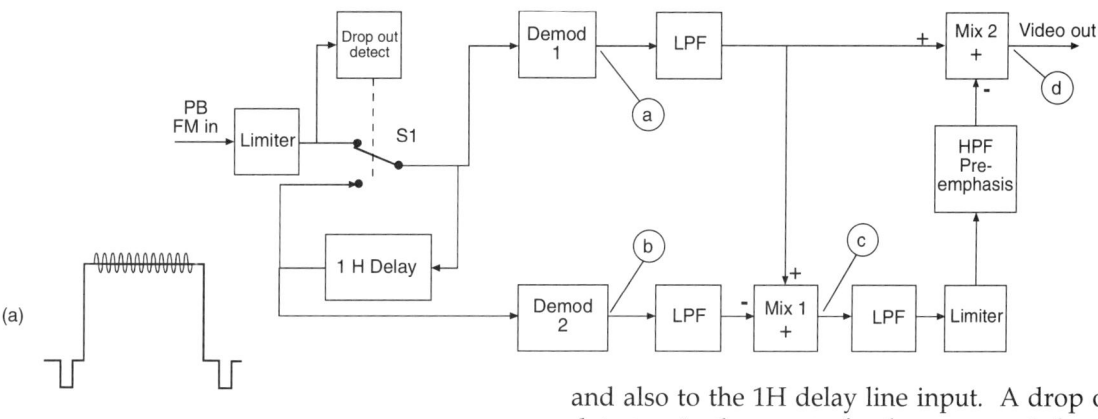

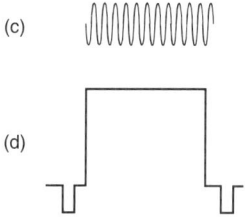

Figure 3.29 *Replay drop out compensator and comb filter*

In this example the wanted or main carrier components will add together and the crosstalk will cancel. Delayed line (n), added to direct line (n+1), will contain 'in phase' main carrier and anti-phase noise, therefore the crosstalk components will cancel.

The block diagram of a working comb filter is shown in Figure 3.29. In order to reduce manufacturing costs, only a single 64 μs delay line is used, which doubles up as the delay line for the drop out compensator as well as the comb filter. The drop out compensator is a standard technique as detailed earlier in this chapter.

Replayed FM carrier is passed through a limiter and via a switch S1 to the direct demodulator (1) and also to the 1H delay line input. A drop out detector, in the event of a drop out, switches S1 over and the delayed carrier is fed to the direct demodulator.

The comb filter does not operate during a drop out, which is very short, so we will consider the comb filter as it is during a normal replay.

Demodulator (1) is in the direct replay path and demodulator (2) is in the delayed path out of the delay line; they are both identical and are followed by low pass filters. Each of the demodulator outputs contains demodulated video, residual carrier and crosstalk; they are fed to the inputs of a subtractive mixer (1). In mixer (1) the delayed signal is subtracted from the direct signal. Video components and carrier components will be in phase and will, in general, cancel; some HF and transient video components may remain.

Crosstalk is in anti-phase on a line-by-line basis as previously shown; therefore it will be additive inside a subtractive mixer (c-(-c) = 2c). The output of mixer (1) will be mostly crosstalk. Figure 3.29(a) shows the directly demodulated signal and 3.29(b) shows the delayed and demodulated signal. Note that the video signal is 'in phase' and that the crosstalk upon it is in anti-phase. For clarity the residual carrier and transient video components have been omitted. When these two signals are subtracted the video cancels, leaving only crosstalk as shown in Figure 3.29(c).

If any two consecutive TV lines are compared it will be seen that they do not contain the same signal content. Indeed, many high frequency signals and transient peaks do not appear on two sequential lines. This results in a difference

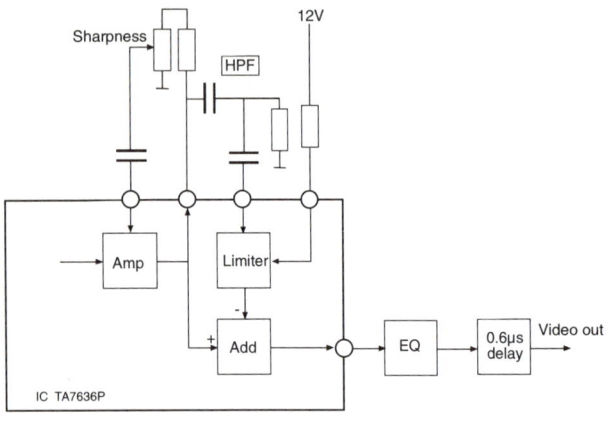

Figure 3.30 *Noise canceller and sharpness circuit*

component of the signal being produced in the subtractive mixer and appearing at the output of mixer (1). If this video component within the crosstalk were allowed to be subtracted from the directly demodulated video in mixer (2), then a severe reduction in vertical resolution would occur.

To prevent any detrimental effect upon the video signal the output of the subtractive mixer (1) is then passed through a low pass filter, which is also a de-emphasis circuit to reduce transients. It then passes through a limiter, which clips the high level signal component to the same level as the crosstalk; this will then reduce its adverse effect on vertical resolution.

Finally the output from the limiter is then fed through a high pass filter to effectively pre-emphasise it again in order to match it to the directly demodulated signal which is still in a state of pre-emphasis, before subtraction takes place.

Mixer (2) is then fed with two signals; the directly demodulated signal still containing crosstalk and the subtractive input which is crosstalk signal with only a small amount of differential video content. Mixer (2) is also a subtractive mixer and the crosstalk input is subtracted from the directly demodulated input. In Figure 3.29, (c) is subtracted from (a) leaving (d) at the output, thus removing the crosstalk from the video signal.

There is still random noise left in the signal and this is removed in a further noise canceller as shown in Figure 3.30.

A subtractive mixer is used again and it is fed with direct video and also the video signal via a high pass filter and limiter. The high pass filter is used to extract all of the high frequencies from the video signal, which will be both noise and HF transients, and the limiter clips the transients to the same level as the noise so that its output is basically noise. The noise is then subtracted from the video signal and cancels itself. R356 adjusts the amount of signal to the limiter and provides a range of adjustment from a soft picture to one that is very sharp and has overshoots and 'ghosting'. The delay line shown (0.6 µs) is just to equalise the luminance and chrominance delays in the signal processing.

Figure 3.31 illustrates the application of the 0.5 fh shift to the FM carrier. It is shown as part of a HA11724 FM modulator IC. The luminance signal, suitably clamped and clipped, is fed internally to the FM modulator. Sync tip carrier frequency of 3.8 MHz is set by the 1K potentiometer and temperature-compensated against drift. The drum flip-flop square wave is applied here via a potential divider of 1000:1 so its level at the FM modulator is very small. Its effect is to increase FM oscillator frequency by 7.8 kHz for Ch.2 head while having no effect on Ch.1 head frequency. If a preset potentiometer is provided to adjust flip-flop square wave amplitude do not adjust it without access to a spectrum analyser!

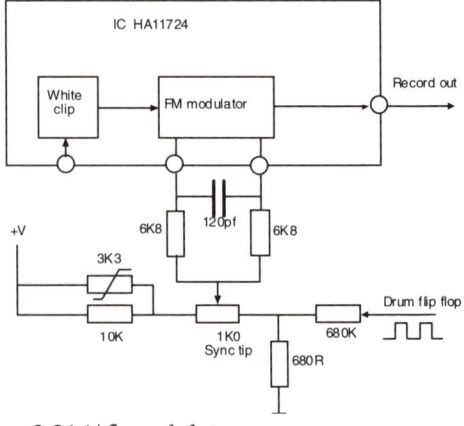

Figure 3.31 *½fh modulator*

Advanced crosstalk Y comb filter

Replay correction techniques have varied somewhat as manufacturers have settled down to a standard or system. Figure 3.32 shows a typical VHS approach that utilises the existing drop out compensator delay line to additionally function in the noise cancellation circuit. In analysing this diagram it is important to remember that the 7.8KHz shift introduced during record has the effect of arranging the crosstalk noise so that it is reversed in phase on each successive line during relay.

Referring to Figure 3.32, the replay FM signal level is stabilised by an AGC network before passing through the drop out switch to an equaliser. (Note that in later dual-speed recorders, long play/standard play compensation is provided at this point).

After equalisation the signal takes two paths, one to the demodulator and the other to the delay line. The delay line output feeds back to the drop out switch in the normal manner for a cyclic drop out compensator, and also feeds a second (delayed) demodulator in order to give two signal paths, direct and delayed. The main path is the direct one, in which the demodulated signal undergoes low-pass filtering and equalisation before passing as waveform (a) into an adder. The one-line-old signal emerging from the delay demodulator is also filtered and equalised, then applied to a differential amplifier as waveform (b). The differential amplifier's second input is waveform (d), the 'cleaned up' output video signal. Waveform (b) is subtracted from waveform (d) to leave just the spurious crosstalk component (c) which is amplified and amplitude limited. The signal emerging from the differential amplifier contains not only the crosstalk noise but also a large amount of unwanted video signal whose level must be reduced to that of the 'pure crosstalk' to prevent it from impairing the wanted direct signal excessively. Since waveform (c) consists of crosstalk only and is inverted with respect to the crosstalk riding on the 'direct' signal the two waveforms (a) and (c) cancel in the adder, whose output now consists of pure video signal (d). This technique is only effective for low frequency crosstalk; HF components are dealt with in the next block (equalisation and de-emphasis), as previously described in Figure 3.30. Where the circuit of Figure 3.32 is used in a dual speed machine a 0.5fh jump circuit is incorporated in the signal output path (see Chapter 8).

A later development on this theme is given in

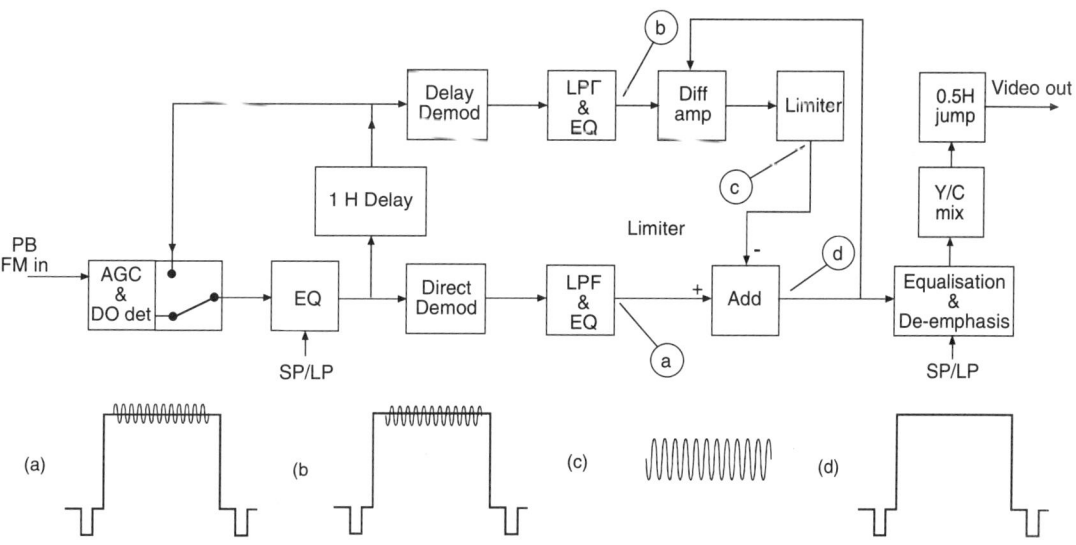

Figure 3.32 *Advanced FM demodulator, drop out compenator and crosstalk cancellation*

Video and Camcorder Servicing and Technology

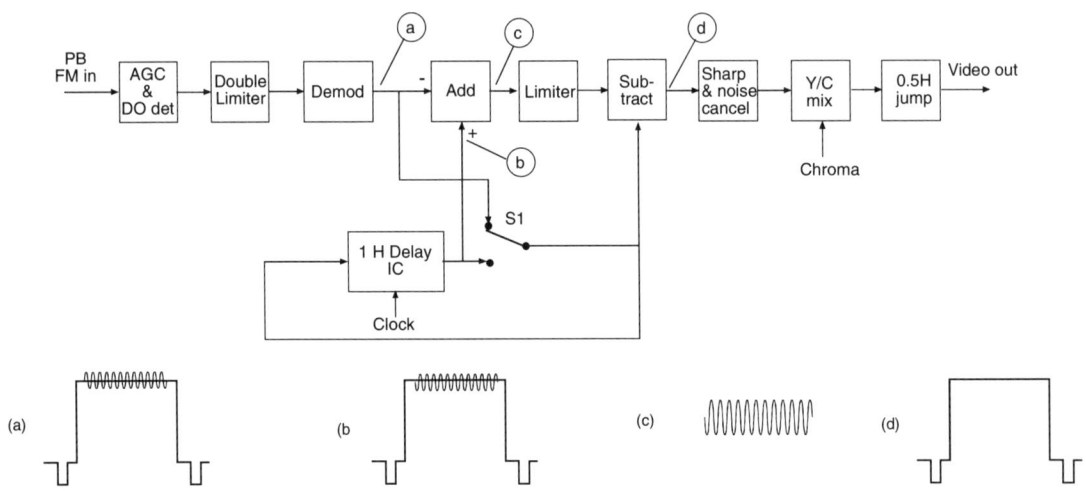

Figure 3.33 *Alternative demodulator, dropout compensator and noise cancel circuit*

Figure 3.33. Here the basic system is deceptively simple, but the circuit is in fact very advanced. After AGC and equalisation the FM carrier passes through a sophisticated double limiter before demodulation to base band composite video.

Let us first examine the drop out compensator. This operates at demodulated video base band, taking a signal from the demodulator via the drop out switch to the adder and to the delay system. The DOC changeover switch is driven by pulses from the drop out detector in the FM AGC section, so that whenever a break in carrier occurs the switch will toggle to 'patch' the hole with a one-line-old delayed signal. You may have noticed that the delay line operates on base band video rather than at FM carrier frequency. This cannot be done in a conventional glass delay line, and the device used here is a CCD (charge coupled device) 'bucket brigade' IC, to be described later.

Noise cancellation for the frequency-interleaved FM carrier also takes place via the delay IC. The 'direct' base band signal (a) and the delayed base band signal (b) are applied to a subtractor the output of which consists of crosstalk, (c). This is amplified and limited for application to a subtractor, wherein the crosstalk component in the main signal path is cancelled to render the clean composite video signal (d). The circuit incorporated in the replay path also reduces tape noise, rendering further noise reduction artifices unnecessary. After addition of the chrominance in the Y/C the CVBS signal is output via the 0.5fh jump circuit, used in LP video search (see Chapter 7, Long Play).

CCD delay line

In this context a charge coupled device is an IC with a certain number of CCD packages, or 'buckets' within it, see Figure 3.34. A clock-controlled gate samples the analogue input signal and the sample level at each clock pulse is passed to the first element. On each subsequent clock pulse the analogue sample is passed on from each bucket to the next on the right until it arrives at the output terminal. In this way the signal is broken down into quantised units, passed along the bucket brigade within the chip at a rate dependent on clock frequency, then reassembled at the output end.

In the 64 µs delay CCD there are 848.5 elements within the IC, and it is clocked at 13.3 MHz; clock frequency is derived as the third harmonic of the 4.433619 MHz colour crystal oscillator. The clock period is thus 75 ns (0.075 µs) x 848.5 = 63.8 µs. A further delay of 200 ns takes place in the filter/equaliser at the CCD output to make up 64 µs, exactly the period of one line.

FM recording and playback

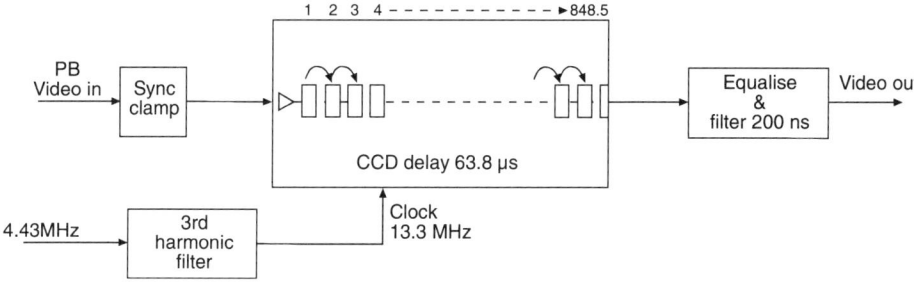

Figure 3.34 *CCD delay line*

VHS HQ technology

The addition of new technology to VHS has allowed the system to develop further to produce high-resolution pictures. This is achieved in three main ways and any manufacturer adopting at least two of them may describe their product as HQ.

The first method to improve resolution is by increasing the white clip level. In video recorders described earlier in this chapter, the white clip level as given in the service manuals was set to 60% above white level, given that sync tip to white level was 100%. In VHS HQ machines this is increased to 80% so the overshoot is 80% of the overall video signal level (i.e. set to a limit of 180%). You will know from this chapter that the white clip is set to prevent overshoots, created by pre-emphasis, from driving the FM modulator above 4.8 MHz and causing replay inversion (black streaks on white lettering). In order to allow the new peak white level of 80% without over modulation, a new compression circuit is incorporated. It is included after the pre-emphasis circuits, or as a non-linear control signal to the main pre-emphasis onto which the peak white clip is connected. The compressed overshoots are then expanded in replay so that the effect is similar to recording and replaying the video signal with peak white levels in excess of a maximum modulation frequency of 4.8 MHz without actually exceeding the maximum frequency.

The second method is to improve the detail of the picture by enhancing low level high frequency components and transitions from black to white and vice versa before the signal is recorded. In some recorders this is referred to as Dynamic Aperture Correction (DYCA).

The third method is to use the Y comb filter noise reduction system, described earlier, more effectively in a process called luminance line interpolation.

Dynamic aperture correction

Dynamic aperture correction is applied to the Y recording signal between the AGC circuits and the FM modulator circuits, in addition to this is the non-linear emphasis and white clip circuits.

Figure 3.35(a) shows the block diagram of the system, formed by a single IC with the exception of equaliser EQ2. Figure 3.35(b) shows waveforms, exaggerated for clarity.

The aim is to emphasise the transient edges of low level high frequency components before recording, such that on replay a signal more closely approximating the original is obtained. In previous models, edge recovery in replay has increased picture detail but at the expense of added edge noise: with DYCA pre-emphasis the same replay correction can be achieved, without edge noise. Note: edge noise shows up as busy little sparkles on all detail edges.

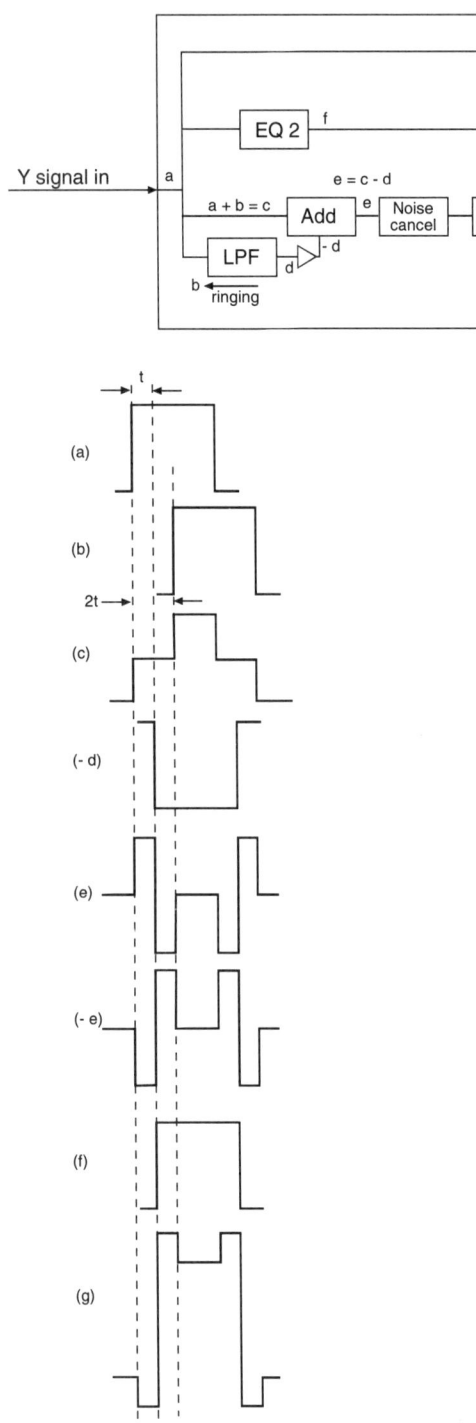

Figure 3.35 *Dynamic horizontal aperture correction and waveforms*

Waveform (a) is a low-level squares wave applied to the DYAC circuit and is split three ways; into an additive mixer, into equaliser EQ2 and into a low pass filter. The filter is not terminated at its output and this will cause input signals to reflect back, effectively creating a 'ringing' delay line of period 't' where t=100 ns. Signals at its output (d) are delayed by 't' whereas reflected signals back to the input (b) are delayed by 2t. The input to the first mixer (c) is, therefore, the sum of input (a) and reflected signal (b).

Output of the low pass filter (d) is inverted and added to (c) by the mixer, so we have (c - d = e).

Mixer output (e) is processed by a noise canceller to remove noise present during high input levels to prevent over-compensation, as the DYAC operates on low signal levels, the circuits have to ensure that high noise levels are not considered to be low level video.

After peak white and dark overshoot clip levels, the amount of dynamic aperture correction is controlled by adjusting the gain of the inverted phase of signal (e), before it is added to (f).

In practice the signal (e) undergoes a delay of 0.5t in the signal processing, making a total delay of 1.5t and, in order to compensate for this delay, signal (f) is passed through an equalising delay of 1.5t (150 ns) in EQ2 thus ensuring (e) and (f) are coincident. The strange looking result (g) is a signal that has pre-compensation. When de-emphasis is applied in replay the picture will look sharp and will not have the 'busy' looking edge noise that was so obvious in earlier models.

FM recording and playback

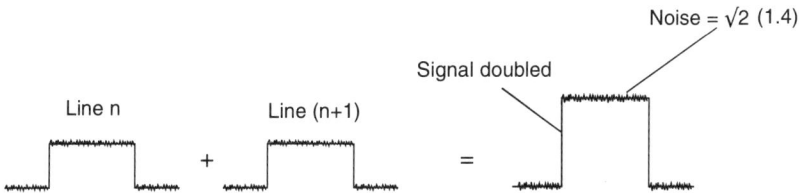

Figure 3.36 *Additive noise correlation*

Luminance line interpolation

The effect of introducing a 0.5*f*h shift into the FM carrier to reduce crosstalk from the lower frequency levels of adjacent track FM carriers, by the use of a Y comb filter is part of the VHS HQ specification. Further reduction of high frequency tape noise can be achieved by the extended use of the Y comb filter, by the addition of two consecutive TV lines, increasing the signal by a factor of 2 while decreasing noise by a factor of √2, as shown in Figure 3.36.

As noise correlation exists between consecutive lines further reduction can be achieved by extending the principle over a number of lines, but adding previous lines on a reducing level basis.

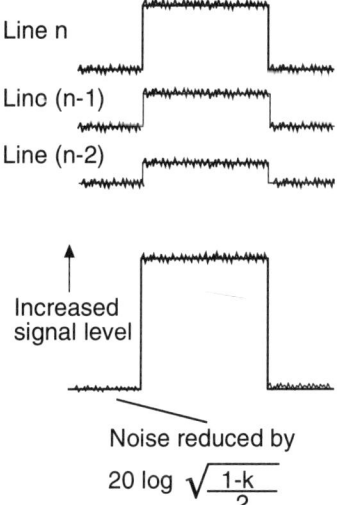

Figure 3.37 *Noise reduction by line interpolation over several lines*

The reduction in noise is then:

$$20 \log \sqrt{\frac{1-k}{2}}$$

Where k is the reduction in level of previous lines, and is 0.5 in VHS HQ.

In Figure 3.37 the first signal is n at full level, n-1 is added to it but at a level reduction of k = 0.5; n-2 is then added but reduced even further by another multiple of 0.5, hence 0.5 × 0.5 = 0.25 of n. Therefore when k=0.5 each delay line reduces the signal by half, so over a number of lines, n the reduction of the added signal is k^n.

The practical approach is to use a circulatory delay line system shown in Figure 3.38 where the output of the delay line is reduced by 0.5 each pass: equivalent to k=0.5. Each line is added into the circulatory system and each time it goes around it is reduced by k, eventually to an insignificant level, so a number of previous lines are being added to the input signal each of a reducing level. The overall effect is that the video components are additive and the random tape noise is subtractive, effectively forming a circulatory filter.

The output of the circulatory filter is a video signal with noise and high frequency components removed. It is also increased in level and so it is reduced in level by the (1-k) network to the same as the input video. In Add2 the 'clean' output of the circulatory filter is subtracted from the input video leaving a resultant signal that is mainly noise and transient video components. As the video transients will have adverse effects they are removed by amplification and limiting leaving a

Video and Camcorder Servicing and Technology

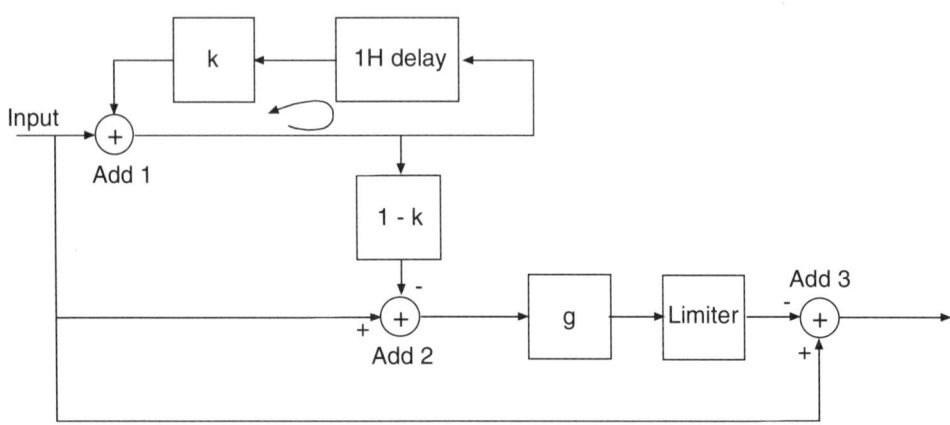

Figure 3.38 *Circulatory line interpolation*

signal that is a sum of random noise. Lastly, the noise is subtracted from the main video signal within Add3 thereby cleaning it.

There is a problem in that this circulatory system would reduce vertical resolution as shown in Figure 3.39 due to the addition of adjacent lines, especially when one line is black and the next contains a full white. In order to minimize loss of vertical resolution in playback the circulatory delay line is brought into use during recording as a vertical pre-emphasis circuit. The pre-emphasis system is shown in Figure 3.40, where the input signal is shown as a TV field signal with line sync pulses, indicating a white square on a black background. The circulatory system along with ADD1 produces the signal (a) with poor vertical resolution and increased level. This is reduced by attenuator ATT2 to the same level (b) as the input signal from which it is subtracted to result in waveform (c). (c) is amplified to (d) and limited to give (e) which is then added to the input wave to produce the pre-emphasised output.

In playback (Figure 3.41), the replay signal (note that its overall shape is not affected by the record/replay process) contains noise. Again the circulatory system produces an increased signal level with the noise removed (b) that is attenuated to (c) and subtracted from the input signal resulting in (d). (d) is then attenuated slightly, (e) and limited to give (f) which is then subtracted from the input signal resulting in an output signal closely resembling that of the original input signal.

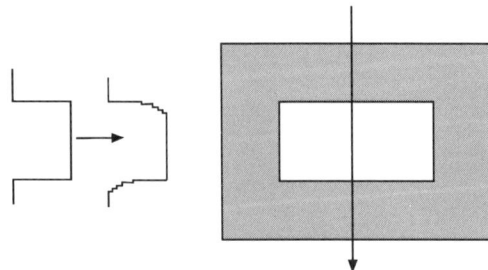

Figure 3.39 *Reduction in vertical resolution by line interpolation*

FM recording and playback

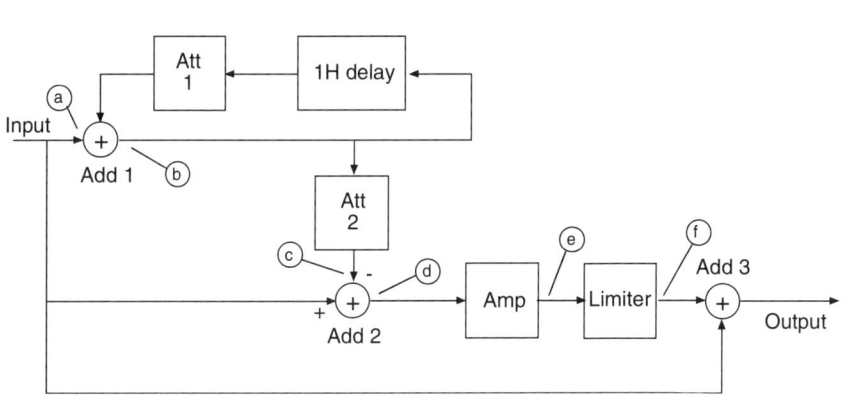
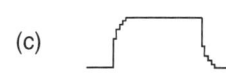

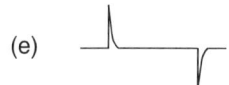
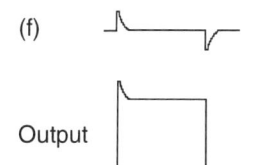
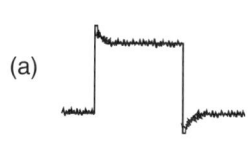

Figure 3.40 *Line interpolation used as a vertical pre-emphasis in record*

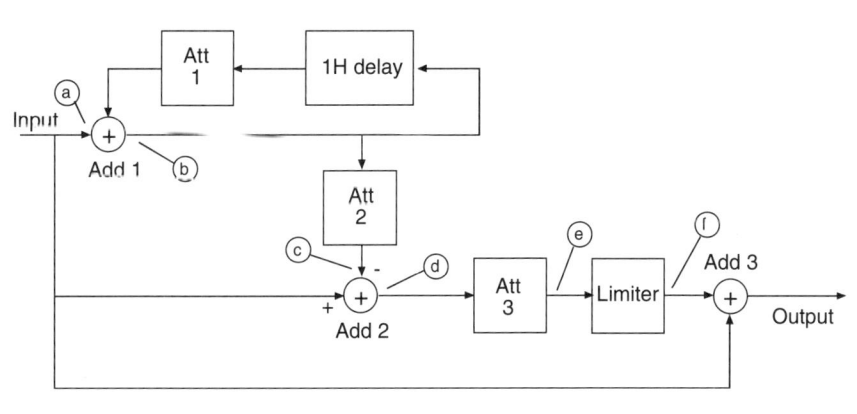
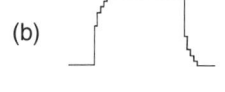
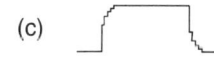
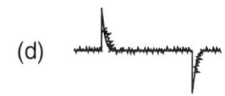
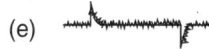

Figure 3.41 *Line interpolation used as a noise reduction circuit in playback*

49

Playback Y detail enhancer

Advances in digital precision delay lines and IC technology have led to the development of circuitry that improves the transient shape of low level, high frequency video signals components without pre-shoots or overshoots on high level, high frequency components.

This is shown in Figure 3.42 along with the waveforms; one row is for small amplitude signals and the other is for high amplitude signals.

In the first 200 ns band pass filter a Y step signal input is converted into a positive and negative component by 'ringing'. A following limiter allows small amplitude signals to pass unaltered but high amplitude signals are clipped (a). Next a 100 ns delay line acts as a differential circuit, low level signals become sinusoidal and high level signals are differentiated to sharp pulses due to the edges left after clipping (c). Via an alternate path waveform (d) is formed, this is of a similar shape for both small and large amplitude inputs.

Within the logic circuit there is signal measurement and selection; (c) and (d) are measured to detect small signals with the same polarity. From this the logic circuit outputs wide signals when the input is small and narrow signals when the input is of high amplitude, signal (e). A further delay circuit differentiates the logic output to produce (g) by subtracting (e) from (f). Note that (g) is wider with more pre-shoot and overshoot for small signals, and narrow negative and positive cycles for large amplitude signals.

Via a third path the original signal is delayed by 400 ns to match timing via the logic path before it is added to the differential signal (g). It can be seen from the Y output waveforms that small amplitude signals have steeper edges with pronounced pre-shoots and overshoots whereas high amplitude signals have a smoother step edge with no pre-shoots or overshoots at all.

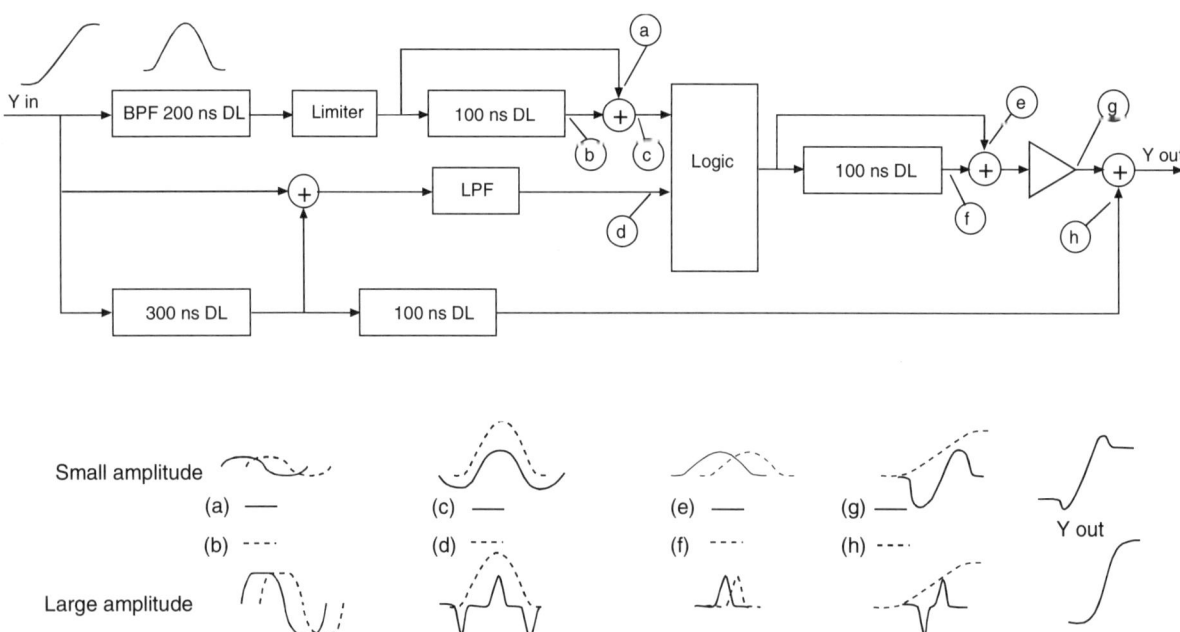

Figure 3.42 *Advanced logic controlled horizontal non linear pre-emphasis*

FM recording and playback

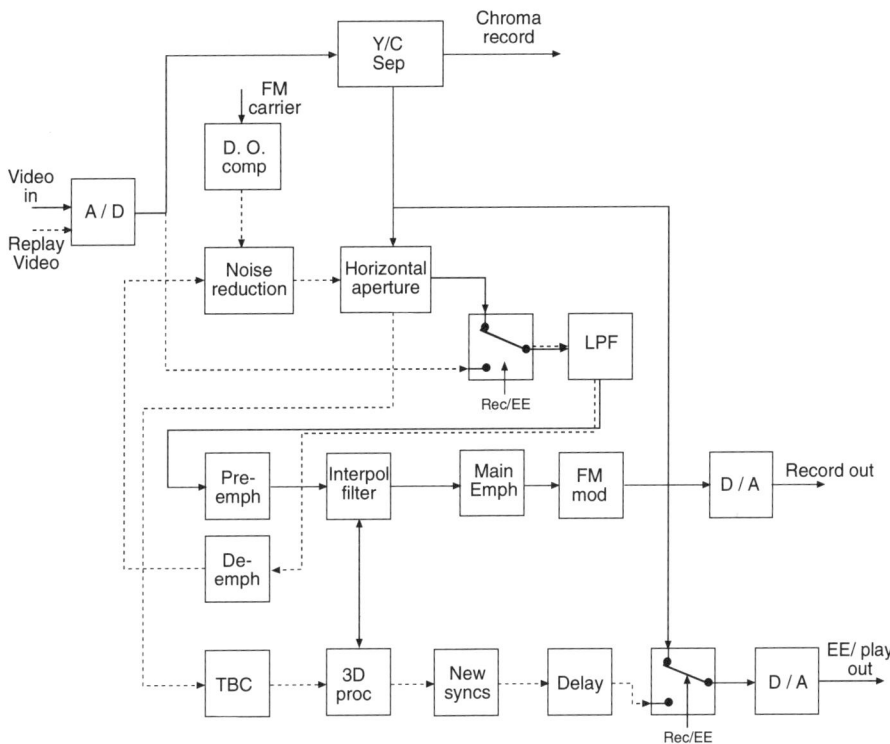

Figure 3.43 *Signal processing carried out in the digital domain*

Digital processing

There are many advantages to be gained by transferring the analogue signal into the digital domain. Many of the analogue signal processing functions, involving filters, delay lines, inductors and capacitors, can be carried out as mathematical calculations when digitised. This gives stability against production tolerances and ageing in the longer term. Production setting-up is reduced from lengthy adjustment of preset controls to a swift software download into an EEPROM memory. Picture quality is enhanced and does not deteriorate during the digital processing and is not subject to additional noise signal components, and it is well suited to dual VHS / S-VHS operation.

All of the record and replay signal processing can be done digitally, as shown in Figure 3.43. After analogue automatic gain control (AGC) to stabilise both amplitude and DC reference levels, the composite signal is digitised in an A/D converter to an 8-bit sample at a sampling rate of 14.2 MHz.

Chrominance and luminance are separated in a three-dimensional, 4-line delay comb filter to maintain bandwidth, minimise cross-talk and residual signal components. Chrominance at this point is still in the 4.433 MHz phase-modulated carrier format, but in the digital domain. It is passed directly to the colour recording and replay section (see p.132).

Recording path

The luminance Y component is first passed to the horizontal aperture processing circuits, and via a selector switch to the output D/A converter for analogue E/E monitoring, see Figure 3.43.

Within the horizontal aperture section the Y

51

signal is subjected to detail enhancement in the range of 1 to 2 MHz. Where the human eye is most sensitive to picture detail, it is not used as an aperture corrector in record. Next the Y signal passes through a low pass filter; it is a four element digital filter with a cut-off frequency of 3 MHz for normal VHS and 5 MHz for S-VHS.

Non-linear pre-emphasis is then applied to the Y signal; again the response is tailored to either normal or S-VHS. In both cases the amount of high frequency boost is both level- and frequency-dependent, more HF boost is applied to the lower level high frequency components and less to high level high frequency components.

A higher clock frequency is used in the main emphasis circuits and the interpolation filter is used to double the number of samples by interpolating two adjacent samples to make a third, and so on. The clock frequency and samples are doubled from 14.2 to 28.4 MHz. As with all magnetic recording the higher frequency components of the signal are boosted in record and reduced in playback, thereby reducing the off tape noise and improving the signal-to-noise ratio.

Frequency modulation takes place in the digital domain and as S-VHS carrier frequencies can extend up to 10 MHz then the clock frequency must be doubled. Normal VHS is sync tip at 3.8 MHz and peak white at 4.8 MHz whereas S-VHS is 5.4 – 7.0 MHz (see p.30 for frequency spectrum) The FM modulator characteristics are also changed between the VHS and S-VHS recording systems.

Playback

Playback signal routing is also shown in Figure 3.43 as a dashed line; the LPF is shared between both record and playback.

In playback, demodulated video Y signal is applied to the A/D converter, as in record the signal has to be stable in both amplitude and DC level. This time the signal is passed through the LPF first and as in record the response is switchable between 3 and 5 MHz.

After de-emphasis the Y signal is passed to a noise reduction circuit, the drop out compensator. Part of the FM carrier is present as an input to the drop out detector to generate pulses as drop out identification in order to fill in any gaps with the delayed signal as required.

Noise reduction is applied as a square root process in the same way as shown in Figure 3.37 earlier in this chapter. Also, the horizontal aperture correction is applied as a standard procedure as shown in Figure 3.35.

Now there are two very special signal processing stages that cannot be done as analogue signals. One is time base correction (TBC) and the other is velocity correction. The integrated circuits that are responsible for the TBC are also part of the three dimensional Y/C separation and digital Y signal noise reduction (YNR) and we shall look at these individually. Two-dimensional filters use line-by-line adaptation and three-dimensional filters use lines from different fields or frames.

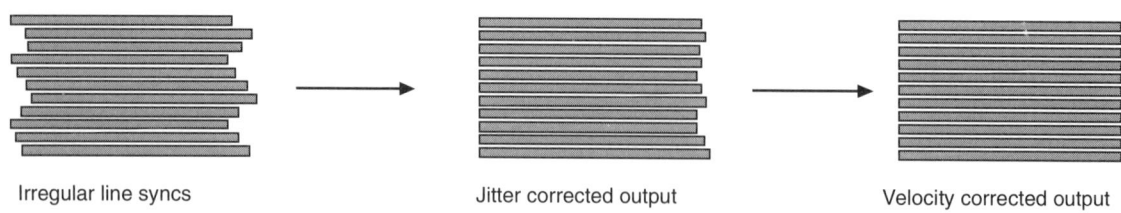

Figure 3.44 *Time base correction at the start and end of jagged TV lines*

FM recording and playback

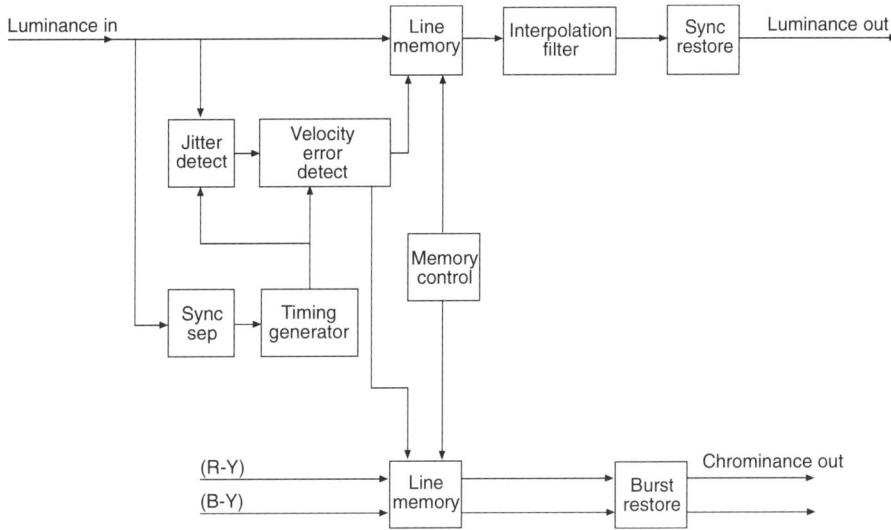

Figure 3.45 *Digital timebase and velocity error correction system*

Time base correction

Time base correction is used to correct line-by-line displacement. It is more common in analogue camcorders to straighten up verticals on the left-hand side of the picture, see Figure 3.44. Dragging the video tape through a mechanical system causes timing errors in line syncs, called jitter, it can be seen as jagged vertical edges due to line-by-line displacement.

Small changes in the velocity of the head drum cause the length of the line of a TV line to change, making the right-hand side of the picture jagged. These line displacements formed by jitter and velocity changes cause jagged vertical edges and can be seen on flag poles, columns or door and window frames. Playback jitter that causes these jagged displacements in the start of TV lines is present in the replayed line sync pulses and these can be used as an indicator to the amount of error.

Horizontal line syncs are stripped off the composite Y signal and used to synchronise the timing generator shown in Figure 3.45, by a phase-locked loop principle. Stable pulses from this timing generator are compared to replayed line syncs in the jitter detector to produce an error component.

Both luminance and chrominance signals are written into line memories for temporary storage.

Jitter error is combined with the result from the velocity error detector. These combined error components are used to read out the stored signals from line memories in a stable timing sequence. The start of the line read out is dependent upon the jitter errors. Velocity compensation is carried out by correcting the timing of each pixel data sample by interpolation over seven samples, three before and four after. If the data sample is in between clock pulses it may be missed and a new sample is constructed as an average from the batch of seven still present in the pixel data memory. As shown in Figure 3.46 the result is that each sample is now correlated with the clock and the line is restored to its correct length, straightening up the right-hand side verticals.

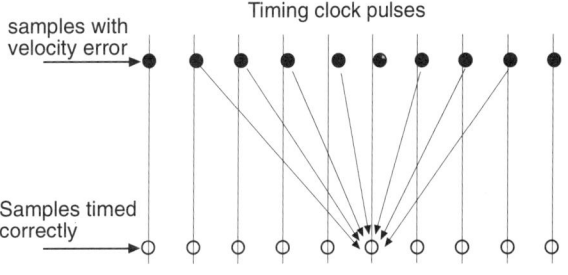

Figure 3.46 *Pixel data velocity correction*

Video and Camcorder Servicing and Technology

2D/3D filter

The Y/C separation block diagram is shown in Figure 3.47. There is a line comb filter, 3D frame memory filter, a 4-line memory, a band pass filter, a field memory, a movement detector and fast switches so that any filter can be used at any instant.

Filtered chrominance from the switch is subtracted from the composite signal leaving just luminance.

In the 2D filter a 4-line delay is used to compare the video content between 2 lines before and 2 lines after the current reference line period. Line (n-2) is compared to a previous line (n-4) and a successive line (n), the current line.

If there are no changes detected between these lines then the comb filter is used as a chroma filter. Where the comparator detects that there are difference between (n-2) and (n) or (n-4) then S1 changes over to filtering via the band pass filter only.

Both switches S1 and S2 are driven at clock or pixel rate. A picture can be optimised for best colour and minimal luma/chroma crosstalk by any of the three filter options being switched into use according to the prevailing conditions detected by the circuit and the high speed switching.

A 3D comb filter is similar to the 2D version except that the 2H delay is substituted by a frame, or field memory. Separation of the chroma signal is carried out using the memory as a delay line in the comb filter. It can only be utilised when there is no movement within the scene between successive frames, otherwise movement distortion occurs.

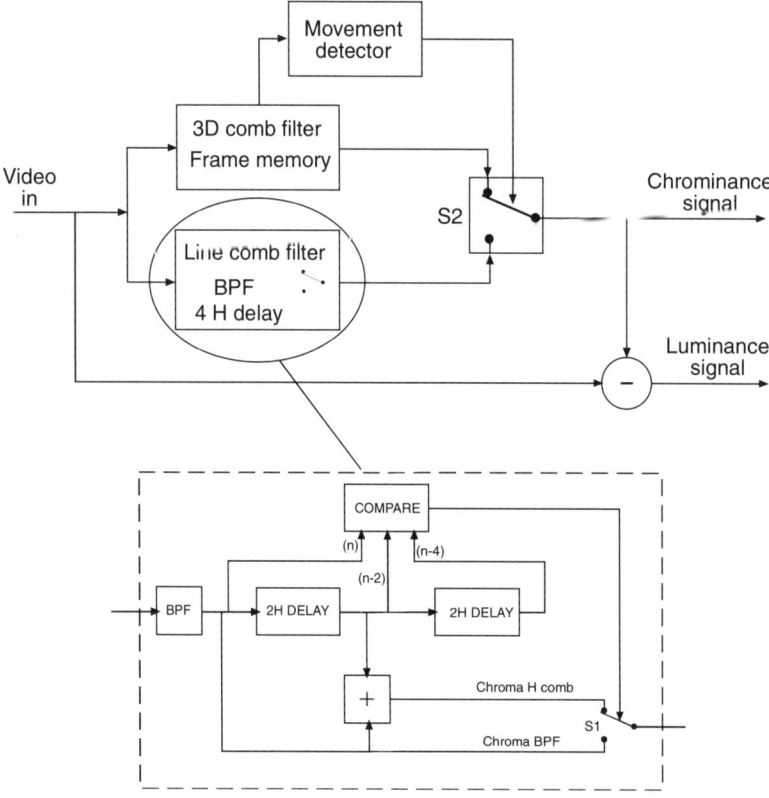

Figure 3.47 *2D/3D Y/C separation filter*

Digital YNR

In subtractor (Sub 1) in Figure 3.48 each line is compared to its counterpart in a previous frame. The difference signal is applied to a Hadamard transfer, which converts the signal from a voltage level to a frequency domain, along with the difference component from a previous line. In the Hadamard transfer section, four samples of each input are combined to eight samples. These are limited to reduce the effect of high frequency signal components within the difference signal, which is mainly noise. A second Hadamard transfer converts the frequencies back to voltage levels and via a level control they are subtracted from the luminance signal (Sub 2). Clean luminance is output and also sent to the frame memory. Consequently a clean signal is subtracted from the noisy signal in subtractor (Sub1), the difference component being noise and some signal HF components. Where there is movement between the two frames then the difference signal from subtractor (1) will contain a large degree of signal components. A movement detector confirms movement and turns the level control down reducing any adverse effects on the picture.

So what does it mean using these complex filters? If a scene has large areas of colour, say red, then a more comprehensive noise reduction takes place to reduce any adverse effects on the picture. Where there is movement or detail within the picture then the noise reduction needs to be modified to clean up edge noise but not to affect the detail in the picture by reducing the noise subtraction level.

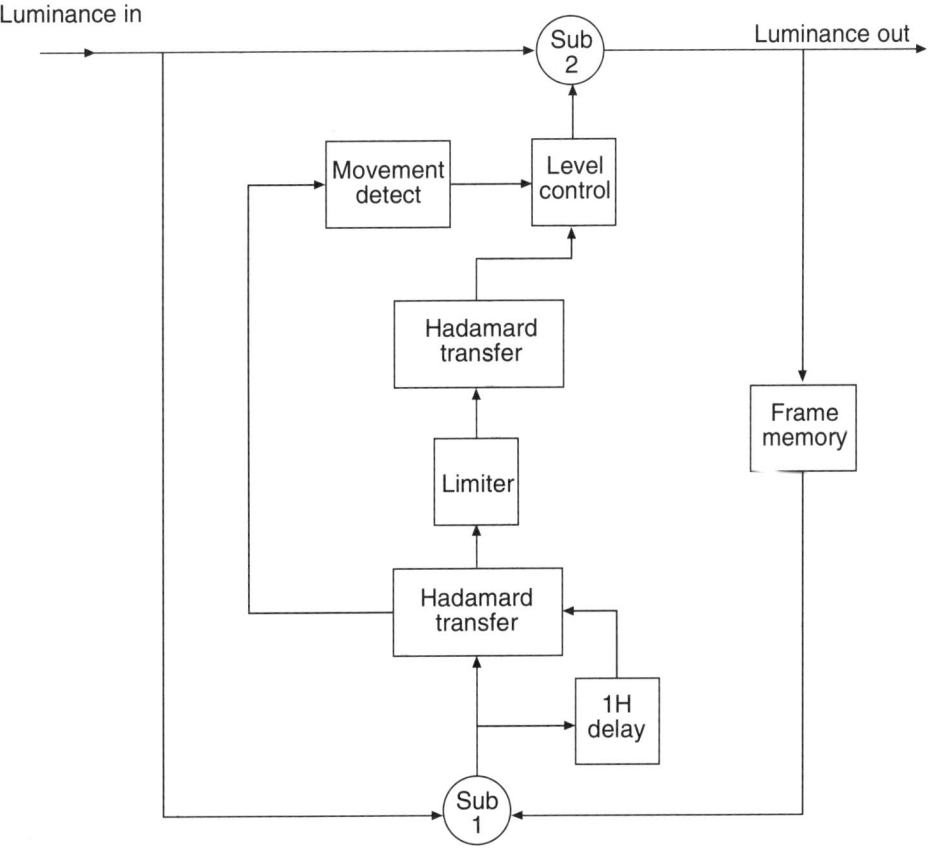

Figure 3.48 *Digital luminance dynamic noise reduction circuit*

Compatibility

The S-VHS recorders are dual VHS standard and will record and playback to both system specifications, standard VHS models in general will not playback the S-VHS specification tapes as the picture looks very black with white speckles. There are some VHS video recorders that will playback S-VHS but not record.

S-VHS ET

An improvement in coercivity of standard VHS tapes allows for wide band S-VHS recordings to be on standard tapes, with some adaptation.

S-VHS ET is possible only on video recorders with S-VHS capability as S-VHS ET is an extension to S-VHS technology. An S-VHS tape has a detection hole that identifies the tape as high coercivity and suitable for S-VHS recordings. S-VHS ET is a recording on tapes that do not have the identification such as standard tapes.

There are a number of changes between S-VHS and S-VHS ET that have to be taken into account. First, that it must be suitable for all tapes, and second, that it will playback on any video recorder with S-VHS playback capability.

Technical changes

1). Changes to the white clip level, that is: VHS +60%, VHS HQ +120%, S-VHS ET +190% and SVHS +210%. These figures are subject to variation between manufacturers and models.

2). Changes to the main emphasis and pre-emphasis characteristics to match the frequency response of VHS tapes.

3). Changes to the FM carrier recording level and chroma recording level. The video recorder may automatically determine this prior to recordings (see Chapter 14, Special features).

Record and replay fault-finding

The most common fault symptom in a video recorder is poor picture quality and it will take some years of experience to define the reason immediately and go to the circuits involved. Most of the time it will be a process of logical elimination of possible fault areas to end up with the correct solution. For 'no picture' read also 'distorted picture'.

Three tapes are required for test purposes. One is a known good tape with some programme material upon it that is known to be of good quality. The second tape is a good quality blank, final testing tape without defects or tape damage. The third is a tape which is worn and has lots of drop out on it to check the drop out compensator circuits.

The main items of test equipment are: a 50 MHz oscilloscope, a test pattern generator, and a manufacturer's high grade reference test tape for final setting up.

If we start from square one (Figure 3.49) we have a video recorder that is powered but has no picture or one that is distorted, either E/E (electric to electric) or replay. Audio may or may not be present in the replay. This is because some TVs mute the audio in the absence of line sync pulse or there may not actually be any audio at all!

First the power supply rails must be checked in both record and playback, as some of them are switched on in E/E whilst others come on only in record after threading is completed. If the power switching supplies are correct then the oscilloscope is applied to the video output socket. If there is no E/E video, return to the input video and AGC or to the IF and tuner. If there is video on the output connector, proceed to the RF modulator and its power supply. Check if audio is present. If it is not, then go back to the tuner and IF. If audio is present then proceed to the video input of the luminance processing circuits.

Second, check the RF modulator. Check for video and audio signals on its inputs and the power rail. Replace the RF modulator if all signals are correct on its inputs but there is no picture on the TV and the tuning is correct.

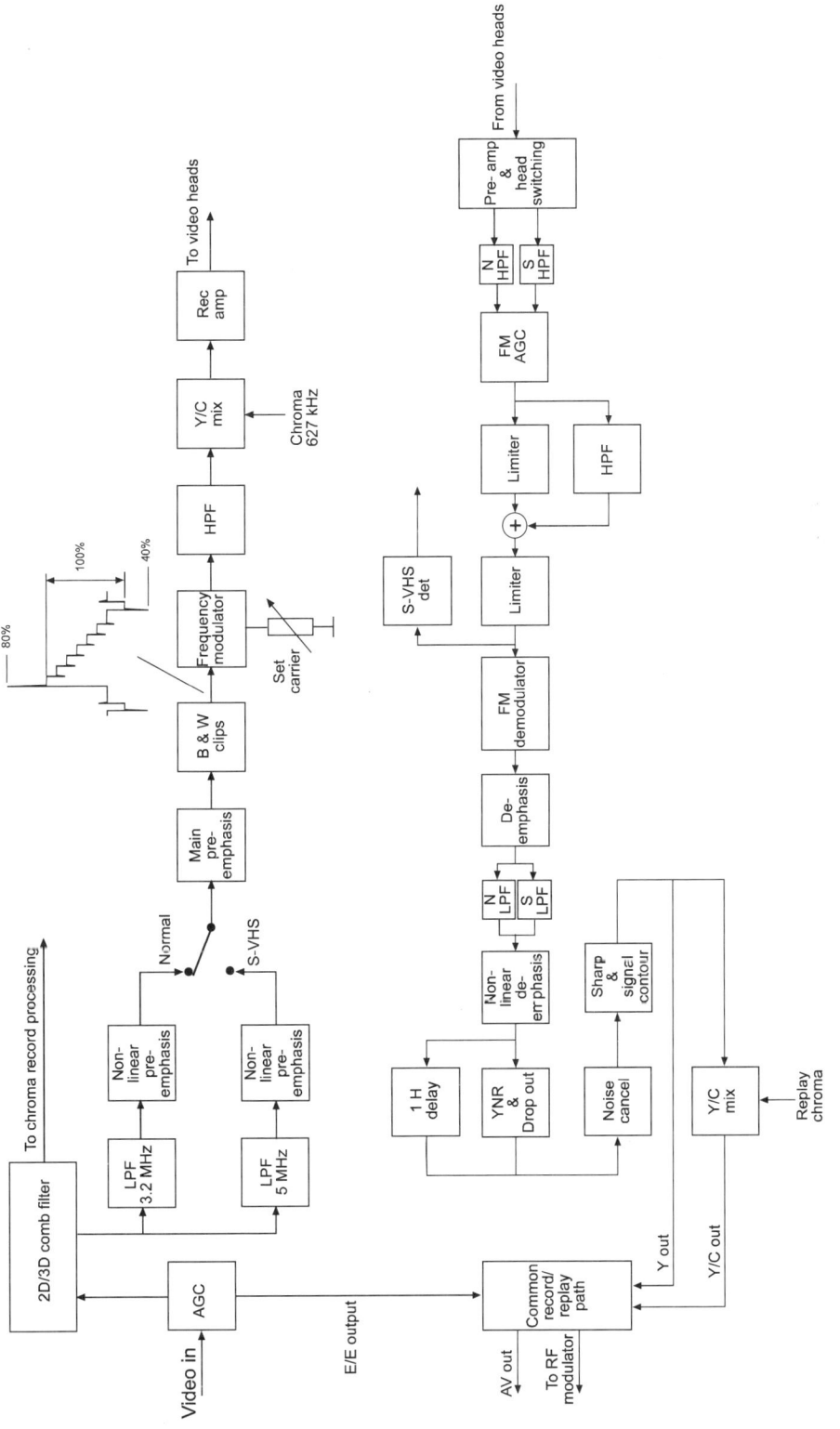

Figure 3.49 Typical S-VHS record and playback diagram

Next, check the tuner and IF. If video and/or audio signals are not present proceed to check these circuits as if they were a TV.

Where E/E signals are present check the replay with a known good tape, a good replay may indicate a recording fault. Determine if the fault is in the record or replay circuits, or if it is a video head problem.

The absence of E/E and replay pictures will confirm a fault in the video output processing circuits. A replay fault can be cross-checked if a recording is made and then confirmed by a good replay on a second video recorder.

Poor replay of the recorder's own recordings whilst a known good tape replays reasonably well indicates head-to-tape problems. Remember that the transfer of signals is head to tape and then back from tape to heads, giving twice the error for the video recorder's own recordings. This is the case in early stages of video head wear or deck friction/tape tension problems.

Record

A fault in the FM record processing circuits will at best distort the picture and at worst give no picture, whereas replays of a known good tape may be good. Poor pictures will usually lack contrast, be milky in appearance, insipid and smeary. Check first the power rails that come on only in record.

Then check with an oscilloscope through the record processing circuits. If the E/E signal is correct then it is a fair bet that the AGC circuits are operational, except that the AGC level on some machines is also the 'deviation' control. Do not twiddle this control; just monitor the video signal level. Then confirm with the service manual that the expected signal level is approximately correct and then proceed with checking the remaining record circuits. Also do not adjust the 'set carrier' control unless you either suspect foul play by the phantom twiddler or you have enough experience and equipment to set them correctly.

Using the oscilloscope follow the signal through the 2D/3D circuits. They may be covered by metal screening so check only the input and output signals, then the next set of low pass filters and the non-linear pre-emphasis. A fault here can affect either VHS or S-VHS, not necessarily both. The signal level may drop here to a few hundred millivolts. Continue to follow the signal through the main emphasis up to the peak white and black clip test point, upon which the service manual will provide an indication of the signal level.

The clip circuits can be easily set by adjusting the gain of the scope vertical calibration so that the difference between sync tip (or sometimes black level check with manual) and peak white is 100% as a reference (two graticules). Peak white clip is set to clip at around 80% (nearly two squares) of this level above peak white so that the clip level is 180% above black level. Black or dark clip level is set at 40% (nearly one square) below the sync tip, note that this will vary so check the actual values in the service manual, as they do change slightly between manufacturers and are very different for S-VHS.

The next check point is out of the FM modulator, through a high pass filter and recording amplifiers to the video heads. Drive to the video heads can be expected to be 2–3V p/p and can be set from the service manual instructions.

It is possible to see the video signal modulation pattern within the FM carrier at field rate to confirm that the FM modulator is working, as there are minor amplitude variations corresponding to field syncs and peak whites. If all the above are correct then recording should be taking place, unless the video heads or rotary transformer are faulty.

Replay

A fault in the replay circuits will degrade the picture with respect to background noise, noisy or 'busy' active parts on verticals, white spots, black spots, very small herringbone patterns, or an orange-peel effect of small wiggles. Large horizontal noise bars within the picture area consisting of picture break-up to black and white streaks will be due to mechanical/tape path errors.

Picture jumping vertically is due to loss of field syncs in the TV. This is due to 'off screen' tracking errors at the start or end of video head scan and indicates tape path mis-alignment, in this case the FM replay signal must be checked for rectangular symmetry.

White spots with comet tails to the right all over the picture are more likely to be due to static, possibly the discharge brush contacts either above the video head drum or beneath the deck not making proper earthing to ground. An increase in random flashes around the picture area could be as much due to head wear as failure of the drop out compensator or magnetised oxide particles on the pinch roller, particularly if they increase in quantity for every pass of the tape.

In any replay situation the quality of the FM carrier out of the preamp should be checked. A test point is provided for this. The oscilloscope should be triggered by the RF 25Hz head switching signal (flip-flop) for a stable display; it will be high for one head and low for the other. Compare the outputs of the two heads. Each should be rectangular and the start and finish edges for any head should rise and fall in parallel with each other when the tracking control is swung around its centre point. Failure of the signal to correspond to this test will be due to mechanical adjustments of the tape path. Irregular drop off in level at the start of a video head FM output will be due to the entry tape guide adjustment, whereas a problem with the level at the finish side will be due to the exit guide.

A low level for one head which does not rise to the level of the other when the tracking control is moved accompanied by ragged and irregular top and bottom edges is due to head failure or incorrect preamp adjustment. Do not adjust the preamp 'Q' preset capacitor or resistor unless you suspect that they have been incorrectly adjusted, or you have a manufacturer's test tape with an RF sweep on it and the experience to set the controls correctly. Incorrect adjustment of the 'Q' circuits will cause 25 Hz flashing on the picture. Specifically it can cause flashing on different frequency gratings on a test pattern or the verticals of crosshatch sections. Some video recorders do not have these controls as they are set by fixed components, making life easier.

Follow the FM replay carrier through the limiter circuits. It will become a bolder signal with less ragged edges at the top and bottom. Some video recorders will have the drop out compensator around the FM demodulator. Other later and higher quality models use the Y noise reduction circuit delay line for drop out compensation. The FM carrier can be measured in and out of the delay line. There is a fall in level out of a glass delay line but not quite so if a CCD delay line is used. With a good oscilloscope, fast drop out switching pulses from the drop out detector can be checked. Visible drop outs can occur if a tape is very worn and there is too much drop out for the compensator to cope with. It is possible with some video recorders to inhibit the drop out compensator by temporarily shorting out the input to the delay line. Ensure that other components cannot be damaged before doing this! Then check if the drop out increases, particularly around the head crossover point. It the drop out compensator is working the number of drop out spots will increase considerably on your worn test tape.

The FM signal is very fuzzy out of the demodulator, and do not be surprised if it is upside down on some models. After the de-emphasis and low pass filter a reasonable video signal waveform can be seen. It can be tracked through the noise reduction circuits to the video out test point, and can be confirmed that the replay path is operational. If the picture is monochrome, check that the Y/C switch is in the composite position on S-VHS recorders, otherwise a colour fault exists.

In S-VHS video recorders there may be a separate de-emphasis circuit and a failure here affects only S-VHS playback.

Figure 3.49 is a generalised record/ playback path to assist with an over-view for logical signal tracing.

4
Motor control and servomechanisms

The DC motor

A DC motor will run at a constant speed provided that it has a constant load. If the load is increased then the motor will slow down. Power supplied to a motor is limited by the value of the applied voltage, whereas the current consumed is dependent upon motor losses and the mechanical load. Motor losses, friction and magnetic losses are small compared to that of an applied load. If the mechanical load is increased then the motor current will increase but the speed falls. A variable mechanical load will result in variations in both the motor speed and power supply current.

A graph drawn for the load/speed characteristic of a DC motor can be shown to be a sloping straight line. An increase in the load, as shown in Figure 4.1, will cause a reduction in motor speed. A change in the load on the horizontal axis will cause a corresponding change in speed on the vertical axis, due to the slope of the load/speed graph. An ideal characteristic would be a horizontal straight line where the speed is constant irrespective of changes in the load. Increasing the voltage to the motor under load will increase the power to the motor. Consequently, both the speed and current consumption will also increase. The motor can be compensated for the reduction in speed due to a constant mechanical load, by increasing the supply voltage until the original speed is restored. A variable mechanical load would require a varying supply voltage to compensate. To do this some measurement must take place of the applied load or its effect. It is possible to determine that if the speed, the back emf, or the current consumed can be measured then the result of this measurement can be used to control the power applied to the motor. It would then be possible to compensate for the applied load and provide the means to run the motor at a constant speed. This will closely approximate the horizontal line graph of constant speed in Figure 4.1.

Methods of measurement can be: optical or magnetic for the speed of rotation; direct measurement of the voltage across the motor for back emf; and the voltage across a low value load resistor for the current consumed.

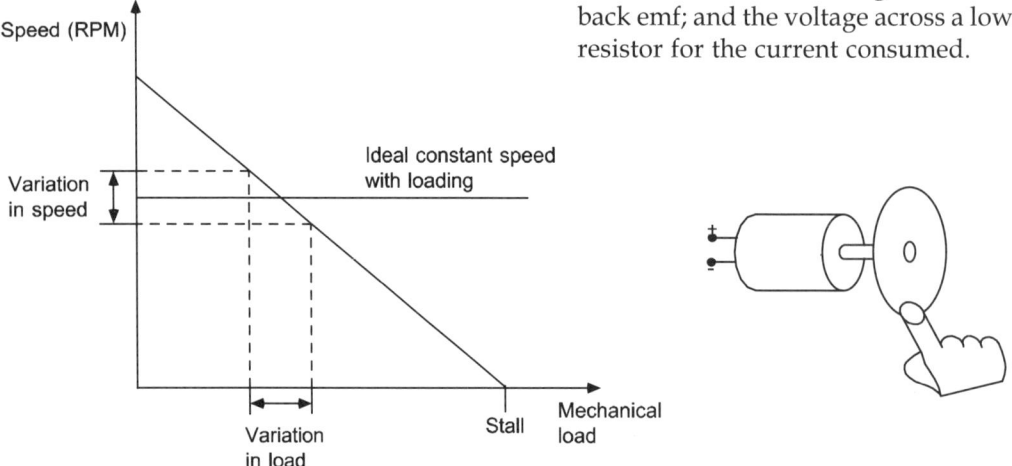

Figure 4.1 *Graph of the variation of motor speed against load*

Feedback speed control

Figure 4.2 shows a simple arrangement that can measure the motor rotation speed and convert it to a control voltage. This is a basic frequency to voltage converter system (f-v), as the motor speed is measured by the number (frequency) of rotations of the flywheel in a given period. A magnet mounted upon the flywheel produces a pulse within the sensor as it passes by. As the signal (F) is a train of pulses, the quantity of these pulses occurring per second is the frequency of rotation. A variable voltage (A) that is produced at the output of the (f-v) converter is proportional to the motor flywheel rotation speed. Within the comparator circuit this signal voltage from the motor flywheel (f-v) conversion is compared to a fixed reference voltage. The motor speed voltage varying, up or down, against the fixed reference voltage will cause corresponding output voltage changes from the comparator. This difference voltage (B) from the comparator is amplified to control the drive voltage to the motor and, being in opposite phase, it will compensate for the original speed variation.

Operation of this servo loop system is very simple. If a load is applied to the motor the speed slows. In consequence the voltage output of the (f-v) circuit (A) will fall against the fixed voltage in the comparator. This voltage fall will cause the output of the comparator (B) to increase, as it is an inverter. An increase in drive voltage to the motor from the comparator will cause the motor to increase its speed, in compensation of the fall off in speed due to the applied load. So we have a constant speed motor drive system using negative feedback that can compensate for a change due to a constant applied load.

We live in a real world, the load will not be constant, and it will vary. A consideration has to be given to rate of change of the mechanical load, whether it varies rapidly or slowly. A load that varies slowly has a greater effect upon the motor speed than one that varies rapidly. It is because of the inertia of the motor that rapid speed changes are 'smoothed out'. It is particularly so when the motor is fitted with a large and heavy flywheel; high frequency speed variations are reduced considerably by the inertia of a flywheel. A disadvantage is that the motor may be too sluggish. It will take a relatively long time to gain speed when started and it is difficult to stop. Mechanical parts, such as the shaft and bearings are subject to higher wear stresses with heavy flywheels and... the motor not very energy efficient. A light flywheel is normally fitted and inertia or damping control is applied electronically as we shall see.

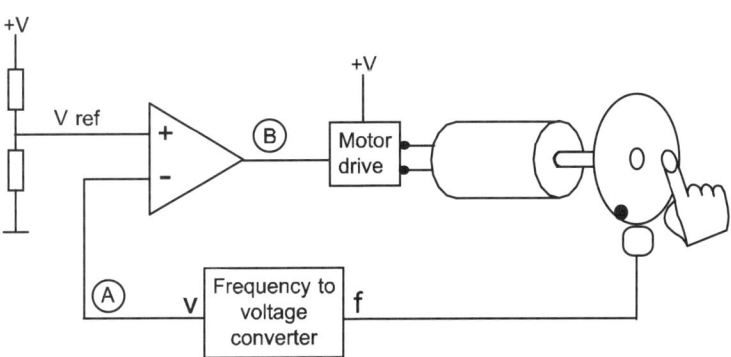

Figure 4.2 *Circuit of speed control of a motor using negative feedback*

Video and Camcorder Servicing and Technology

Characteristics of feedback servo control

In this example a single load is applied to the motor for a very short time, in other words it is a pulse load. A timing diagram is illustrated in Figure 4.3. This type of load is synonymous with driving over a bump in the road. It allows the 'step response' of the servo to be analysed. When the load is applied to the motor there is an immediate drop in speed, and recovery takes a bit longer. A reduction in speed causes the output (A) of the f-v converter to fall rapidly. The motor power is increased by a complementary rise in drive voltage (B) to the motor. As the motor picks up speed the pulse load is removed. A gradual decrease in speed back to the original value takes place. In response to the motor speed slowing the drive voltage is reduced in the form of a slope.

In a closed loop, such as we have just seen, the gain of the comparator amplifier is very important. If the gain were to be increased then the drive voltage to the motor would be too high. An effect of this is to over compensate the motor by over driving to such an extent, that instead of gradually returning to its original speed the motor has been given too much power and overshoots. Drive then has to be further reduced to compensate for this extra power and this gives rise to over compensation. This shows up as a 'bump' in the curve below the original drive voltage level, as illustrated in Figure 4.4.

By increasing the gain even further the servo loop would oscillate following the 'step load'. These oscillations cause the motor speed to vary rapidly out of control before gradually dying down, as shown in Figure 4.5. This is termed 'hunting'. Note that the oscillations from the (f-v) converter are in antiphase to the motor drive voltage. If these were 'in phase' then the drive

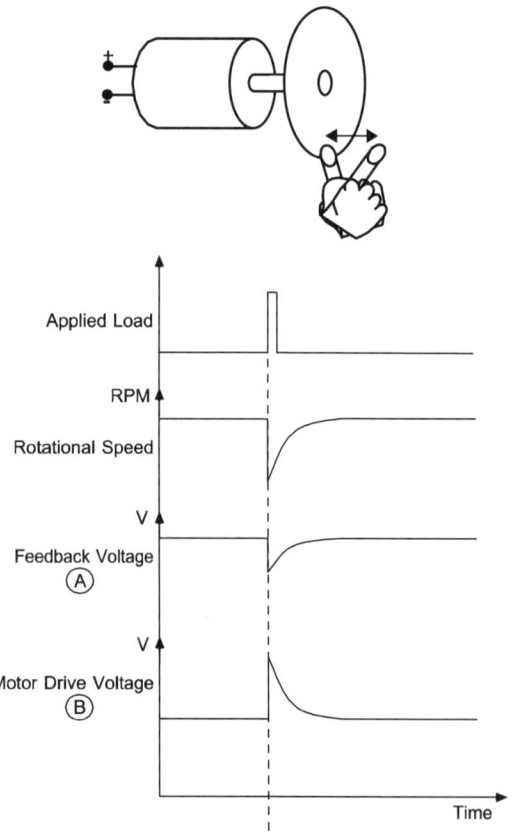

Figure 4.3 *Pulse response of a motor speed control circuit*

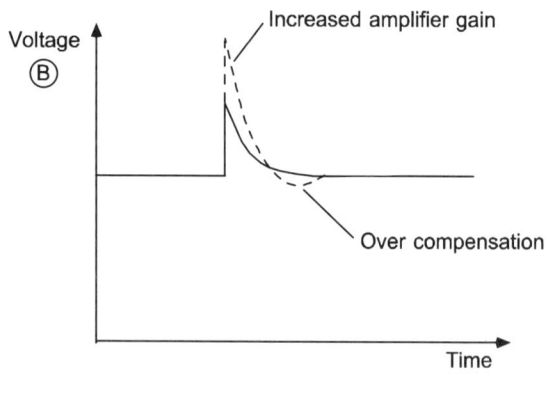

Figure 4.4 *Effect of increased amplifier gain on pulse response*

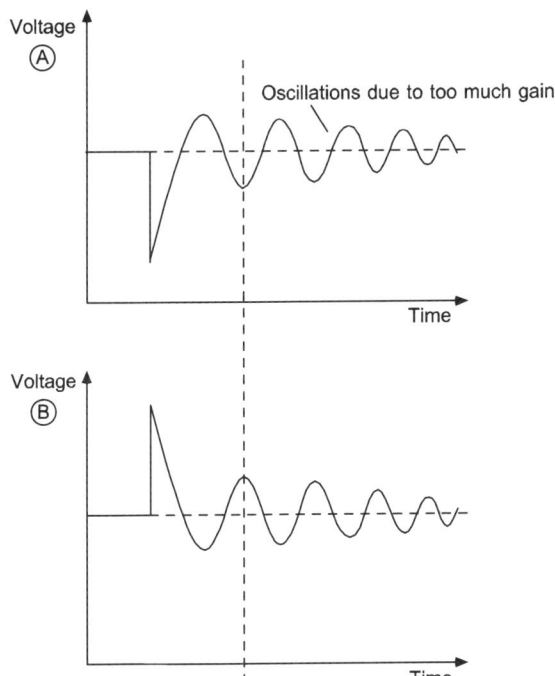

Figure 4.5 *Oscillations due to excessive circuit gain following a pulse load*

voltage (B). A consequence of this phase shift is to introduce a delay between the motor slowing down and the extra power to correct for it, with the overall effect of reducing loop control. It can be seen from the additional waveform in Figure 4.6 that at higher error frequencies (t) is the equivalent of 180° delay. The drive voltage is inverted to be 'in phase' with the speed error, resulting in the drive to the motor aiding the error and not correcting it. In other words the drive voltage (B) becomes positive feedback so that the servo becomes unstable. Gain/frequency characteristics have to be added to the error amplifier to reduce the loop gain to zero at the frequencies where loop phase shift becomes 180°. This prevents unwanted oscillation of the servo loop.

For example, consider a motor running at 1000 rpm. The speed of the motor varies between 950 rpm and 1050 rpm five times in a second, this means that the servo has a dynamic oscillation of 5 Hz. Inserting a 5 Hz filter, as a high pass filter in the error amplifier will prevent this happening and stabilise the motor at 1000 rpm.

would become 'positive feedback' and drive the servo further out of control. Servo loop gain cannot be increased to improve response without the danger of hunting occurring.

The degree of correction in a servo discussed so far is more effective at low frequency error rates than high frequency errors. Flywheel inertia is more effective for stabilising high frequency errors along with mechanical accuracy of bearings to reduce vibrations. A heavy flywheel cannot always be utilised, particularly in a servo where rapid changes are required; for example when a motor is required to accelerate from stop to running speed in less than a second or so.

A servo loop has an inherent time lag. An (f-v) converter has a time lag between input and output. The amplifier has a small delay and the time difference between increased drive voltage and subsequent increased speed of a motor becomes significant. All of these factors accumulate to a loop delay of (t), as shown in Figure 4.6. At low error rates (t) delays the correction in phase between the motor speed change and the correction

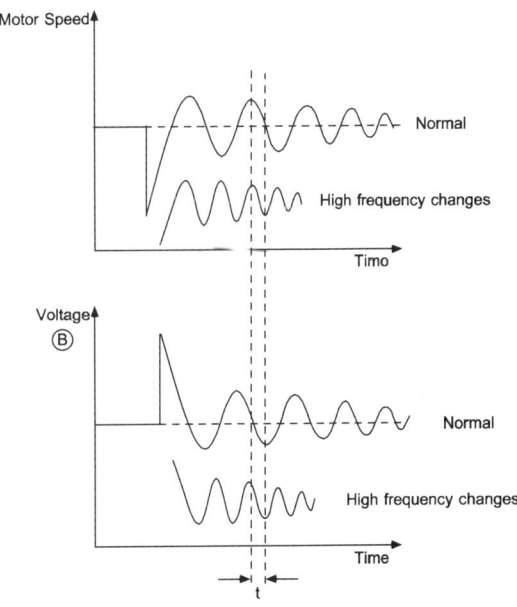

Figure 4.6 *Phase shift to positive feedback at higher frequencies*

Bode diagram and phase correction

A Bode diagram shows the relationship between servo loop gain and loop phase characteristics over the frequency response range. Two graphs relate the gain/frequency characteristic and the phase/frequency characteristic. By comparing the two graphs at corresponding frequencies both the gain and phase can be evaluated.

Figure 4.7(a) illustrates a Bode diagram of the servo in Figure 4.2. At a frequency (f1) where phase is delayed by -180° due to the accumulated 'lag' the corresponding gain is greater than zero. If the servo is disrupted by a pulse load, instability results in the form of 'hunting'. It can be corrected by adding a 'phase lag' circuit, better known as a low pass filter. The low pass filter reduces loop gain to less than zero at (f) with only a small amount of effect on loop phase, as shown by the dotted lines.

Another way is to add a phase lead circuit to compensate for the phase lag at a given frequency. The effect of such a circuit is to raise the frequencies, at which -180° phase lag occurs, beyond the point where comparable gain is below zero. This is shown in Figure 4.7(b). Without a high pass filter -180° phase shift occurs at (f1) where the corresponding gain is greater than zero. By introducing the high pass filter into the servo loop the frequency response is extended to (f2) where gain is less than zero when the phase lag is -180°; the loop phase response is also extended to (f2) and the servo is stabilised.

Damping

Within the DC control of the servo loop, after the comparator and before the motor drive circuits, are various phase/frequency compensation components. These form either high or low pass filters. High pass filters have a very low frequency cut-off to compensate for hunting; they cut out low frequencies that are just a few hertz. At the higher error frequencies rapid speed changes are

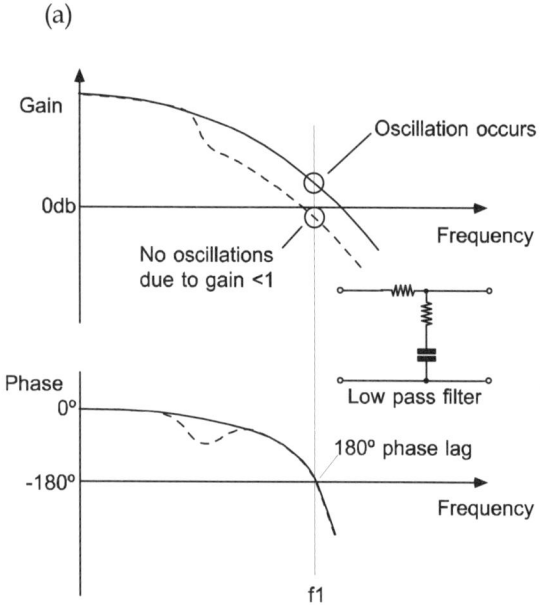

Figure 4.7(a) *Stabilisation of gain/phase response with a low pass filter*

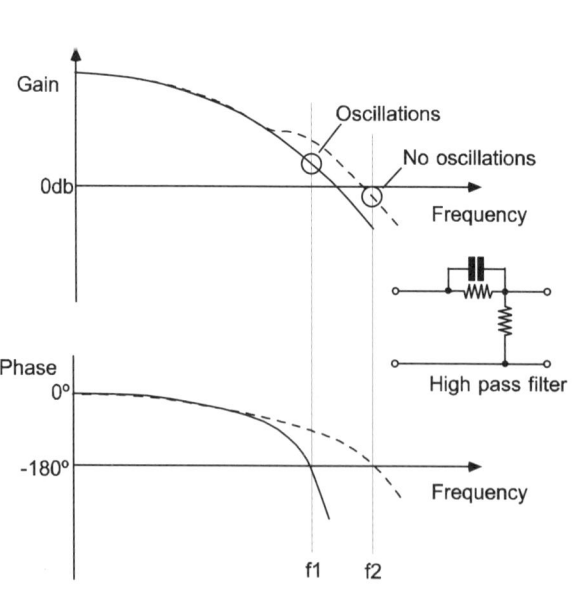

(b) *Stabilisation of gain/phase response with a high pass filter* in the form of jitter and low pass filters are utilised to cut these out. The complete response shaping is a mixture of several filter types and the nature of these is determined by the type of motor and its inertia.

Stability of the servo loop is indicated by its reaction to a pulse load. Figure 4.8 shows the four variations of possible response to a heavy load of short duration, which is similar to the start up response.

(a). This is a servo that has no damping at all, loop gain is high and there is so much over correction that dynamic oscillation takes place; the motor speed varies greatly.

(b). Here the servo has had some phase correction applied to it but there is insufficient damping. The servo hunts for a comparatively long time before it settles down.

(c). There is too much damping applied and the response is sluggish; it takes too long for the servo to reach its running speed and phase control cannot easily be applied. (For phase control see later.)

(d). For a servo to operate with the best stability then the response is that of **critical damping**. In a critical damping response the control graph over-swing is; (1) large, (2) small, (3) very small, followed by stable speed.

The degree of damping can be varied according to the level of error. If the motor is just starting then the servo gain is very high and rapidly accelerates the motor. As the nominal speed is approached then the servo gain reduces and damping is increased. This is dynamic control and is usually carried out by a microcomputer-based servo control IC.

Frequency to voltage converter

A major element in the speed control circuit is the frequency to voltage converter and it is usually an inverter. That is to say that the DC error output voltage varies inversely with the input frequency. Although the error voltage is nominally DC, it is a variable DC voltage. This is a term describing a DC control voltage that is varying, rather than

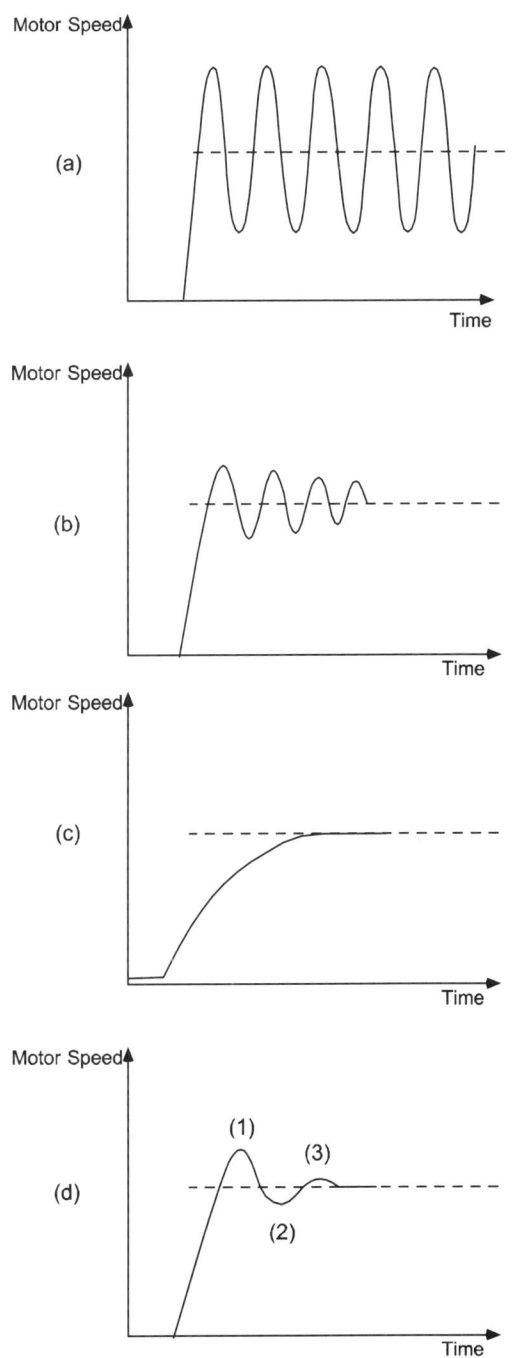

Figure 4.8 *The approach to ideal critical damping*

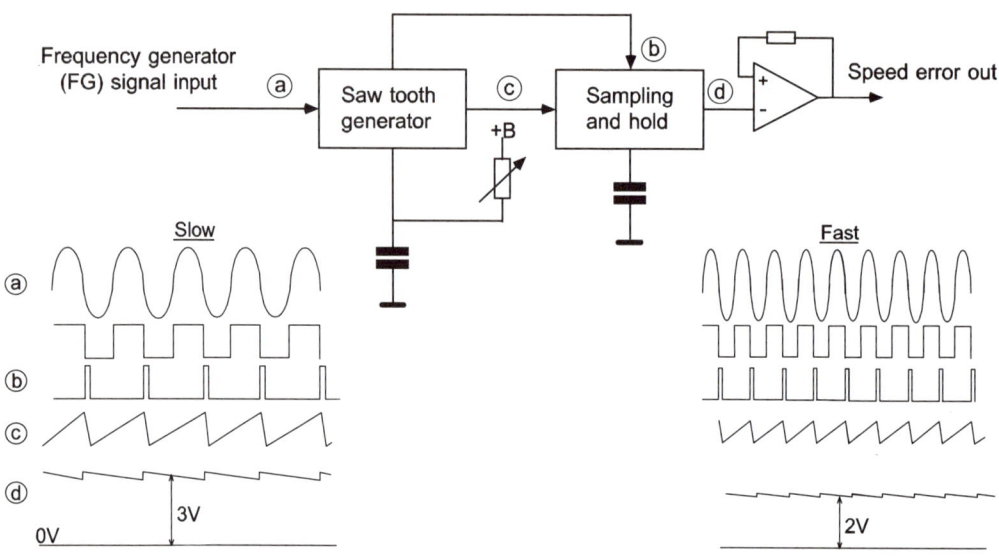

Figure 4.9 *Speed control (f-v) circuit and waveforms*

using the term AC and it will change in level at a rate determined by the servo loop correction control and damping. Feedback from the motor, or its flywheel, for speed control purposes is a sinusoidal waveform from a frequency generator (FG). From the motor it may only be 2–3 mV so this signal is amplified before being used within the servo. Frequency range is between 100 Hz and 2 kHz depending upon the individual video recorder design criteria. It is generally expected to be 250–1500 Hz.

Frequency generator input to the (f-v) circuit is around 2–3V p/p, after amplification, see (a) in Figure 4.9. A squaring circuit, usually a Schmitt trigger type circuit, converts the sinusoidal waveform to a square wave. From the falling edge of the square wave a reset pulse (d) is formed and fed to the sample and hold section. Simultaneously, a ramp, or sawtooth waveform (c) starts to rise; it is started by the falling edge of the reset pulse. A variable preset sets up the slope, or rise time, of the ramp. Irrespective of the input frequency the rate of change of the slope of this ramp is constant. Once the slope is preset, the peak level that it reaches before falling to zero will be dependent upon the period between the reset pulses (b); which in turn is determined by the input frequency.

If the input frequency is low then the period between reset pulses is longer. So is the time that is allowed the constant rate of change of the ramp which attains a higher peak value. The peak level of the ramp is switched into the hold circuit by the same pulse that resets the ramp. It follows that the voltage stored in the sample and hold circuit (d) will be higher for a low frequency input than that for a high frequency input.

Should the motor speed increase then so will the input frequency. In this case the ramp has less time to increase before being stored and reset; consequently the stored voltage is lower.

To recap: if the motor is slow then the control voltage is high, if the motor is fast then the control voltage is low. This is why the (f-v) converter is normally an inverter.

It is possible to set the speed at which the motor rotates by setting the slope of the ramp by a preset control, this sets up a narrow speed 'window' ready for the phase control section to take over and is set up in the record mode.

Phase control

So far we have looked at the control of a motor in terms of speed control and the next step is the concept of phase control. A fairly good way to get the difference into perspective is to consider the speed control (f-v) as coarse control and the phase control as fine control. Phase control is a fine degree of 'adjustment' and we must accept that phase control cannot be applied unless the motor is running at the correct speed. That is to say that phase control is not possible without some prior degree of speed control.

Let us consider the arrangement of a servo with the motor fixed to a flywheel via a shaft with a rotational speed of 1500 rpm; this is equivalent to 25 revolutions per second or one revolution in 40 milliseconds. A marker that is mounted upon the perimeter of the flywheel will be seen as a blur. Speed control can stabilise the spin of the motor/flywheel to a speed of 1500 rpm. Phase control then takes over from the speed control to give fine regulation within **one revolution** at that speed.

To illustrate the meaning of this idea we can look at the motor and flywheel system as shown in Figure 4.10. The situation is that the motor is speed controlled at 1500 rpm (25 rps) by the (f-v) feedback signal. An additional strobe light is triggered by an external trigger reference of 25 Hz.

Therefore, the strobe is flashing at approximately the same rate as one revolution of the flywheel. Under these conditions the flywheel may appear to be either going around very slowly or may even seem to be stationary, depending upon the accuracy of the speed control system. If we now connect the phase control loop by closing the switch SW1 the marker upon the cylinder will now appear stationary. This is because the phase control section is synchronised to a reference which is the same 25 Hz trigger signal as the strobe light.

A practical use of this example is the way a car mechanic will set engine timing. A strobe light that is triggered off the ignition is used to make the crank shaft timing marks appear stationary.

Speed control is obtained by using the (f-v) converter to measure the speed of rotation so that the motor speed is the required 1500 rpm (25 rps). This is done using an impregnated magnet forming the circumference of the flywheel, whereby a continuous sine wave is generated in a pick-up sensor. The magnet mounted upon the perimeter of the cylinder generates a single pulse. It is utilised to phase lock the motor to the reference 25 Hz trigger pulse.

Now that phase control has been achieved the marker on the flywheel appears stationary under the influence of the strobe light; it will appear to sit anywhere around the circumference.

A variable phase delay between the feedback

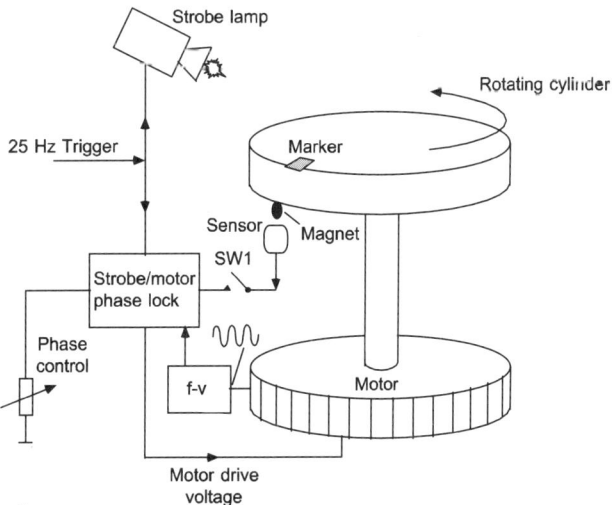

Figure 4.10 *Phase control of a rotating cylinder motor*

Video and Camcorder Servicing and Technology

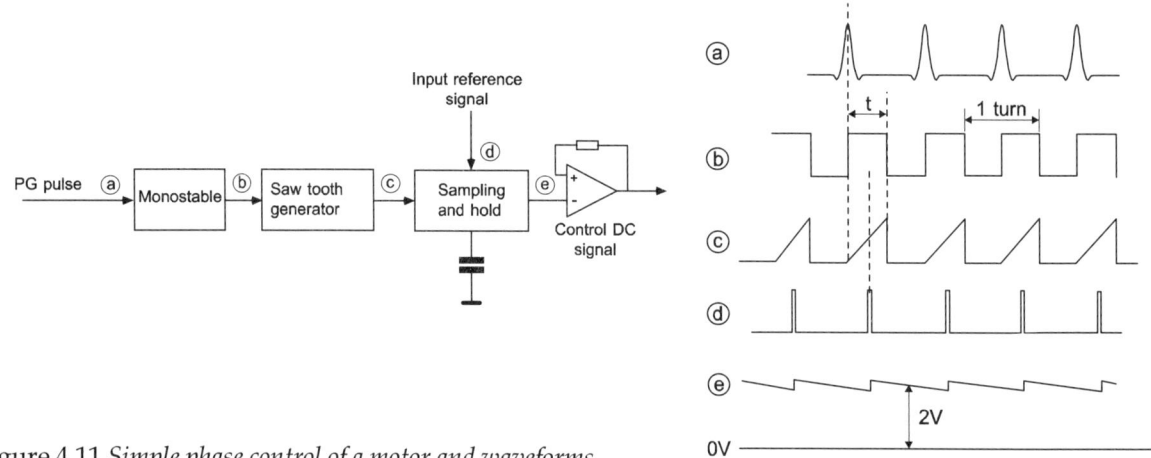

Figure 4.11 *Simple phase control of a motor and waveforms*

sensor pulse and the reference strobe trigger pulse makes it possible to move the position of the marker. If the phase delay is capable of 360° then the marker can be set to be 'stationary' anywhere around the circumference. This is an example of the very fine degree of phase control.

To sum up, speed control sets the motor/flywheel to a given speed. Phase control stabilises rotation even more accurately to an external reference.

In the terms of a video recorder servo, phase control allows the marker (i.e. video heads) to be phase locked to an external video signal.

Basic phase control circuit

Operation of the phase control is not too dissimilar to that of the speed control in Figure 4.9, except for a second reference input; the phase control circuit is shown in Figure 4.11 along with the corresponding waveform diagram.

For phase control the flywheel signal is converted into a square wave with a period equal to that of one full rotation. A pulse (a) is generated by the action of the magnet mounted on the perimeter of the rotating cylinder passing by a pick up coil sensor. This is called the pulse generator, its output signal is referred to as the PG pulse. The pulse is processed and sharpened in order to trigger a monostable with a period (t) being approximately half of the time that it takes the cylinder to rotate. This produces a square wave as shown in (b).

One half of the square wave (t) is converted into a ramp (c) by a sawtooth generator; the slope is fixed.

A reference signal input is transformed into a short pulse (d) for sampling purposes. It samples the centre of the ramp, with the whole ramp being its control range or operating 'window'; correction is applied either side of the ramp centre value. The ramp is sampled at its centre by a 'sample and hold' circuit and an error control voltage is produced, indicative of the voltage value of the ramp at the time of sampling. In this case a 4 V ramp that is sampled halfway will result in a control voltage of 2 V that can, in theory, vary between 0 and 4 V.

If the drum motor tends to slow down then the PG pulse will lag behind the input reference. At the 'sample and hold' stage the ramp is now later than the sample pulse and sampling takes place lower down the ramp. At (e) the output voltage falls. A high impedance buffer stage inverts the error voltage and so the control error signal from the circuit increases, increasing the motor drive voltage, speeding up the motor and advancing the PG pulse timing back to the centre of the ramp.

Speed control and phase control circuits produce variable correction voltages for the motor drive amplifier. These are added together to form coarse and fine control.

For the phase control circuit to work the motor speed must return a square wave that has the same frequency as the reference signal, which means that the motor must be running at the correct speed. If it is not the correct speed then the sample pulse cannot regularly sample the ramps; it may miss a few and the servo will be unstable and will not 'lock up'. Coarse control is the speed control, it must be correct before the fine phase control can be effective.

Video recorder head drum servomechanism

In order to fully understand the operational aims of phase control in a video recorder video head drum servomechanism, the structure of the information on the tape must be examined.

An illustration of the structure of the magnetic video track is given in Figure 4.12(a). This is taken as the tracks on a manufacturers test tape and shows the position of the TV field sync pulses that must be written on each and every subsequent track.

It is worth noting that the view of the tape is that of the magnetised side from a point in the centre of the head drum. Many years ago it was possible to see this pattern by the use of a small sealed dish containing ferrite powder in a suspension fluid. After shaking the dish it was placed over the magnetised side of the tape and the ferrite was allowed to settle, as it did so it was affected by the magnetic structure upon the tape and took on the magnetic pattern shown in Figure 4.12(a).

Figure 4.12(b) shows the equivalent video signal with dotted lines for the start and finish of each track, showing the overlap period where the duplicated signal is recorded and video head switching takes place. Tape path contruction is such that the video heads make contact with the tape at the bottom edge. They travel across the tape in the same direction that the tape is moving. A1 and B1 are the reference head switching points for the field sync pulse position. This position is given as a period of 6½ TV lines prior to the beginning of the field sync period where the half-line pulses start.

In the recording mode of the servo there has to be a phase relationship between the incoming video to be recorded and the video head switching points. The field pulses of the incoming video signal and the head switching waveform have a phase shift of 6½ TV lines in order to put the field sync pulse on the tape in the correct position. There is a small variable delay circuit in the PG path to allow for fine adjustment of the 6½-line period; it is set up in playback on a manufacturer's reference tape.

It is the objective of a head drum servo to ensure that during recording the video heads are phase related to the incoming signal, such that the video tracks are recorded as laid down by the system specification. This is achieved by the servo phase comparator working between incoming field syncs and the PG signal from the video head drum or its derived head switching (flip flop) waveform designated HSW.

There are two video heads mounted upon the head drum and spaced 180° apart so that each one records and replays a **TV field**; both heads in a single drum rotation record or replay a **TV frame**. The video head drum is rotating at 25 Hz (1500 rpm) this being equivalent to 40 ms per revolution, therefore the reference field pulses at 50 Hz must be divided by two down to 25 Hz to match the PG pulses in the phase comparator.

As previously shown in the phase control system of Figure 4.10, it is possible to run the cylinder motor to speed and then apply phase

Video and Camcorder Servicing and Technology

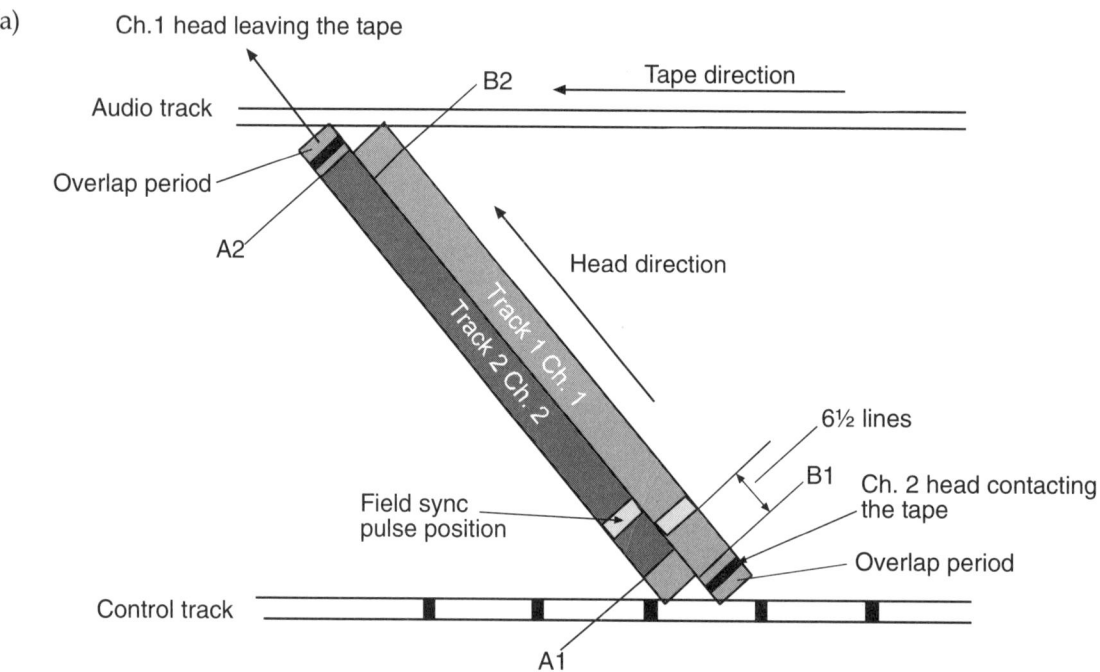

Figure 4.12(a) *VHS video track layout showing position of field pulses*

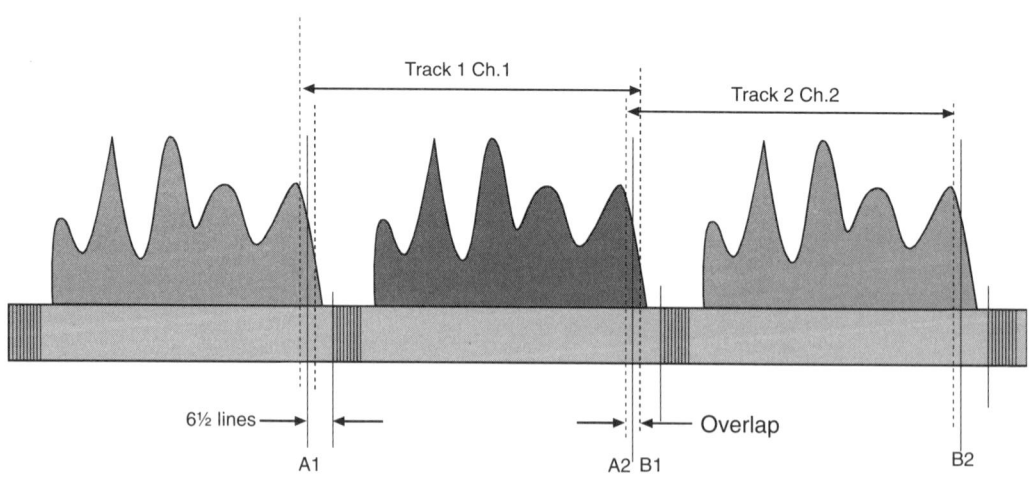

(b) *The relationship of the tracks to the video signal*

Motor control and servomechanisms

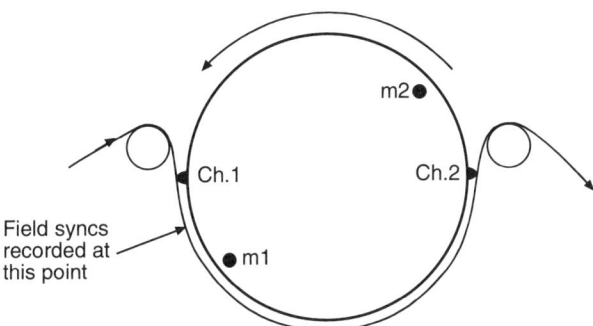

Figure 4.13 *Video head wrap around showing both heads in contact with the tape during the overlap period*

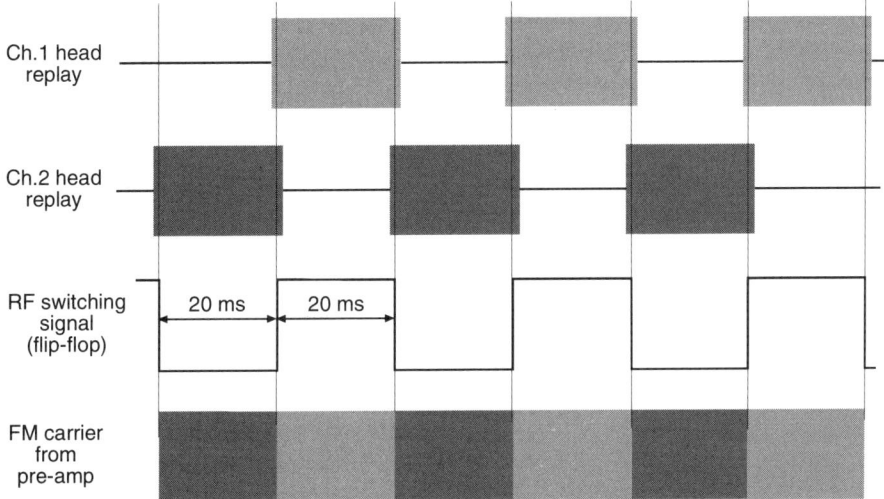

Figure 4.14 *Video head switching during playback ensuring contiguous signal by use of the overlap period*

control from a reference input signal to determine the position of the video head around the circumference and design this to be in the right place at the right time. This means that when the video head contacts the tape in the record mode it is just before the head switching signal and 6½ lines before the arrival of field syncs in the incoming video signal that is to be recorded. The phase control section of the servo ensures that this is so.

In practice the tape wrap around the head drum is greater than 180° and both heads can be in contact with the tape at the same time, shown at the end of track 1 and the start of track 2. Two margin or overlap sections are generated by the extra wrap around and are used for head switching in replay to avoid signal interruption.

The tape wrap is shown in Figure 4.13 along with the relative positions of the video heads during the overlap margins just prior to the head changeover point and field syncs. Figure 4.14 illustrates the replay head switching waveforms. Replay for each head exceeds the 20 ms period of the square wave head switching signal by a few milliseconds each side. This is due to extra wrap around allowing for the overlap margins. It can be seen that switching transitions occur at a time period when both Ch.1 and Ch.2 heads are producing an output; this results in a continuous FM carrier with no gaps between each head.

Video and Camcorder Servicing and Technology

Head switching does not take place in record as both heads are driven continuously with the recording signal in order to record the extra overlap margin.

Basic analogue drum servo

The example of a basic drum servo to illustrate the concept of servo phase control is shown in Figure 4.15 with its accompanying waveforms in Figure 4.16. There is no speed control or FG loop as such but there are circuit designs that precede the addition of an FG speed control loop which are found in later versions of servomechanisms shown later in this chapter.

In the recording mode the objective is to write the video tracks so that the recorded field pulse is in the correct position. Phasing of Ch.1/Ch.2 heads must be maintained with respect to the incoming video signal, represented by the field pulse. Ch.1 head is shown in the correct position for recording the field sync section of the video signal. Magnet M2 has just passed by the PG pick-up head having induced a negative PG pulse that sets the head switching signal from high to low.

PG pulse shaping

Each PG pulse is a very small induced signal blip, either positive or negative, from opposing magnets M1 and M2. M1 is a N/S magnet, which induces a positive pulse whereas M2 is a S/N magnet inducing a negative pulse.

PG pulses are amplified and squared off to give sharp triggering edges. Each PG pulse is delayed by about 0.2 ms (about 4 TV lines) in a variable monostable delay. This allows for adjustment for each head in replay of the 6½-line offset between the head switching point and the field pulse.

From the negative transition of the head switching pulse a ramp is formed. The slope of the ramp will determine the 'gain' of the servo; for

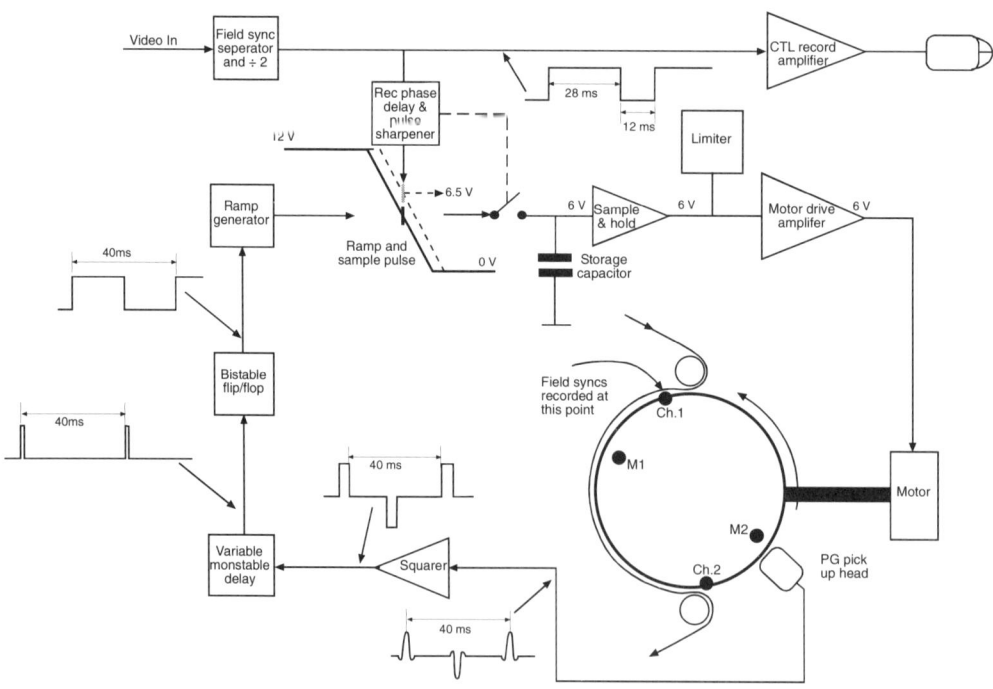

Figure 4.15 *Basic drum servo utilising phase control to accurately write the track*

Motor control and servomechanisms

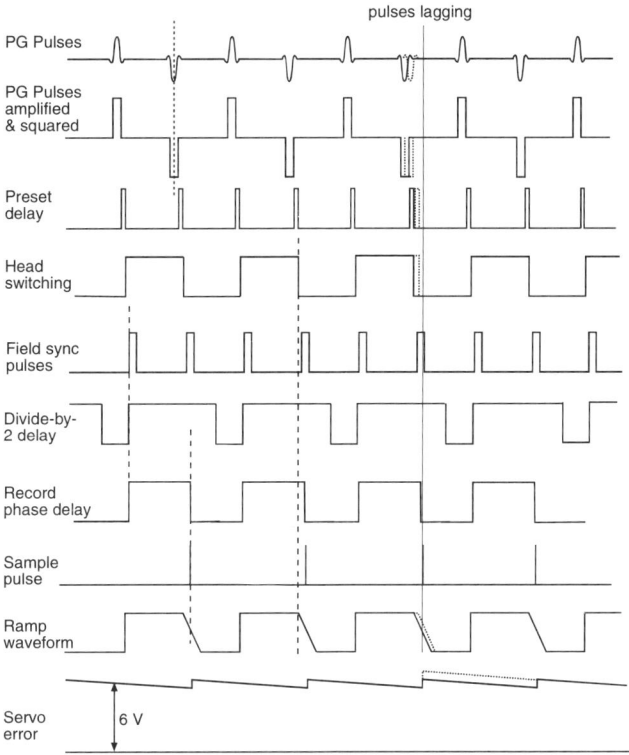

Figure 4.16 *Drum servo waveforms*

high gain the ramp is short and steep, for low gain the slope is more gradual and longer.

Record phase

Incoming video is processed to filter out and separate field sync pulses. These are then divided by 2, using a 28 ms delay monostable; the resulting 28/12 ms waveform is further utilised as the recording signal for the control track.

From the trailing edge of this 28/12 ms (mark/space ratio) waveform a variable delay monostable is triggered, this is the record phase delay.

Having first adjusted PG1 and PG2 controls in playback to set up the 6½ line period between the head switching signal and field syncs using a manufacturer's test tape, the record phase can then be set. The record phase control sets the phase position of the video heads in record. Referring to Figure 4.12(b) it sets the exact position of Ch.1 head to record field syncs 6½ lines after the head switching signal transition.

Sampling

After the record phase delay a very short sampling pulse is produced to sample the ramp derived from the head switching waveform. In effect the pulse momentarily closes a switch into a very high impedance circuit with a storage capacitor. Whatever the voltage that the ramp is at the time the switch closes is transferred to the storage capacitor. Consequently the capacitor is updated every 40 ms and the voltage across it is the servo error signal. In this example the servo error signal is normally 6 V given that the 12 V ramp is sampled halfway, as it should be. Via a motor drive amplifier this 6 V error signal becomes the drum motor power supply control.

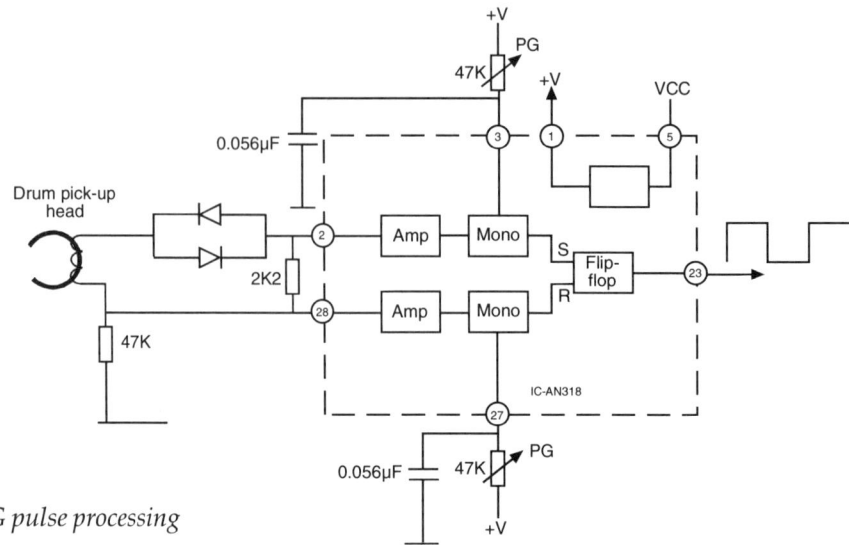

Figure 4.17 *PG pulse processing*

Servo control

In order to illustrate how the control circuit functions it is assumed that the head drum slows very slightly. An effect that may be produced if there were additional friction between the head drum cylinder and the tape.

Slowing of the drum rotation due to additional friction has the effect that the PG pulses are produced over a longer period, meaning that each would be 'later' than the previous one. An oscilloscope display would show the pulse train stretching to the right-hand side of the display. In servo terms the PG pulses would be lagging. Consequently transitions of the head switching signal would be lagging and so the ramp arrives late; this is shown as the dashed line in Figure 4.15. As the field pulse sample is always on time with respect to the ramp being late sampling takes place further up the slope. Now the error signal rises (to 6.5 V) across the motor producing more power and increasing the drive to overcome the additional friction applied. In terms of phase, the motor does not reduce in speed measurably with increased loading, it simply lags in phase. If uncorrected the phase lag would move the Ch.1 head a few degrees clockwise, recording the field pulse lower down the video track in the wrong place. Ramp and sample phase control maintains the phase relationship between the video heads and field syncs, set by PG1, PG2 and record phase controls.

PG circuit

Figure 4.17 shows the circuit diagram of the analogue PG pulse processing section of the AN318 dedicated servo IC. While this and the following examples are of early discrete circuits, they do serve the purpose of exemplifying the more subtle circuit techniques required to operate and stabilise a video recorder servomechanism.

As previously mentioned, a magnet passing by the drum pickup head induces either a positive or negative pulse, depending upon the polarity of the magnet.

Two parallel diodes employ the forward voltage characteristics of diodes in that signals below the forward voltage do not turn them on and they maintain high impedance. Noise generated by the rotating flywheel is therefore blocked and only the PG pulses, which are of a level high enough to turn on the diodes, pass through.

A positive PG pulse through the diode pair arrives at pin 2, and via pre-amplifier, triggers a short delay monostable. When the monostable resets, a following flip-flop is set high. A negative PG pulse has no effect on pin 2, but across the 47K resistor this becomes a positive pulse to pin 28. After amplification either monostable is triggered for a short period and off the back edge the flip-flop is either set or reset. In this way the output head switching signal is constructed.

For the monstables:
$$t = CR$$
$$22K \times 0.056\mu F = 22 \times 10^3 \times 0.056 \times 10^{-6}$$
$$= 1.232 \text{ ms.}$$

Sample and hold

Another section of the AN318 is used for the sample and hold processing part of the servomechanism.

The drum switching signal inputs at pin 22 to a delay monostable with voltage control of its timing components. This is followed by an inverter, and then a trapezoid timing circuit for the construction of the ramp. Timing components on pins 16 and 17 determine the slope of the ramp and influence the overall gain of the servo loop. Circuitry around the sample and hold section has high impedance. For the servo error voltage storage capacitor on pin 13, an 8M2 resistor biases pin 13 to 6 V for soft start-up control. This helps to get the drum motor up to speed, without influencing servo control because of the 8M2 resistor.

A frame sync pulse, which generates a short sample pulse using the record phase monostable on pin 14, is used to switch the sample and hold. It is further sharpened within the IC. Servo error voltage at around 6 V is buffered out of the IC on pin 2.

In replay, triggering of this monostable is by replayed control track pulses with phase shift provided by the tracking control, to adjust for noise free tracking.

In this basic system the capstan motor servo runs from a crystal reference in both record and playback. Later in this chapter the capstan servo can be seen to take an active role in replay tracking.

Part of the servo error signal is fed back to the monostable. It can rise to its nominal level of 6 V but the lower level is limited by a diode and divider to about 3 V. Feedback control of the monostable is used to aid the start up sequence and reduce the 'lock in' time. At the time of start-up for the servo there are few ramps to sample,

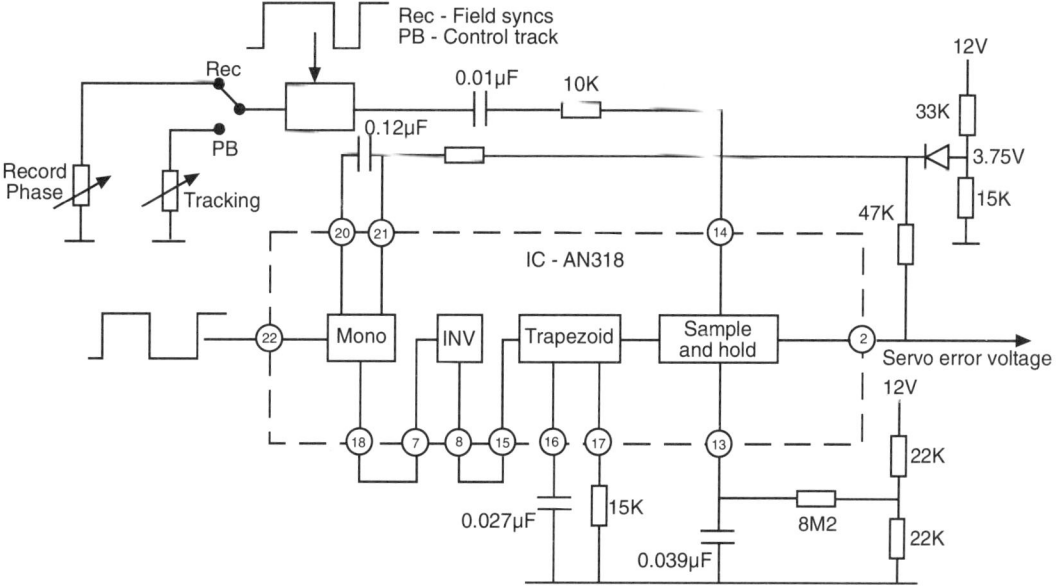

Figure 4.18 *Sample and hold circuit*

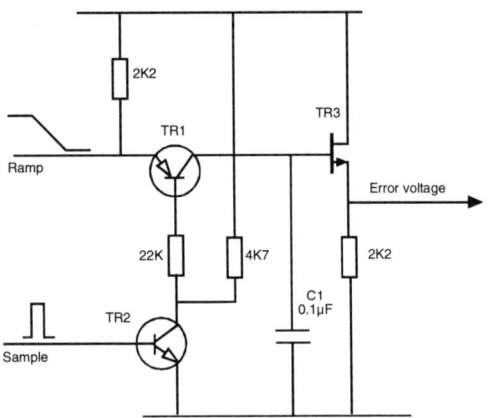

Figure 4.19 *Detailed sample and hold descrete circuit*

and the number of ramps also increases. This then gives the sample pulses chance to hit more ramps and the servo starts to stabilise until lock is finally achieved and each ramp is sequentially sampled.

Figure 4.19 illustrates the high impedance ramp and sample operation. A ramp voltage is applied to TR1 and during the ramp period a very short pulse is applied to TR2. TR1 switches on momentarily and C1 charges up very quickly through the low impedance path of TR1. TR3 is a high impedance FET transistor configured as a source follower to buffer the error voltage without any significant drain on C1 between samples.

and as the motor is running slow the samples are either sampling the top or bottom of the waveform. In order to accelerate the motor, the average number of high samples must be greater than that of the low samples. Feedback control to the monostable reduces the time period of the low state so that the mark/space ratio of the ramp waveform (see Figure 4.16) is affected and there are longer periods of high level. As the motor speeds up the monstable time is increased and the ramp waveform approaches a 1:1 mark/space

Error voltage limiter

To understand this part of the circuit in Figure 4.20 we have to consider the signal components that make up the servo error voltage. There is the basic 6 V DC but there are AC components upon the DC signal. These are normal servo AC control signals that are required for servo control and 'noise' that is unwanted and must not be allowed to affect the motor. During normal phase correction in both record or playback the 6 V DC error voltage varies in level at around 25 Hz and this

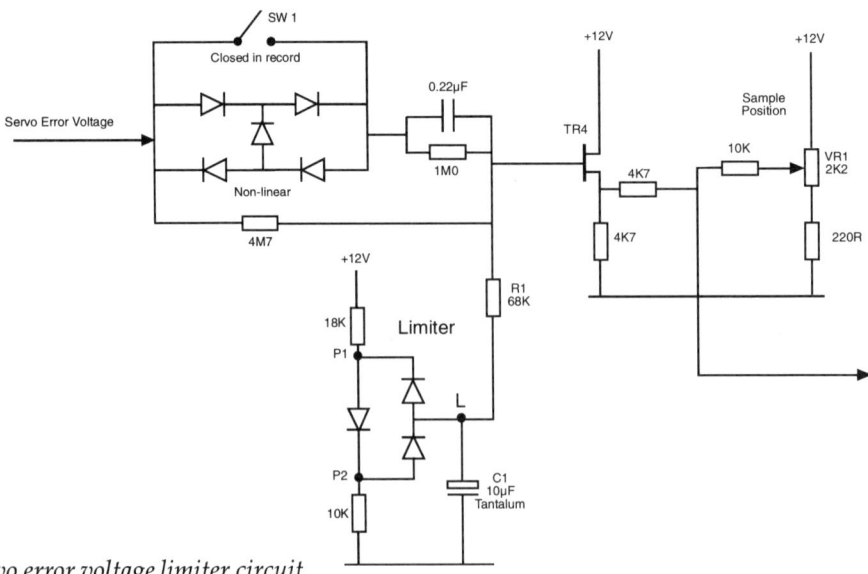

Figure 4.20 *Servo error voltage limiter circuit*

can be seen as a low frequency AC component. When the servo is unlocked the AC ripple increases in frequency and level, and can be used to bring the servo back into the locked condition.

During both the recording and playback of the control track the tape varies in tension, this causes small phase errors, or position errors in the control track pulse. In a phase control system the resulting AC component is considered as 'noise' on the error voltage and must be eliminated otherwise the motor will be adversely affected.

In record the switch SW1 is closed and the error voltage path is via 4M7 in parallel with 1M0 and 0.22µF. The presence of the capacitor gives extra gain for AC components. TR4 buffers the error signal from the filter section and a fixed DC is added. It is adjusted by VR1 and sets the pulse position to be centred on the ramp in record mode.

Non-linear network

In playback any small AC components, due to tape path friction noise are filtered out. As sampling is done by the control track pulses in replay any tape path jitter varies the phase of the replayed pulse and an AC noise component is generated. These must not be allowed to affect the servo operation. As the noise is of a relatively low level it cannot pass through the diode non-linear network as SW1 is open. The diodes are not forward biased and remain at high impedance. Although there is leakage path via the 4M7 the level is very much reduced.

Limiter

Another section of the circuit is the limiter. Point P1 is set to about 5 V and P2 to 4.3 V Should the servo error signal at point 'L' rise to high it will be clamped at 5.7 V as the diode becomes forward biased to P1. If the error signal falls in level it will be clamped at 3.6 V by a second diode to P2. A control window between 3.6 and 5.7 V is used to limit the control range. Errors outside this window are inhibited to prevent abnormal motor speeds. Within the window upper and lower limits both clamping diodes are of high impedance and limiter circuit has no effect upon the servo control.

The components 0.22µF/1M0/4M7 form a high pass filter and components R1 (68K) and C1 (10µF) are a low pass filter to shape the servo AC gain and phase to keep the servo within the Bode diagram parameters, see p.64.

Motor drive amplifier

In order to control the drum motor, in this case a DC brush motor, a discriminator amplifier is used. It will control and stabilise the motor characteristics so that any changes in the motor do not influence the overall servo operation. Both the motor drive amplifier and discriminator are combined in one circuit. It is important that the gain of the amplifier is set to unity and VR1 is preset for this adjustment.

R1 is used to monitor the motor current during normal operation. This current can vary with temperature, brush wear, or extra loading due to friction. An increase in current due to temperature changes, will cause a corresponding voltage change -dV across the motor and to the inverting input of the amplifier. As the gain is unity then there is a subsequent change of +dV at the output which is applied to the other side of R1 to compensate. In effect the change in voltage across the motor is cancelled out, preventing any change in motor speed that would adversely affect the servo.

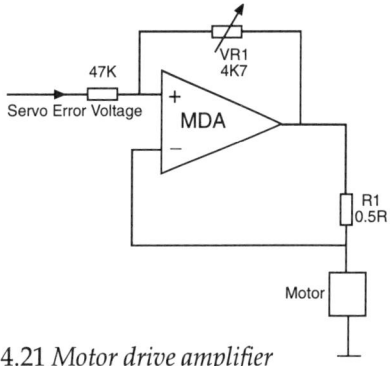

Figure 4.21 *Motor drive amplifier*

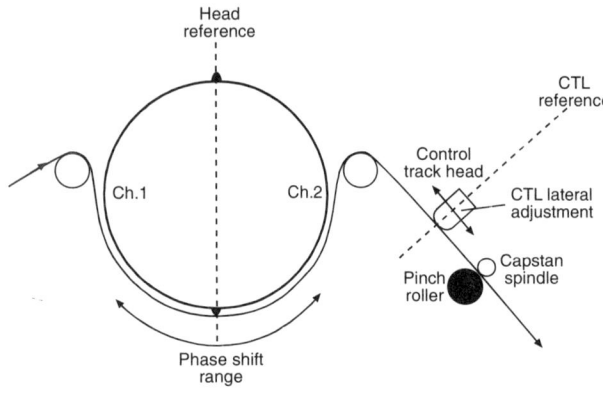

Figure 4.22 *Tracking reference points and CTL head lateral adjustment*

Tracking

In an ideal world any video recorder that records and then replays its own tapes does not need a tracking control. This is because the tape transport mechanism is the same for both recording and replay, there are no mechanical or electrical differences. Where tapes are interchanged or pre-recorded tapes are played then some degree of tracking optimisation is needed; this may be manual or automatic.

Initially video recorders had the tracking control as part of the head drum servo, now it is unusual to come across this arrangement as it caused the picture to weave sideways when the tracking control was operated.

Tracking control is a phase shift in the capstan servo to shift the phase of the capstan motor with respect to the replayed tape, or more to the point the replayed control track pulses. For the capstan phase shift to be effective upon the video tracks then the head drum servo must be running at constant speed and phase during replay.

Referring to Figure 4.22 two reference points are shown: one is the video head currently scanning a track about halfway through; the other is the fixed position of the control track head (CTL head) replaying control track pulses.

In the capstan servo the tracking control is a phase shift control, the range is about 180–200°. It is a phase shift between the replayed control track pulses and the capstan motor, and the directly driven tape. When the tracking control is adjusted it advances or retards the tape position with respect to the video head reference point in Figure 4.22. By adjusting the control track phase shift, the video track that the video head is currently scanning can be replaced by an adjacent one, see Figure 4.23. This produces a lateral shift in tape drive phase.

Altering the lateral position of the control track head can produce a similar affect. When setting up a video recorder CTL head tape path the tracking control is set in its central position (with auto tracking it is switched off), then a manufacturer's test tape is replayed. Why? Well it has an accurate magnetic tape pattern where control track pulses are positioned correctly with respect to the video tracks, as is the field pulse period. The control track head is then adjusted laterally to give the best tracking, and a maximum FM output level as measured at the video pre-amplifier output test point.

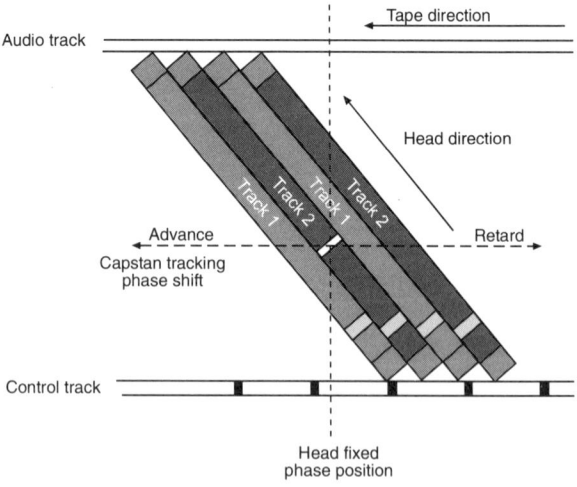

Figure 4.23 *The affect of capstan phase shift upon the video head and tracks*

Motor control and servomechanisms

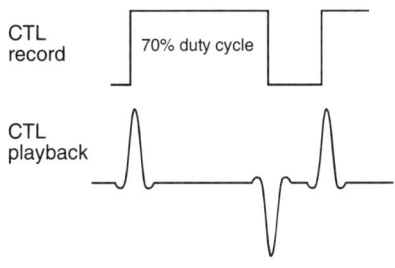

Figure 4.24 *Differentiation of the rectangular control waveform when recorded onto magnetic tape*

Recording tracking pulses

A 25-Hz rectangular waveform with a mark space ratio of 28 ms/12 ms is recorded onto the tape as the control track. It is recorded by the lower section of the CTL head as a series of pulses on the lower edge of the tape.

A magnetic head can only record onto tape a current that is varying and for a square waveform it records the differential of that waveform, as shown in figure 4.24. More information on recording rectangular waveforms is given in Chapter 12, p.255.

The original standard of control track waveform 28ms/12 ms is termed a 70% duty cycle as 28 ms is 70% of the total period of 40 ms.

CTL developments

When tape indexing and title archiving were introduced, a standard for CTL duty cycle was drawn up to ensure that there would not be any interference with the control track that would impair playback and maintain interchangeability. For a standard control track without any special operations, the duty cycle was agreed as 60%. A standard was then drawn up for extending the control track for indexing recordings and data codes applicable to archiving.

For data usage the standard duty cycle is tightened up to 60 ± 5% and is data '0' and a duty cycle of 27.5 ± 2.5% is given as data '1'.

Now the control track can carry data that can be used to identify specific tapes and carry information on what is recorded on that tape directly or by a code that can refer to data stored in a memory.

Indexing

A specification was also drawn up for the index signal coding of the control track. It consists of a 63 bit data word, bearing in mind that there is one control track pulse for each frame, which is a total of 126 video tracks. Identification of the index signal is specified as a 63 bit word starting and ending with '0' and with 61 data '1's in between. This is then identified by the system control microcomputer as an index signal when searching in fast forward or rewind. An index signal can be erased by replacing the word with a control track pulse having a 60% duty cycle. See also Editing duty cycle, p.279.

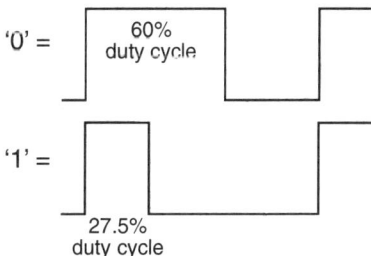

Figure 4.25 *Arrangement of control track duty cycle for data 0s and 1s*

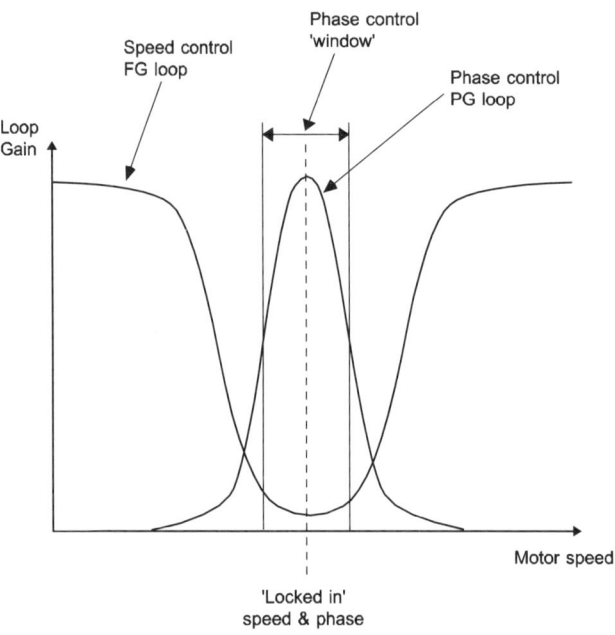

Figure 4.26 Theorectical inter-control of the servo FG and PG loops

FG/PG control

When servicing a video recorder that may have a servo fault it can be very difficult to evaluate which of the two loops may be giving the problem and to decide which has control. The phase control diagram in Figure 4.26 helps to illustrate how the two loops work together.

In a dual loop servo the two loops work together; the frequency generator (FG) loop sets the motor speed to a value within which the pulse generator (PG) can phase lock.

Once the FG speed section of the servo has driven the motor to its correct speed its influence on the servo reduces to a minumum; the FG loop gain can be considered to be low within the phase control window area. Within the window area the phase section has a high loop gain and has full control of the motor to maintain the phase locked loop. Each of the two servo loops could be considered as coarse (FG) and fine (PG) control, although this would be an over simplification. Dual loop servos aid faster servo lock-up from a standing start and tighter control over the motor, which is required for modern high performance video recorders.

This allows motor designers to reduce the weight of the head drum cylinder and capstan motor flywheel and reduce overall power consumption.

Lighter motors allow faster acceleration, and faster 'lock-up' times improve performance in the time between play/pause/play and editing functions.

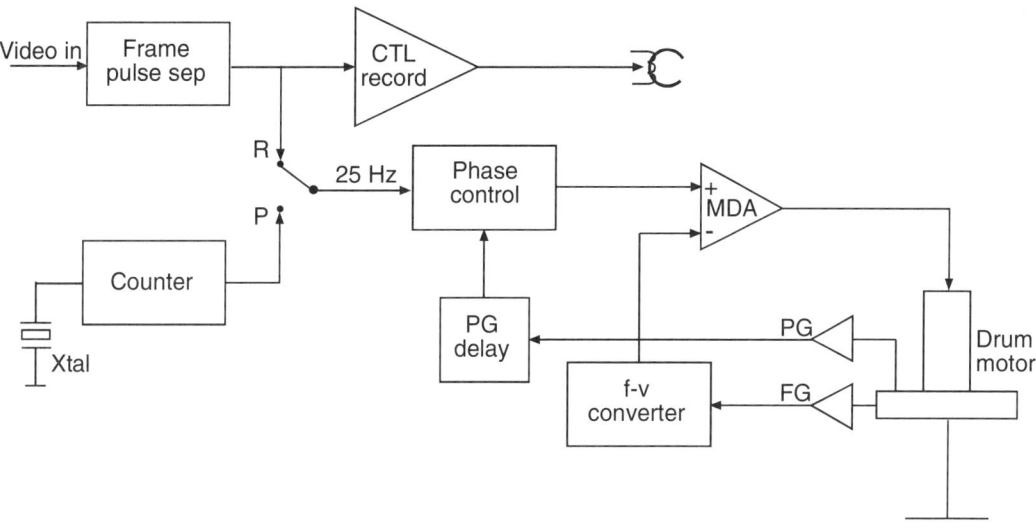

Figure 4.29 *Drum motor servo system*

Drum motor servo

While there are variations in the circuits of drum motor servos between earlier and later technology or between manufacturer's individual models, they all follow the basic structure of Figure 4.27.

Motor speed is set by the FG loop and phase lock by the PG loop. Both are combined into a composite motor drive amplifier (MDA). Note that as the (f-v) converter is an inverter it is connected into the -ve input of the MDA and is re-inverted back to the correct phase. If the motor slows down under load the output voltage of the (f-v) will fall and as a consequence the output of the MDA will rise. This provides extra power to correct for original loading effect.

As described on p.72 the phase control section compares the PG pulse, converted to a ramp voltage, to the frame input pulse. If the motor tends to slow, or lag, then the output drive voltage increases slightly to compensate.

A modern VCR servo in record, references to the incoming video signal in the form of the 25 Hz frame pulses and utilises the 25 Hz PG pulses for phase control.

In playback the drum is stabilised in a fixed phase to a reference crystal oscillator, suitably counted down to 25 Hz. The only variable in playback is tracking and this is taken care of in the capstan motor servo.

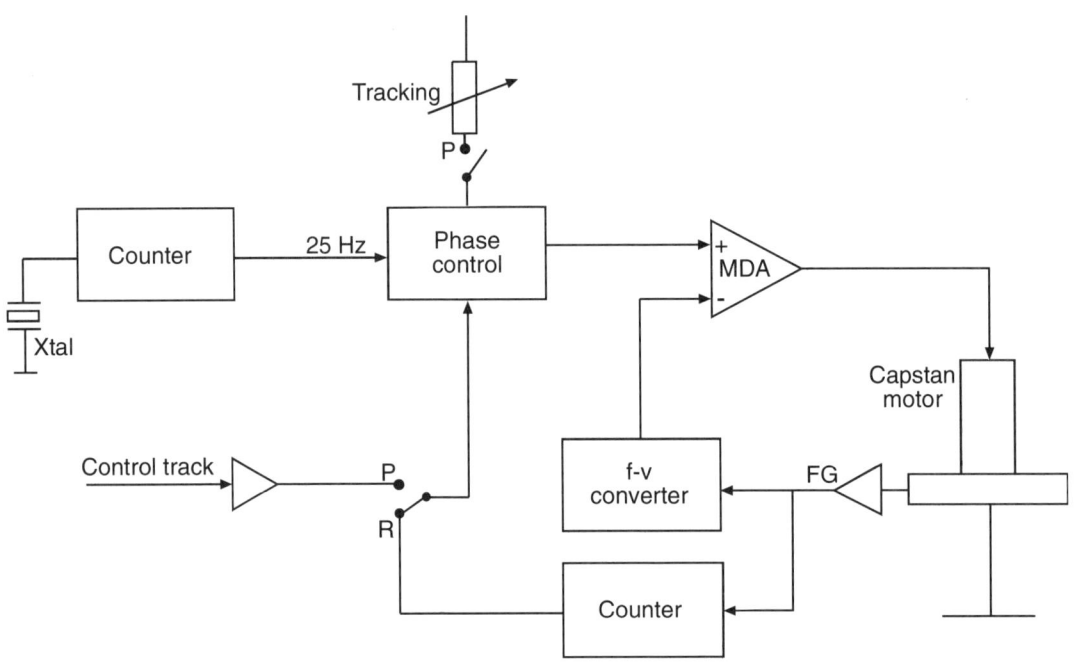

Figure 4.28 *Capstan motor servo system*

Capstan motor servo

In record, the role of the capstan motor servo, shown in Figure 4.28, is to run the tape at constant speed and phase, and so that the servo is locked to a crystal oscillator reference. Speed control by the FG loop is the same as for the drum motor servo. Phase control is achieved by counting down the FG frequency to 25 Hz and phase-locking it to the crystal 25 Hz reference output of a counter. It is not necessary to vary the phase of the capstan motor in record. It is kept at constant phase in order to avoid any unwanted phase variations being recorded, as these would show up as instability in any subsequent replay.

In playback the motor is directly responsible for driving the tape with a degree of phase shift for tracking. Off-tape control track pulses are phase locked to the crystal reference to stabilise the tape speed and phase. There is then an additional phase shift control for tracking adjustment purposes. This phase shift is equivalent to a 180° shift of the drum servo, enough to change the video heads over by one track in the tape lateral direction.

(a)

Servo	Reference	Comparison or Feedback	FG
Drum Record	V-sync/2 = 25 Hz	Drum pulse PG - 25 Hz	N/A
Drum Playback	Replayed Control Track = 25 Hz	Drum pulse PG - 25 Hz	N/A
Capstan Record	V-sync/2 = 25 Hz	FG 225 Hz /9 =25 Hz	225 Hz
Capstan Playback	Crystal 32 kHz/1280 = 25 Hz	FG 225 Hz /9 =25 Hz	225 Hz

(b)

Servo	Reference	Comparison or Feedback	FG
Drum Record	V-sync/2 = 25 Hz	Drum pulse PG - 25 Hz	1500 Hz
Drum Playback	Crystal 32 kHz/1280 = 25 Hz	Drum pulse PG - 25 Hz	1500 Hz
Capstan Record	Crystal 32 kHz/1523 =21 Hz	FG 126 Hz /6 =21 Hz	126 Hz
Capstan Playback	Crystal 32 kHz/1280 = 25 Hz	Replayed Control Track = 25 Hz	126 Hz

(c)

Servo	Reference	Comparison or Feedback	FG
Drum Record	V-sync/2 = 25 Hz	Drum pulse PG - 25 Hz	1500 Hz
Drum Playback	Crystal 32.768 kHz/1311 = 25 Hz	Drum pulse PG - 25 Hz	1500 Hz
Capstan Record	Crystal 32.768 kHz/1311 = 25 Hz	FG 505 Hz /20.2 =25 Hz	505 Hz
Capstan Playback	Crystal 32.768 kHz/1311 = 25 Hz	Replayed Control Track = 25 Hz	505 Hz

(d)

Servo	Reference	Comparison or Feedback	FG
Drum Record	V-sync/2 = 25 Hz	Drum pulse PG - 25 Hz	1600 Hz
Drum Playback	Colour reference 4.443 MHz/1773 = 25 Hz	Drum pulse PG - 25 Hz	1600 Hz
Capstan Record	Colour reference 4.443 MHz/1773 = 25 Hz	FG 200 Hz /16 =25 Hz	200 Hz
Capstan Playback	Colour reference 4.443 MHz/1773 = 25 Hz	Replayed Control Track = 25 Hz	200 Hz

Figure 4.29 *Some examples of servo FG and PG frequencies and references*

Examples of servo operation frequencies

Figure 4.29 illustrates some variations in servo operation signal frequencies. Figure 4.29(a) is an example of an early video recorder that had no drum frequency generator. It is the same system as shown in Figure 4.15. Record operation is fairly standard, the reference input is frame syncs and the feedback is the PG for phase lock.

In playback the drum PG is compared to replayed control track pulses with phase shift via a user tracking control. The capstan motor runs at constant speed and phase in both record and playback, referenced from a 32 kHz crystal counted down to 25 Hz.

In all of the other examples of drum cylinder and capstan servo, speed control is stabilised by FG frequencies around 1500 Hz. All drum servo phase comparators have a frame pulse at 25 Hz and a PG pulse at 25 Hz in record. Playback requires that the drum cylinder runs at constant speed and phase, a 25 Hz reference is produced from a crystal oscillator.

Capstan motors run at constant speed and phase from a crystal reference in record. Phase control is obtained by counting down the FG signal to 25 Hz for comparison to 25 Hz from a crystal source. It is not necessary to have a PG pulse magnet and sensor on the capstan flywheel as positional control is not required, only constant phase.

Capstan motor phase control in playback uses the control track pulse off of the tape. It provides capstan motor phase errors indirectly as the tape is driven directly by the motor. Any slippage or irregularities in the pinch roller has an adverse affect on the capstan servo in replay.

It can be seen from the chart of Figure 4.29(d) that improvements in counter technology allow the colour reference oscillator at 4.433 MHz to be used in place of a separate reference crystal.

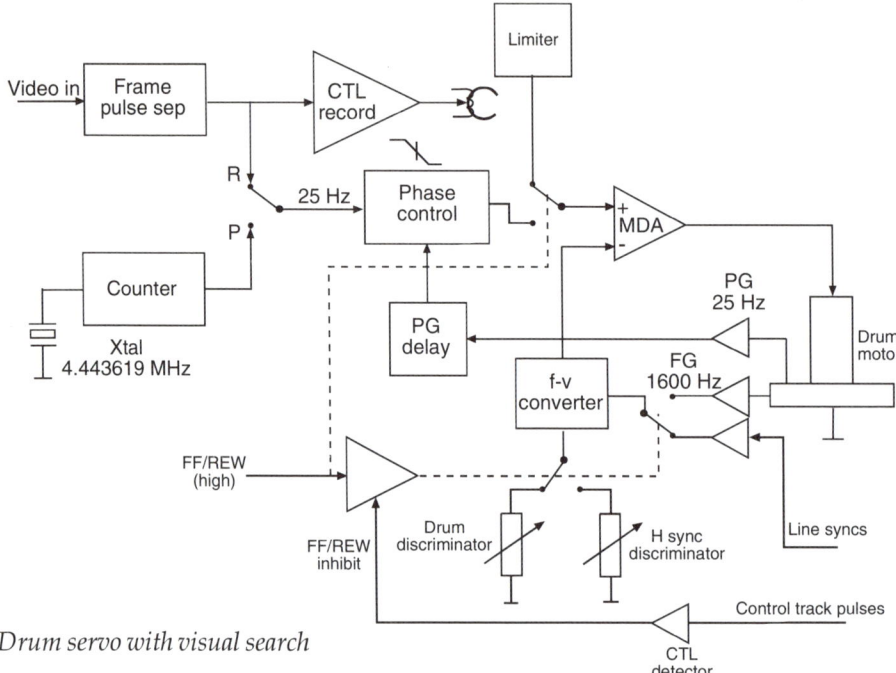

Figure 4.30 *Drum servo with visual search*

Drum servo application

Figure 4.30 shows a more advanced drum servo system. In record mode the reference input is of course field sync, then suitably divided by two to become frame syncs. The feedback signal consists of drum PG pulses; this is conventional practice for a drum servo. The speed control loop is based on an (f-v) converter, taking a signal at 1600 Hz from the drum FG. The only preset control (drum discriminator) enables the ramp sample pulse position to be correctly set.

Although the drum discriminator preset is within the FG loop its effect is to set the pulse position on the ramp. Consider that the drum sample pulse sits too high upon the ramp. Adjustment of the drum discriminator preset control decreases (f-v) converter output voltage to the inverting input of the motor drive amplifier. The MDA output voltage rises and as a result there is more power to the drum motor. This tends to increase the speed and advance the phase of the ramp with the effect that the sample pulse is lowered down the ramp, thereby reducing the power out of the phase control loop. In other words a small increase in power by adjustment of the FG loop is compensated by an equivalent reduction in power by the PG loop. This allows for adjustment of the sample pulse to correctly centre it upon the ramp.

An analogy can be seen in the phase-locked loop of a TV flywheel line sync circuit, where small adjustments of the horizontal hold control will shift the picture laterally. If too large an offset is attempted control is lost at the point where the loop unlocks and all synchronisation disappears.

The playback reference is a 4.433619 MHz crystal, counted down to 25 Hz. The drum locks to this to maintain constant speed and phase in playback.

Visual search compensation.

An increase in lateral tape speed in the forward visual search direction (cue) will result in a reduction in replayed line sync frequency causing picture disturbance on the TV, and possibly loss of horizontal hold. This is because the tape is travelling in the same direction as the head drum;

this can be seen in Figure 4.12(a). Consequently in reverse visual search (review), the line sync frequency will increase. One way to compensate for this and stabilise the picture is to control the speed of the head drum using replayed line sync pulses and aim for a consistent 15.627 kHz.

When visual (picture) search is selected, either forward or reverse, several switching functions take place. The phase control output is inhibited and replaced by a fixed voltage from the limiter (see Figure 4.20 for limiter). The input to the (f-v) converter is switched to off-tape line syncs, and the drum discriminator preset is replaced by a 'horizontal discriminator' control. Since we are anticipating a change of FG tone from 1.6 to 15.625 kHz (a factor of ten) it is reasonable to expect the horizontal discriminator preset to present a resistance of about one-tenth that of the drum discriminator pot. In search modes, only the FG loop is operational, working to hold the drum speed at a point where off-tape line syncs have the correct frequency of 15.625 kHz.

Effectively the drum speed is increased by about 6% in cue and reduced by about 7% in review. These figures are dependent upon the visual search speed x8, x10 etc.

A safety function is also incorporated, i.e. monitoring control track pulses, if they disappear it will be because a blank section of tape is passing. In these circumstances the input to the (f-v) converter reverts to the FG signal as a temporary reference until pictures reappear along with the control track.

Capstan servo application

The principles described in the last section are also applicable to a capstan servo, and the similarity between Figures 4.30 and 4.31 is immediately obvious. Starting at the top left-hand corner of Figure 4.31 (capstan servo) the reference in both record and playback is the counted-down output of a 4.443 MHz crystal, usually obtained from the colour circuits. During record, feedback from the rotating device comes in the form of a

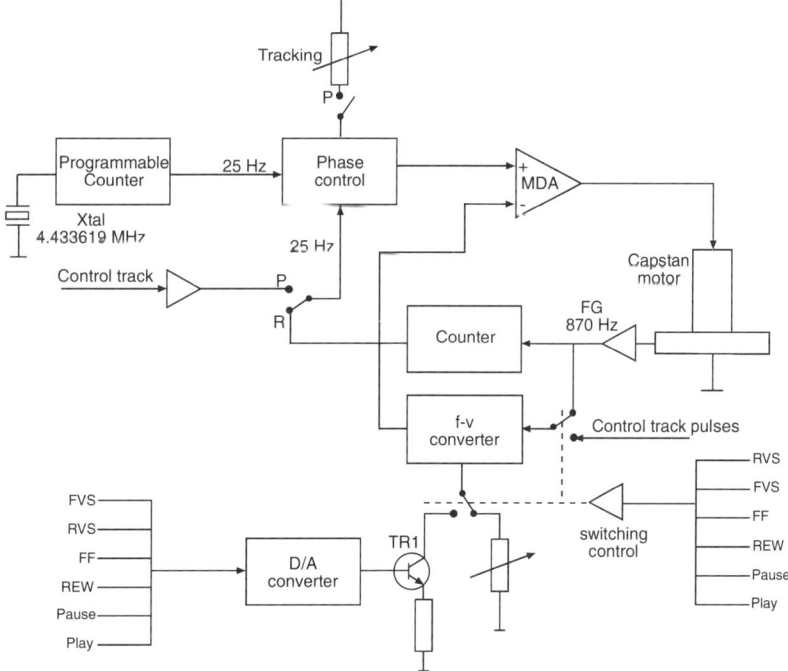

Figure 4.31 *Capstan servo with visual search*

870 Hz tone from a flywheel FG, divided to 25 Hz in a counter, and matching the 25 Hz output from the 'programmable' counter.

The capstan servo (f-v) converter plays a major role, primarily for speed control as described earlier. An 870 Hz FG signal from within the motor provides the input to the frequency discriminator. If visual search is selected during playback the (f-v) input is switched to control track replay, and will operate to maintain an input frequency of approx 250 Hz, corresponding to x10 search; 10 times the normal control pulse rate of 25 Hz. By referring back to the (f-v) converter (c) in Figure 4.9, it can be seen that a steeper slope in the ramp generator within the (f-v) converter will increase the output voltage. In this way by altering the slope of the ramp the 'gain' characteristics of the (f-v) converter can be altered and the loop can be set to specific motor speeds. This is 'programmed' by the components towards the bottom of Figure 4.31. A D/A (digital to analogue) converter examines data from the system control microprocessor to produce a specific output voltage for each transport mode. This is applied to the base of constant-current generator TR1 whose collector current sets the ramp slope to the appropriate degree. Taking the visual search example given above when 'cue' is selected. A high binary number will be presented to the D/A converter and its correspondingly high output voltage is presented to TR1 base, giving rise to a high collector current. This increases the charging current of the ramp-forming capacitor to generate the required steeper slope.

The use of frequency discriminators in advanced servos enables the designer to have safer control over tape handling, and to have more control over the various conditions, which are to be found in multiple function video recorders.

Digital servo

Analogue servo functions can be transferred into the digital domain; where there was a ramp, there is now a counter, and where there was a sample, there is a latch circuit.

A digital binary counter can be used as the equivalent of a voltage ramp as used in an analogue ramp and sample servo system. A voltage ramp is dependent upon the charging or discharging of a capacitor and there are component variables that, due to tolerances, will require time spent in setting up adjustments, in production, or as ageing takes place. Capacitor value, supply voltages and resistor tolerances all play a part in the stability of the servo in the long term. A binary counter is clocked by pulses from a crystal source. Use of this counter as a servo ramp improves short- and long-term stability and removes the need for many adjustments. It is not subject to the drift and tolerance problems of the discrete timing components found in the analogue circuits. A digital ramp circuit consists of four basic elements: a counter, a comparator, a latch, and a pulse-width modulation bistable.

Sampling of an analogue ramp takes place once every 40 ms. This is too slow a rate for a pulse width modulation bistable, and filtering to a DC error voltage is not practicable. A solution is to clock the bistable at a much higher rate, 1 kHz or more, in order to integrate the PWM signal into an equivalent DC level by an integration filter.

The servo system must therefore be capable of running the output pulse-width modulation at a high clock rate whilst responding to servo errors produced at a rate of 25 Hz. This is achieved by the use of a latch circuit, a counter memory, a comparator, and two ramp counters.

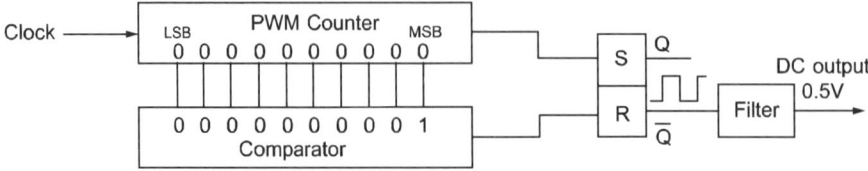

Figure 4.32 *Digital ramp and sample to form a square wave output*

Digital ramp

A digital version of the voltage ramp is a binary counter. A 10-bit counter has a count of 2^{10} which is 1024 steps compared to a voltage ramp as shown in Figure 4.33. It is not too difficult to work out that the halfway count, where a ramp would be sampled halfway, is a count of 512. As the counter is driven continuously by clock pulses it will count from 000000000 to 1111111111 and back to 0000000000. Each time the counter is full of 0s it 'sets' the output bistable, so if nothing else happens output Q would remain permanently high. But this is not the case. A comparator is connected in parallel with the counter binary output and is programmed to a count of 512 for the purposes of this example, that is 1000000000. (Note that the actual counter in the diagram is laid out as 0000000001 where the most significant bit is on the right.) When the count of 1000000000 is reached in the main counter it matches the value in the comparator. At this point when there is a match an output pulse from the comparator 'resets' the bistable. So in this example the bistable will have a symmetrical mark/space ratio as it is **set** at the start of the count and **reset** halfway by the comparator. The integrated output from the following D/A filter will be half of the supply voltage. Where the comparator count is less than 512 then the mark/space ratio of the bistable is high for a shorter period and the integrated DC voltage will fall. If the comparator is higher then the DC voltage will rise. This is equivalent to sampling an analogue ramp, halfway down, with a 'sample and hold' circuit. To complete the picture more processing is required in the form of a second 10-bit ramp counter and a 10-bit latch required to match the lower frequency sampling part of the servo system to the higher bistable frequency.

Digital drum servo operation

Clock pulses for the servo, in this example of a video head drum motor servo, are derived from a colour reference crystal. A divide-by-4 counter reduces 4.433619 MHz to 1.1 MHz to clock both the pulse-width modulation and ramp counters. In this example the PWM counter is only 9 bits as a clock pulse is used as bit 0 for the comparator. Also an output from the PWM counter first stage (divide by 2) at 554 kHz is used for the clock to the ramp counter.

A 9-bit counter has a count of 512. With a 1.1 MHz clock the count period is 1/1.1 MHz x 512 = 465 µs or a frequency of 2.14 kHz. The output frequency of the bistable to the D/A filter is approximately 2 kHz; being a set/reset bistable it does not divide by 2.

At the bottom of Figure 4.34 is the ramp counter clocked at 554 kHz. It is a 10-bit counter and therefore the count is 1024 and the count period can be calculated from: 1/554 kHz x 1024 = 1.85 ms. This is equivalent to the average analogue ramp waveform period.

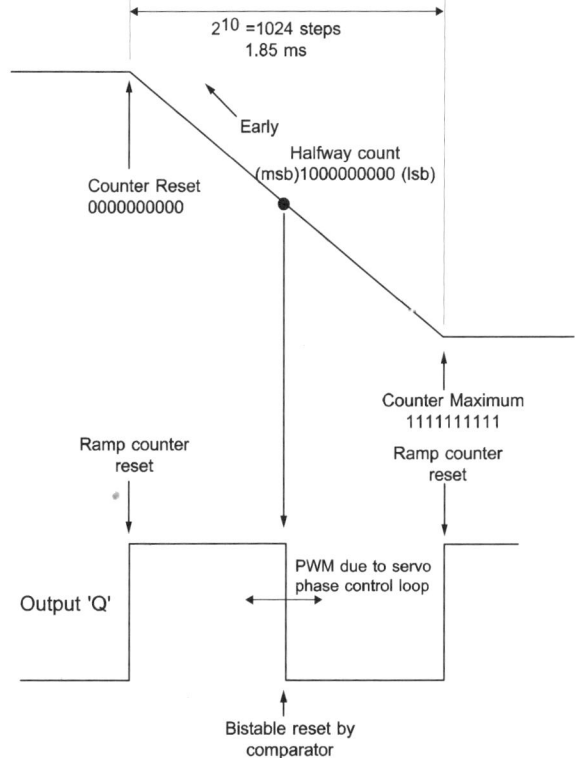

Figure 4.33 *Ramp counter and bistable relationship*

Video and Camcorder Servicing and Technology

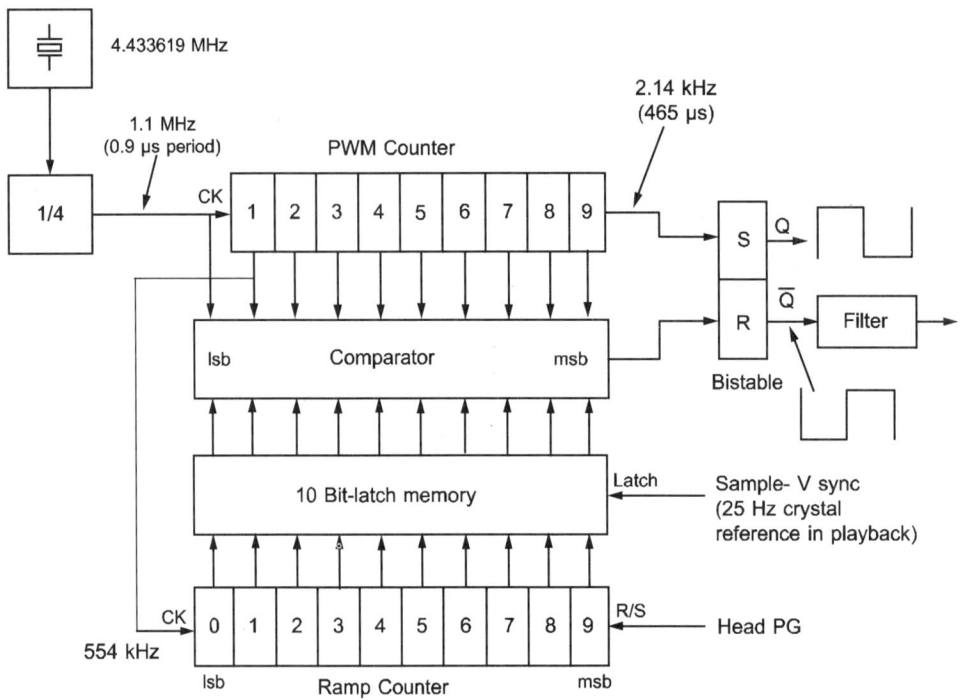

Figure 4.34 *Digital drum servo circuit*

PG pulses are sent to the ramp counter once every 40 ms and reset the ramp counter to zero; the count then commences driven by the 554 kHz clock pulses. During the count period of 1.85 ms (normally at the halfway count at 0.92 ms or 1000000000) a sample pulse arrives at the trigger input of the 10-bit latch; this sample pulse is derived from either V-sync in record or control track in playback. When the sample pulse arrives, the 10-bit data representing the current ramp counter value is transferred into the latch memory. This is equivalent to the sample pulse transferring the ramp voltage into the analogue 'sample and hold' capacitor; this capacitor has been replaced by a latch memory.

Now the latch memory becomes the fixed reference into one side of the 10-bit comparator. It is only refreshed once every 40ms, and until it is refreshed again it remains a constant count value.

After setting the bistable so that Q is high at the start of count '000000000' the PWM counter commences to count up, driven by the 1.1 MHz clock pulses. At sometime during the count period the parallel output to the comparator will match the count stored in the latch.

When both inputs to the comparator are matched then it outputs a 'reset pulse' to the bistable and Q goes low until the PWM counter reaches zero once again. In this way the PWM counter is 'sampled' proportionally to the ramp counter but at a much higher repetition frequency. The sampling rate for the PWM counter is almost 100 times faster than that of the ramp counter.

Digital phase correction

Phase correction in this digital servo may be difficult to follow, this is because any phase variation due to drum motor rotational changes affects the PG pulse that resets the ramp counter, rather than the sample pulse which is fixed in phase by vertical syncs. For an example let us consider that the drum motor slows down so that it lags in phase. This means that the ramp counter reset is late. When the V-sync pulse is on time and the ramp count reset is late then the ramp count value is less than normal at the time the V-sync pulse triggers the latch. Via the comparator the reduced count is applied to the PWM counter with the effect that the bistable set period is reduced in width. The bistable inverted output that is used for the D/A integration filter is low for a shorter period than it is high; the integrated voltage therefore rises proportionally.

Conversely, if the drum motor speed increases, then the period between the PG pulse and the latching signal increases, as does the count. This time the PWM count between set and reset increases, increasing the bistable low period and reducing the integrated voltage.

The affect of a low value ramp count is to produce a higher servo DC error voltage. This corresponds to a low mark/space ratio from the bistable, whereas a high count produces a high mark/space ratio and a lower error voltage, as shown in Figure 4.35.

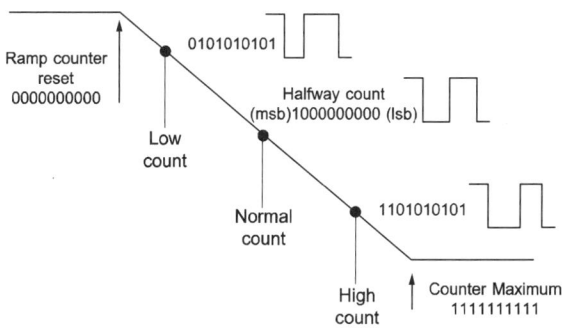

Figure 4.35 *Relationship between ramp counter and bistable PWM mark/space ratio*

Digital speed control

The basic circuit for the digital speed control is similar to that of the phase control, the difference being that both the counter reset and latch control are the same signal, the FG pulse. The basic circuit is illustrated in Figure 4.36.

The overall operation is identical to the analogue speed control shown on p.66.

After the sinusoidal FG signal has been squared in a Schmit trigger type of circuit it is applied to the speed control circuit. Although the processing is digital it is still essentially a frequency-to-voltage converter.

An FG counter is reset to zero by the rising edge of the FG pulse. It is only a small 3- or 4-bit counter and the count period is short and acts as a delay monostable. After the count period, the FG counter resets the saw-tooth counter, which then starts counting. When the falling edge of the FG pulse occurs it triggers the latching part of the 10-bit latch memory and stores the count present at that time in the memory where it forms a reference input to the comparator.

As it was in the phase control system, the PWM counter sets the output bistable at the start of count and a match in the comparator resets the bistable. In this way the mark/space ratio of the bistable is varied according to the count value held in the comparator. In turn this count value is dependent upon the time period between the FG rising edge and falling edge.

If the motor is running slow then the FG pulse period is longer so the saw-tooth count value is greater, and so is the value in the comparator. Consequently the PWM count is greater and the bistable mark/space ratio is greater.

In this example bistable output is from the non-inverted port, and therefore the high period is longer in raising the value of the integrated control voltage from the D/A integration filter and increasing power to the motor.

Conversely if the motor is too fast then the count is shorter and the control voltage falls, reducing power to the motor and slowing it down. This circuit can apply as a speed control FG loop to both drum and capstan motors.

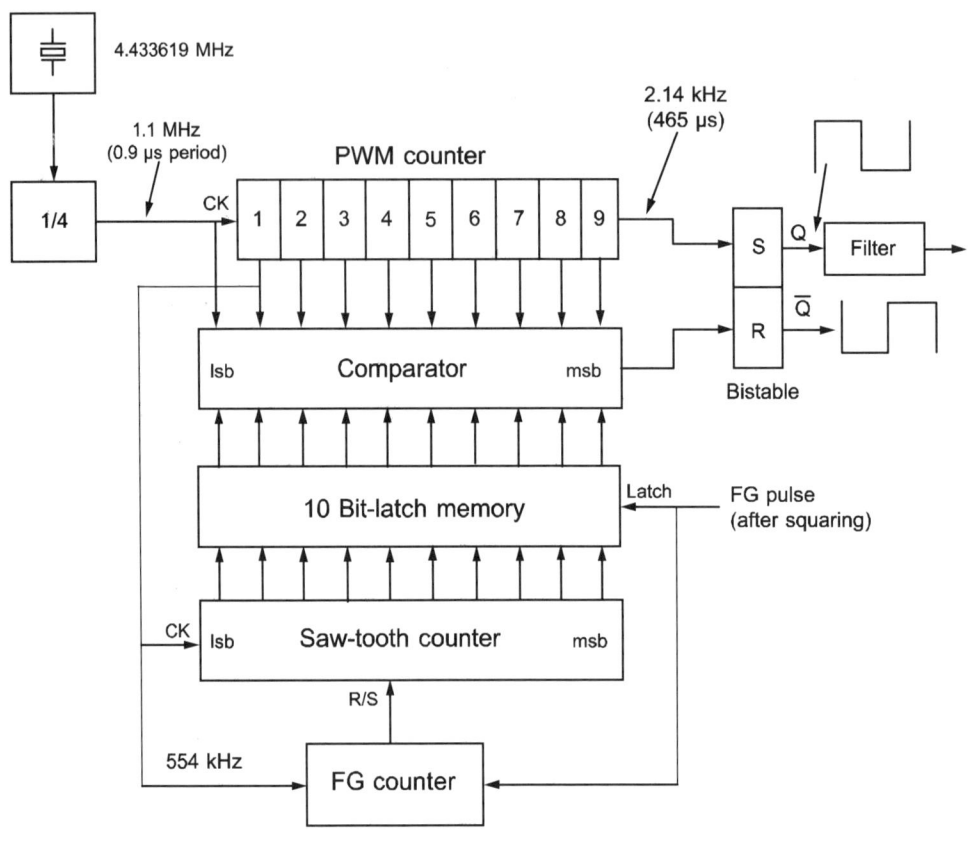

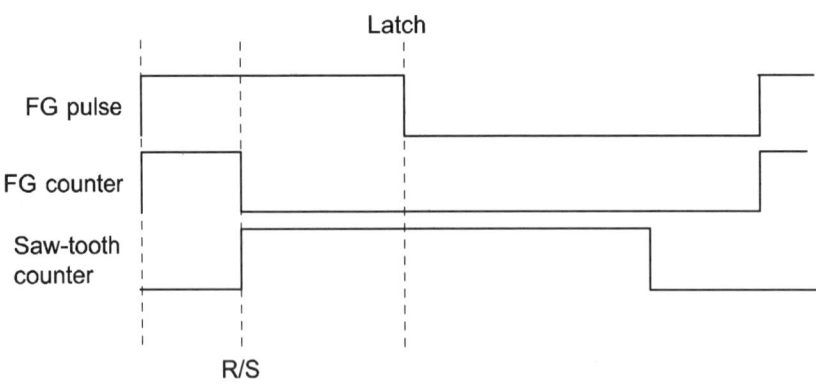

Figure 4.36 *Digital speed control circuit and FG timing waveforms*

Digital control track

So far we have looked at the control track recording signal as a rectangular waveform generated by a monostable timed at 28 ms to effectively divide the incoming 50 Hz vertical syncs, by 2 to 25 Hz; also indexing and tape identification require that the control track has different mark/space ratios for data 0s and 1s. By using precision programmable counters instead of capacitor-based monostable delays then the control track mark/space ratio characteristics can be adapted to suit any requirement.

Confirmation of the monostable approach can be seen in Figure 4.16 between the field sync pulse and the 'divide by 2 delay' waveform.

Once a monostable is triggered by a field sync pulse it is unaffected by a following pulse whilst it is still in the triggered mode; it can only be triggered again after it has reset. As long as the timing period is greater than the period between two successive field pulses (20 ms) then it will divide the field pulses by two.

Figure 4.37 shows a counter with a bistable and an And gate; the count is set to 51 (1+2+16+32). A count period of 24 ms is achieved from the input clock of just over 2 kHz (465 μs) from 51 × 465 μs = 23.715 ms. This is within the specification of a 60% duty cycle for a 40-ms period as shown in the accompanying waveform.

When a field sync pulse arrives it triggers the bistable, output Q goes high and resets the counter to zero. Driven by clock pulses with a period of 465 μs, counting commences up to 51, detected by the And gate. An output from the gate then resets the bistable and Q goes low. As with the monostable, once the bistable is set it ignores any subsequent pulses until it has been reset once again. As the count duty cycle may be less than 50% then the vertical sync pulse has to be divided by 2 to 25 Hz before the control track duty cycle counter. The And gate can be programmed by external logic to any count, so if the control track mark/space ratio has to be altered to 27.5% duty cycle (27.5% × 40 ms = 11 ms) then the count has to be altered to 24 (11 ms/465 μs =23.65) to be within the specification for data '1'.

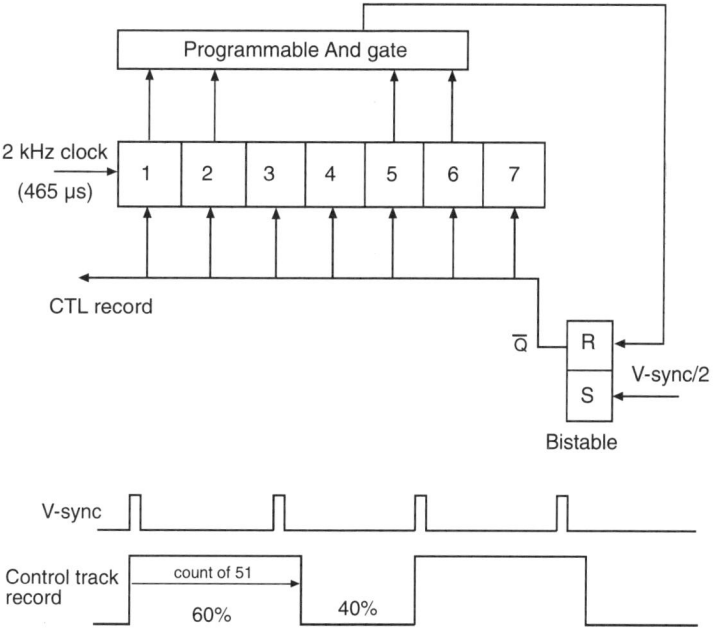

Figure 4.37 *Control track duty cycle counter*

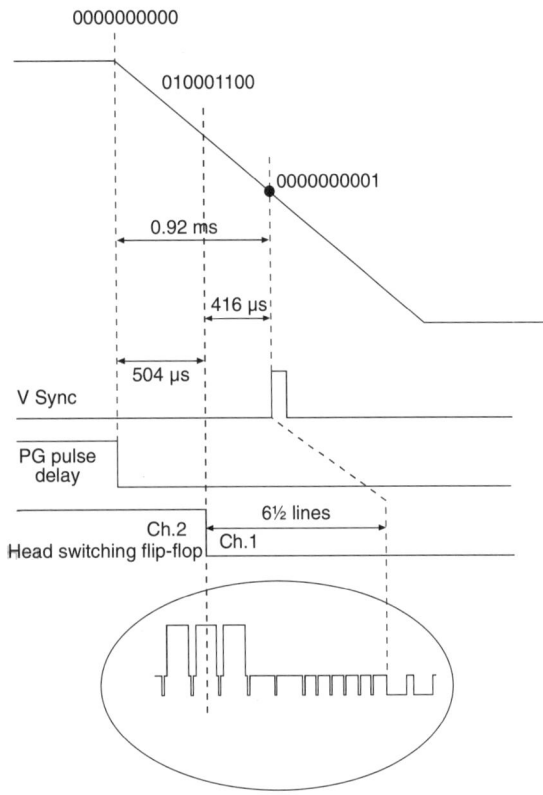

Figure 4.38 *Automatic record phase timing*

Digital record phase

With precision counters, digital control circuits can eliminate many set-up functions by automating the system. The servo is designed to set the field sync pulse, in record, in the correct position where it should be on the track, as seen in figure 4.12(a). First, the servo has to be set up in playback using a precision manufacturer's test tape, as the field sync pulse on this tape has been recorded in the correct position. The head switching signal is set to be displaced from this pulse by 6½ lines, or whatever the manufacturer's service manual gives; in some cases this can be up to 8½ lines. Hence, step one is to fix the head switching signal with respect to the field sync pulse on the test tape.

Step two is to set the record field sync with respect to the head switching signal by adjusting the record phase control, unless there isn't one and it is automatic!

In a digital servo it is possible to fix the record phase control by locking the drum switching (flip-flop) signal to field syncs by 6½ lines, using digital counters. Figure 4.38 shows the waveforms to do this, it is important to note that the head switching signal in replay is now assumed to have been correctly set.

A virtual ramp counter is reset to start at a count of 0000000000 at the transition from Ch.2 head to Ch.1 head by a PG pulse from the drum motor. As the ramp count takes 1.8 ms and the sample pulse occurs halfway through the count at 0.92 ms, or 920 µs; this represents the time period between a PG pulse and field sync.

A 6½ line period is 6.5 x 64 µs, which is 416 µs, so if a delay of 504 µs is introduced between the PG pulse and the head switching flip-flop then the fixed period of 416 µs is defined between the head switching signal and the vertical sync pulse.

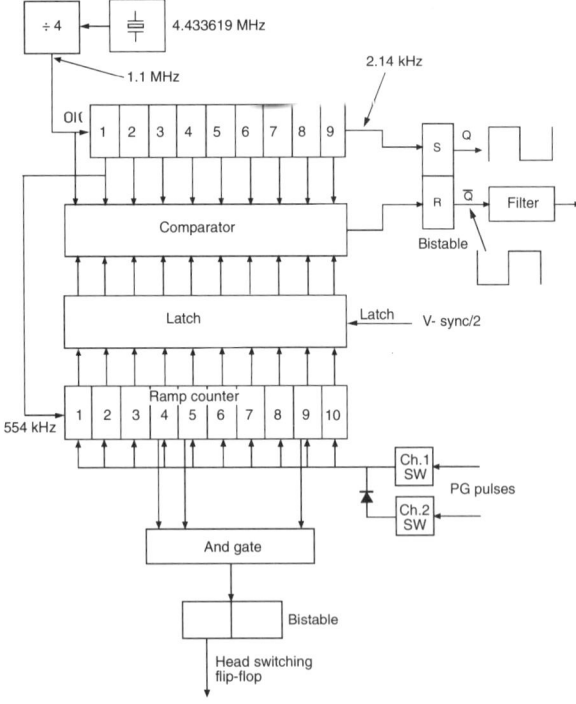

Figure 4.39 *Automatic record phase timing circuit*

Motor control and servomechanisms

This concept does take some thinking about, but it does rely upon two important factors. One is that the 504-µs delay is present in the playback mode when the head switching signal is fixed with respect to the test tape; then in record the vertical sync signal is fixed by 416 µs with respect to the head switching signal. Consequently the vertical sync pulse is recorded onto the tape in the correct position.

Figure 4.39 is the circuit diagram showing how the PG pluses via Ch.1SW and Ch.2SW delays reset the ramp counter. They are both set up in playback, often as a single setting, to what are essentially, dual programmable delay counters. Either PG signal will reset the ramp counter and the head switching waveform is formed at a count of 010001100 (280). Only the Ch.2 to Ch.1 transition is used as the ramp; Ch.1 to Ch.2 is required only for head switching waveform symmetry.

Clock pulses into the ramp counter are at 554 kHz or 1.8 µs period. The output to the And gate is set to (8 + 16 + 256) which is a count of 280, (280 x 1.8 µs = 504 µs). Each time the count of 280 is reached the bistable is toggled over, producing the head switching flip-flop symmetrical waveform with the inbuilt 504-µs delay between the PG delay counters and the bistable transistions. This leaves the remaining 416 µs, or 6½ lines, between the head-switching waveform Ch.2/Ch.1 transistion and vertical sync.

LSI servo IC

Rotational information from the video head drum in Figure 4.40 is sent to the drum pickup input (PG) pin 22 as a 25-Hz square-wave signal; the drum PG pulses are amplified and squared in a Schmitt trigger circuit externally to the IC. Drum PG pulses from pin 22 set a monostable with a variable delay connected to pin 20 which is used to set the 6½ H switch point position in playback. A set/reset bistable is used to obtain the drum flip-flop on pin 17, there is an inbuilt 2H delay for dual speed versions to compensate for the 2-line shift between LP and SP due to the mounting offset of the video heads (see also p.173). Internally, the flip-flop signal is used on the drum phase detector, and the drum phase error is on pin 42 as

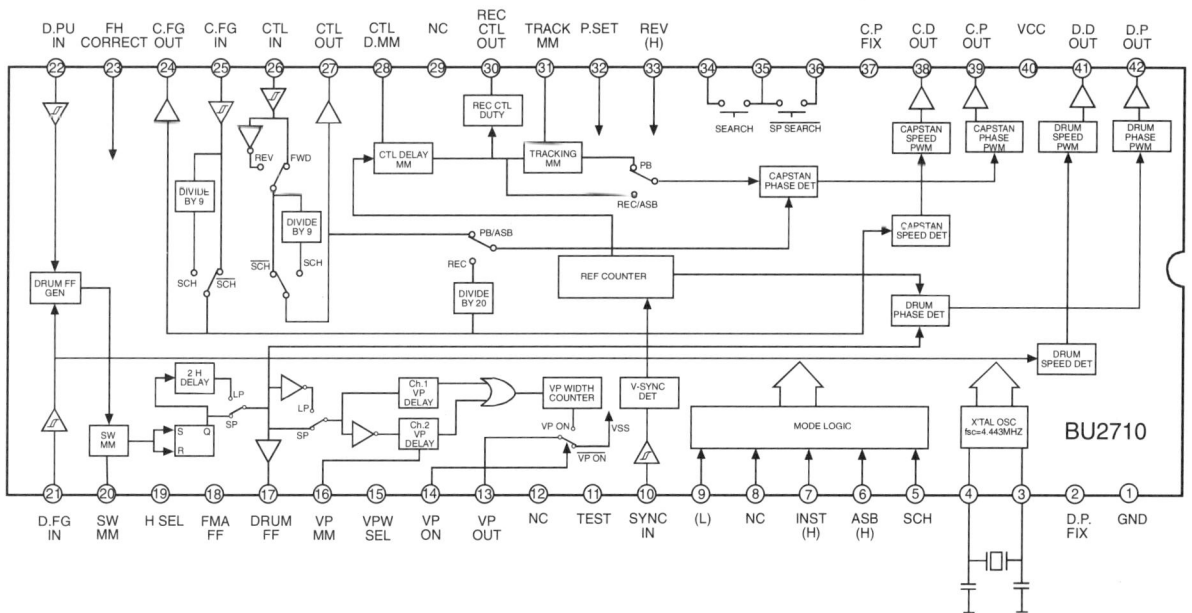

Figure 4.40 *Typical large scale integration servo IC with some digital processing*

a 1.1 MHz pulse-width modulated signal.

On pin 21 there is the drum FG signal for speed control entering the IC at a frequency of 1500 Hz. This is frequency-to-voltage converted in the drum speed detector to produce the drum speed error signal for the pulse-width modulator on pin 41.

Pin 19 is a head select input for 2- or 4-head use and is set to a fixed mode on a video recorder that does not have separate LP heads.

Pin 17 is the output of the drum flip-flop signal to the system control and colour circuits, and the phase-shifted FM audio head-switching flip-flop is on pin 18.

In record, composite syncs are applied to pin 10, horizontal syncs are removed and vertical syncs are divided by 2 to 25 Hz; these are used to reset the reference counter. In the absence of V sync, such as playback, the reference counter outputs 25-Hz pulses counted down from the reference oscillator on pins 4 and 3. This is shown as a crystal oscillator, but it can often be a 4.43 MHz subcarrier signal from the colour circuits.

The output of the reference counter is used as the drum servo phase comparator reference whilst the drum flip-flop provides the second, variable, input.

Both the drum speed and phase pulse-width modulated signals are integrated into DC error signals and combined as the drum error signal in an adding amplifier, before being applied to the drum motor. Their combined range of operation is shown in Figure 4.27.

Vertical pulse generator

This part of the circuit within the IC produces the synthesised vertical sync pulse, which is used in still, slow and visual search modes. An inverter is used between LP and SP modes to compensate for the 180° phase shift between LP and SP video heads, which are mounted opposite each other (see Figure 8.1, p.154).

Two delay monostables form the synthesised vertical syncs with one monostable being adjustable for vertical jitter, as the vertical lock control on pin 16. A vertical width counter determines the actual width of the synthesised vertical sync pulse that exits on pin 13.

Record CTL track

An output from the reference counter is used to construct the control track recording signal. After delay by the CTL delay monostable the record duty cycle counter determines the duty cycle mark/space ratio for indexing or data recording out of pin 30.

Capstan servo

The reference counter provides a 25-Hz clock for the capstan phase comparator, with the motor feedback coming from the capstan FG signal on pin 25 divided by 20. As the FG signal is 500 Hz division by 20 results in a 25 Hz match for the sample component. Capstan FG signal on pin 25 is also frequency-to-voltage converted in the capstan speed detector for speed control. Similarly to the drum, pin 24 is a buffered output to systems control for monitoring the capstan motor.

In replay the drum phase control consists of the drum PG pulses and a 25-Hz crystal reference, counted down in the reference counter from the 4.43 MHz oscillator.

The capstan phase detector input is the replayed control track from pin 26 via the switching and tracking monostable on pin 31. The reference for the phase detector is from the reference counter output at 25 Hz, counted down from the 4.43 MHz oscillator input on pins 3 and 4.

The special facilities accommodated within the IC are phase-locked picture search and assembly editing.

For picture search operation at x9 play speed, the capstan FG input is divided by 9 to run the capstan speed at x9 play. Loose phase lock is achieved by comparing the replayed control track, also divided by 9, to the reference counter with an extra division by 9 inbuilt into the programmable counter, this stabilises the noise bars on the TV screen.

Assembly editing is achieved by the ASB switch positions where it is required to lock the capstan servo to the incoming signal for a short while (approximately 500 ms; see assembly editing). It enables the capstan phase detector to reference from incoming syncs (REC/ASB) and replayed control track (PB/ASB) for the short run-in period before switching over to full record mode.

Pins 34–36 control the capstan servo gain for visual search while mode control logic from the main microcomputer controls all of the functions on pins 5–9. Pins 2 and 37 fix the drum and capstan discriminators for test purposes, and pin 33 controls the capstan motor reverse operation.

In this IC some functions are digital while others still use capacitor-controlled monostable delays. These could easily be substituted by digital counters.

General fault finding on servos

In order to determine a servo fault, close inspection of the picture and measurement by an oscilloscope is necessary. A test meter is of little use except for continuity of circuits or motor windings.

The first step is to decide if the fault is in the drum or the capstan (tape) servo, or indeed in both. A complication may arise in camcorders and advanced VCRs wherein if the drum servo has large errors then the capstan servo is inhibited to prevent tape damage.

In order to decide which servo is at fault two differing symptoms can be considered. Speed changes or errors in the drum will cause increases or decreases in the 'frequency' of the replayed video. This particularly affects the horizontal sync frequency. Speed changes or errors in the capstan servo will cause changes in the pitch of the normal linear audio — sometimes causing a warble, which sounds like underwater speech.

To check a servo, a 'known good tape' is required to analyse the replay signal. If a perfect replay is in evidence then there may be just a recording fault. Good playback of a known good tape with consequential irregular playback of its own recordings indicates a recording fault. This is fairly rare for servo faults as most affect both recording and play-back and can be diagnosed in replay.

Systems for checking where a good replay and a faulty recording is confirmed

A faulty recording is confirmed when replay of own recording is a sequential display of a good picture interrupted by a break-up to spots covering the whole screen, with additional change in audio pitch (linear sound track) or muting (Hi-Fi track).

For control track recording circuits check that a square wave is present at the servo output and follow it up to the control track head. Note that it will be very spiky with considerable overshoots across the head due to its inductance. Check that the control track head coil earth end is at ground, as it may be switched by a semiconductor from logic. If the drive waveform is evident on both sides of the winding then the earth return is open circuit. Check for the presence of the capstan servo reference, which may form the drum servo reference or a crystal clock reference. A dirty control track head can prevent a good CTL recording and playback. Inefficient head-to-tape contact can adversely affect both recording and playback doubling the error, whereas interchanging the tape with another video recorder may not indicate a problem.

Head switching point rolling, uncontrolled up or down the picture

Check that the head switching point is evident on the replay of the recorder's own recordings. It is a thin line with horizontal pulling below; it may be moving either up or down and not stationary. If this is so, then the field sync input reference is not getting through to the drum servo, or it is not being correctly separated from the input video signal. Whatever the cause, it is evident that the drum servo is not locking on to the incoming signal in record mode.

Check that the drum servo reference from input syncs, separated in either the luminance or chrominance circuits, are clean syncs with no shadows of video signal. Confirm this at field rate.

A playback fault will indicate drum or capstan servo by inspection of the replay signal – audio and video. It is a fair bet that if there is a playback problem then the recording will be faulty. Concentrate on solving the problem in playback using the 'known good tape'.

Video

Loss of horizontal syncs in a manner similar to a TV horizontal hold problem, or the whole picture shifting or swaying sideways rhythmically, are both symptoms indicating a drum servo or motor fault.

Audio

Variation in the pitch of the audio will be an indication that the speed of the tape is running fast or slow or changing, either way, when the capstan servo is not locking. This may be accompanied by a regular change in tracking (for example, a clear picture which is regularly degraded to a mass of spots and then back to clear picture). If the guides and tape path compatibility of the test machine are not good, then break-up may commence anywhere on the picture as a band of noise which then spreads up or down, rather than evenly all over the picture, or in horizontal bands.

Drum servo faulty drum running fast

First, consider the f-v circuit: this is a frequency-to-voltage converter. It is usually an inverter, substantiated by going into the negative input of the mixer op-amp DC amplifier. To test this part, slow down the drum and watch the output voltage of the f-v circuit rise in sympathy – if it does, then so far, so good. If not, check for FG waveform from the drum motor and amplifier up to the servo IC.

Checking the phase control section

A magnet passing a coil generates the drum PG signal and the level is usually fairly good for measurement. The PG signal has to trigger a dual monostable in order to form the drum flip-flop square wave. There are usually odd points around this circuit to check with the 'scope that both monostables are duly triggered. Any irregularity in the flip-flop signal can again be caused by low PG level or a monostable mis-triggering, or both.

The drum flip-flop is compared to the main steady reference signal, which may be a crystal, internal IC clock, or the 4.43 MHz colour reference from the Y/C PCB.

Check for the presence of the reference signal or its presence at the phase comparator. The DC output voltage may be rapidly varying up and down if the servo is not locked, and this variation may be followed through to the MDA. A difficult fault to rationalise is when all the signals and variations are present but the servo still does not lock. As the phase control and speed control join in an op-amp prior to the MDA, it is necessary to establish which is at fault. Both may vary with all inputs present and be correct if the motor itself is the cause of the problem.

Checking the servo waveforms

The signal levels given are general, so refer to the particular model service manual for absolute values. Figure 4.41 shows a basic drum and capstan servo.

The first job in the fault-finding sequence is to establish which parts of the servo are functional, and this can be commenced by slowing the drum to see if a picture can be restored. It will indicate

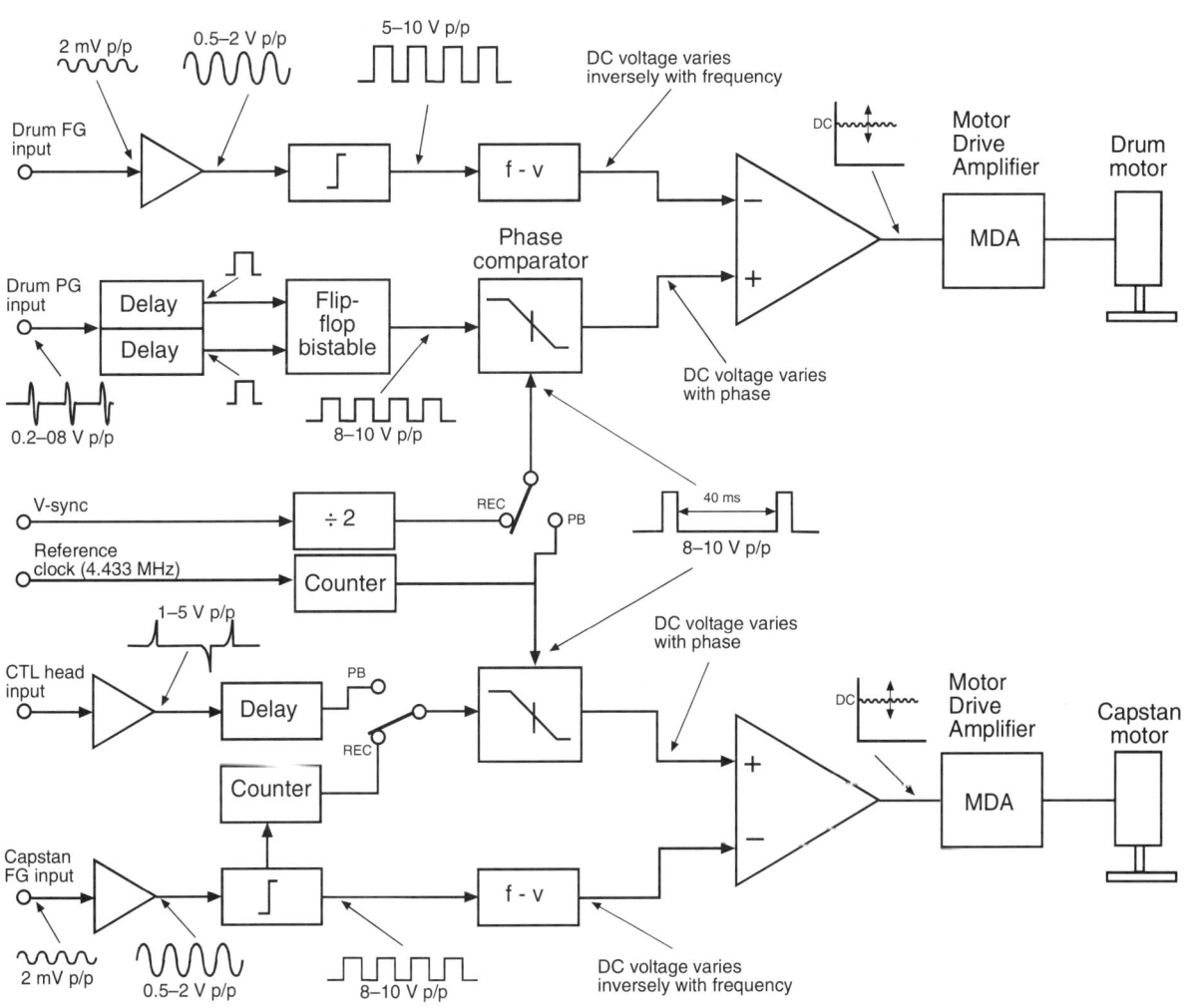

Figure 4.41 *Overall servo system with expected waveforms as an aid to fault finding*

that the drum is running too fast if a picture of sorts can be obtained, and, conversely, too slow if a picture cannot and the lines get worse and increase in number.

An oscilloscope is essential from now on. Next, check if the two drum servo feedback signals, PG (pulse generator) and FG (frequency generator), are present from the motor section, although they may be varying considerably.

The FG circuit

It is not very practical to measure the drum FG signal prior to the first amplifier but out of the amplifier it should be around 1 or 2 V, at least enough to drive the following Schmitt trigger which squares it off. Any irregularity in the mark/space ratio of the square wave out of the squaring circuit would indicate an insufficient FG level from the motor. If this is the case, check the frequency of the reference signal and also the electrolytic capacitors on the DC amplifiers — particularly tantalum types used in DC amplifier feedback components, which are used to shape the AC response of the servo. Leaky tantalums reduce the gain, and hence effectiveness, of the amplifiers. The same electrolytics can dry up and increase their impedance, particularly when they are used to couple a pick-up coil to an amplifier input where the DC voltage is low. In the example shown where the DC voltage derived from the FG signal is mixed with the phase signal on the negative amplifier input, failure of the FG signal processing will result in high speed of either drum or capstan motor with no phase control.

Phase control circuits

As shown in Figure 4.41, the level of the PG signal is around the 0.2-1 V p/p and can be measured off the pick-up coil input to the servo circuit board. The main point is the output of the dual monostables that are triggered by the PG pulses and combined to form the drum flip-flop or RF switching signal. Erratic outputs or wide mark/space ratios will point to mis-triggering due to too low a level. Multiple outputs will point to too high a drive level, and some VCRs critical to this have a PG level control. The square wave should have an even 20 ms mark/space and a 40 ms period, assuming correct drum speed.

The reference input in playback is derived from a clock source and must arrive at the comparator input with a 40-ms period for the drum servo to lock up to it. If the output from the comparator is varying by more than a few hundred millivolts then the servo is not locking. If both inputs are present and samples are running through a ramp waveform then the DC voltage at the output will vary considerably. The fault may be in the motor drive circuits if the FG circuits check out as functional.

Capstan servo

The analysis of the capstan servo is similar to that of the drum, with only one difference. Instead of PG pulses, the signals are derived from replayed control track pulses. Measurement is not practical on the audio/control track head, but the CTL pre-amp levels of the pulses can be expected to be around 0.5 V p/p.

As a final measure, if you are not sure what is happening in a servo that is out of control it is often possible to stabilise the motor by applying an external voltage from a modern variable power supply to the input of the motor drive amplifier. The output of the servo should be disconnected to avoid damage by feeding back the DC voltage. A good starting point is 2-3 V; increase it and measure the drum or capstan FG until the correct frequency as given in the service manual is achieved. This technique is not advisable on VCRs with digital servos, however, as errors will be too great for the IC counters and pulse width outputs.

Motors and motor drive amplifiers

DC brush motors used in early machines suffered from wear on the brushes, which created a ripple on the servo. In the drum servo it would result in a 'soft lock', taking a long time to stabilise with an additional side-to-side sway. In the capstan servo there is a warble on the sound (in the worst cases it sounds like underwater speech). Check for ripple across the DC motor and in both instances replace the motor and associated belts.

Multipole motors suffer from missing drive phases; this makes the motor 'lumpy' and seriously affects servo stability. A cause may be a faulty drive transistor or IC, or a missing Hall sensor output.

Digital servo fault finding

The operation of a digital servo has been covered but fault finding is different to that of the analogue servo. Refer back to Figure 4.40, which is that of a typical LSI servo IC.

Pins 38 and 39 are the capstan speed and capstan phase control outputs, respectively. Pins 41 and 42 are the corresponding drum outputs. All these outputs are a high frequency square wave with variable mark/space ratio for control. In the event of a fault the output will range from fixed 0 V to fixed high output (5–12 V according to the particular supply voltage). In between the fixed high or low levels may be sporadic bursts of square wave with nothing between, or a continuous square wave with sections missing. This indicates that the servo is trying to work and that the IC is operational. Whatever the shapes or breaks in the square wave, the resultant integrated DC level can be checked. It will correspond to the equivalent analogue DC error voltage.

A fixed DC level which cannot be persuaded to change state by judicious manipulation of the motor speed may indicate a faulty servo control IC or an input reference missing. However, in the event of, say, a large error in the speed there will be some changes in the levels of the speed pulse-width modulation outputs, but the phase outputs are clamped high or low, as the servo control is well out of operational range. This may be a normal situation, so be careful in your judgement. This example IC has the required inputs for servo operation that must be checked.

Pins 4 and 3 are the internal clock from which the servo reference is derived. It could also be a 4.43 MHz carrier wave from the colour oscillator tapped off the Y/C circuits. This can be easily checked and measured.

Pin 22 is the drum PG input and pin 21 the drum FG input. In some modern machines these two signals may be mixed in the drum motor as a single PG and FG signal and subsequently separated within the servo IC. FG frequency can be measured, but this is pointless on an uncontrolled servo. Oscilloscope checks are better, as these will show up any variations in level or frequency, which would create faults.

Pin 25 is the capstan FG input. In record it controls both speed and phase circuits. Speed control is derived from an (f-v) speed detector, where it is divided by 20 and compared to the reference signal for phase control. As pin 24 is an output then the FG can be monitored. Triangular waveforms appear on pins 28 and 31 as capacitors form the timing components for the monostables. The recording CTL signal is passed out of pin 27 and can be checked at this point.

In replay the amplified control track signal is fed in on pin 26, where it can be measured. Look for a signal greater than 0.5 Vp/p, as it will have been amplified from the pick-up head by a pre-amp prior to the servo IC input.

Multipole motors

Motors in digital circuits are invariably direct drive brush-less multipole motors. The rotor section is the only moving part and the drive force is magnetic.

Coils are driven by Hall effect devices and they set up a rotating magnetic field that is followed by the impregnated magnet forming the rotor. Part of the rotor is a flywheel and around the circumference of this flywheel is a rim magnet impregnated with many N/S segments for generating the PG and/or FG signals depending upon the application.

The rotating magnetic field pulls the flywheel around by acting upon the rotor section. It is the rotating magnetic segments on the flywheel that switch the Hall effect devices. These in turn switch the drive coils via the motor drive transistors that set up the rotating magnetic field, and rotation continues. Speed is controlled by the value of the current in the drive coils.

Hall effect device

The Hall effect device is not unlike any other semiconductor or magnetic deflection system, as it is a combination of the two. A bias current (Ib) is passed through the slab of semiconductor material, set up by the bias voltage (Vb) across it.

As shown in Figure 4.42, if a current is passed through a piece of semiconductor material, a small magnetic field will be present around the current path in a similar manner to that set up by passing a current through a wire. If an external magnetic field is now passed through the semiconductor material, the two fields will interact and the flow of electrons within the material will be modified and create positive and negative halves across the semiconductor slab at right angles to the bias current. This is called the 'Hall' voltage (Vh) and can be amplified and utilised.

The value and polarity of Vh depends on two variables: the external flux density (B), a change in flux density produces a change in Vh; and the bias current Ib, a change in bias current will affect Vh.

Two other factors affect the value of Vh: the thickness (d) of the semiconductor material and the Hall coefficient (Rh). These are fixed by the material and manufacture of each device.

With any given device, d and Rh will be constants, so we do not concern ourselves with these from a servicing point of view. The value of Vh will depend on the bias current, Ib, and the external flux density, B.

$$V_h = \frac{R_h \times I_b \times B}{d}$$

In practice with a Hall motor, Ib is usually kept at a constant value using a source voltage (Vb) of around 2 V per Hall device, with three Hall effect devices in series. A typical Vh output value is in the region of 0.5 V.

Reversing the polarity of either Ib or B will result in reversal of the polarity of Vh. This is an important factor in the operation of a Hall effect motor.

In a typical Hall motor the brush and commutator arrangement is replaced by three Hall

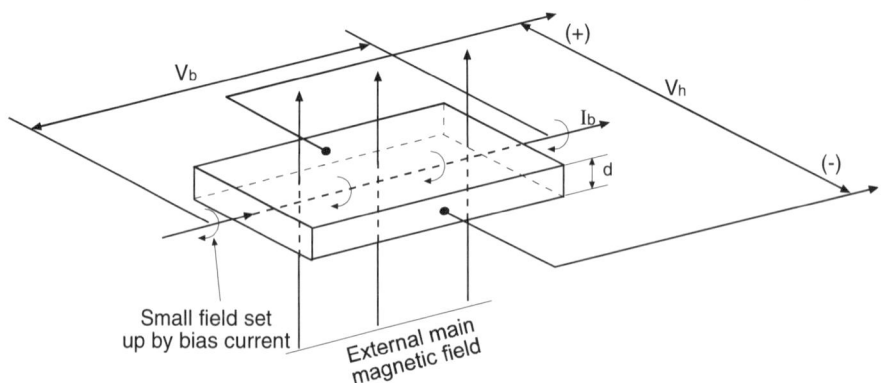

Figure 4.42 *Hall effect device construction and operation*

Motor control and servomechanisms

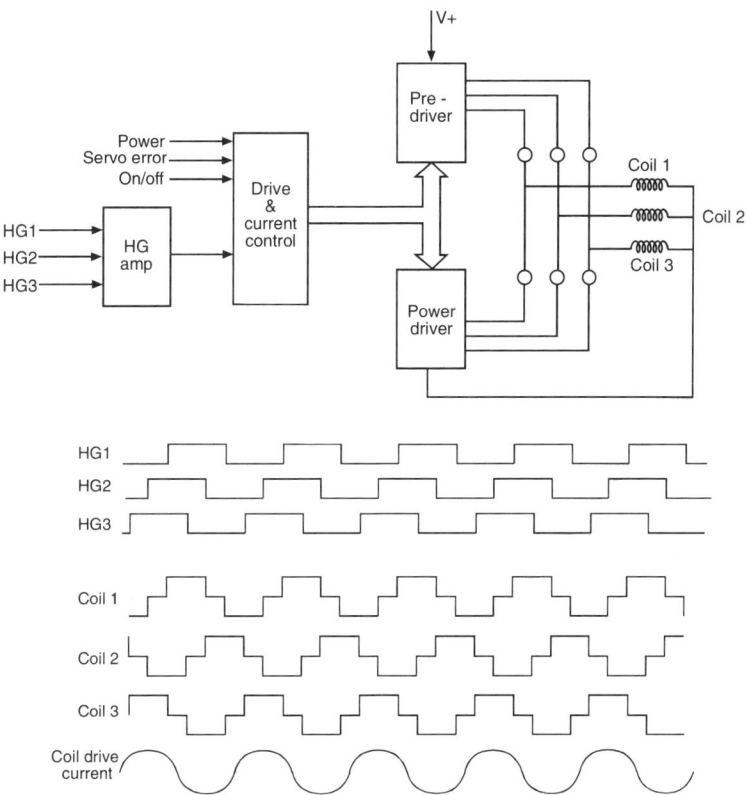

Figure 4.43 *Typical motor circuit and waveforms*

elements that are sometimes referred to as diodes. A series of permanent magnetic sectors pass over these elements and as the motor rotates, the magnets activate the Hall elements that in turn control the motor drive currents that flow through the windings. The traditional mechanical coupling between the motor's rotor and stator is thus replaced by a magnetic coupling, which is the reason for the Hall motor's smooth, noise-free operation and long life.

The rotor consists of a multiple-pole ring magnet, which spins over the Hall elements with a very small clearance of about 0.5 mm.

Adjusting the current that flows through the stator coils can control the motor's speed. In a video recorder the motor is servo controlled by the supply voltage being regulated by the phase and speed output error voltage of the servos. Power to the motor can be supplied via a DC/DC power converter using an output error voltage and increasing it in level and power. What is now a servo-varied power supply is applied to a three-pole transistor bridge network to provide three-phase drive to the motor coils, current in the coils being controlled by the DC/DC converter.

A typical motor diagram and waveforms are shown in Figure 4.43. Hall effect device outputs go into a high gain amplifier to increase the level of the sinusoidal waveform and then clip it to give a square wave with stable switching edges, HG1, HG2 and HG3. Servo inputs, error, stop/start, direction and power, are fed into the controller along with the HG square waveforms. A three phase drive to a switching transistor matrix provides the motor drive. Initially the drives are stepped voltage waveforms as illustrated for coil 1, coil 2 and coil 3 but the inductance of the coils smoothes this out to a sinusoidal current.

Drum motor circuit

Figure 4.44 shows a typical direct-drive head drum motor assembly with a 16-pole rotor and three Hall effect devices. Each Hall effect device is mounted at 120° from the others, giving the motor a three-phase 120° drive. Bias current is supplied from a 5-V supply via a temperature compensating diode with a common return resistor. This results in approximately 1.3V (V_b) across each Hall effect device. As the north and south magnetic sections pass each Hall device positive and negative pulses are generated from the V_h outputs. Due to the magnetic impregnation of the rotor, the V_h waveform is almost sinusoidal and so the waveforms HG1, 2 and 3 are clipped in the 'Hall amp and drive wave mixer' section to give the square waves shown on Figure 4.43.

Drum servo reference is used to develop a drum error voltage to the DC/DC converter to supply power for the current within the drive coils. As the combined drive current passes through a common 0.68R return resistor a check for even, three-phase drive can be made here. Any 'lumps' in the waveform will indicate that one or more of the drives to the coils is absent, or that one of the drives may have an incorrect current value. Measurement of the Hall gate serial bias DC levels will indicate a failed Hall device. HG2 will be around 4.4 V, HG3 about 3 V, and HG1 about 1.7 V.

As there are many variations in motor design, the drum motor may have a separate PG pick-up coil and magnets upon the flywheel circumference, or the PG pulse may be embedded in the FG

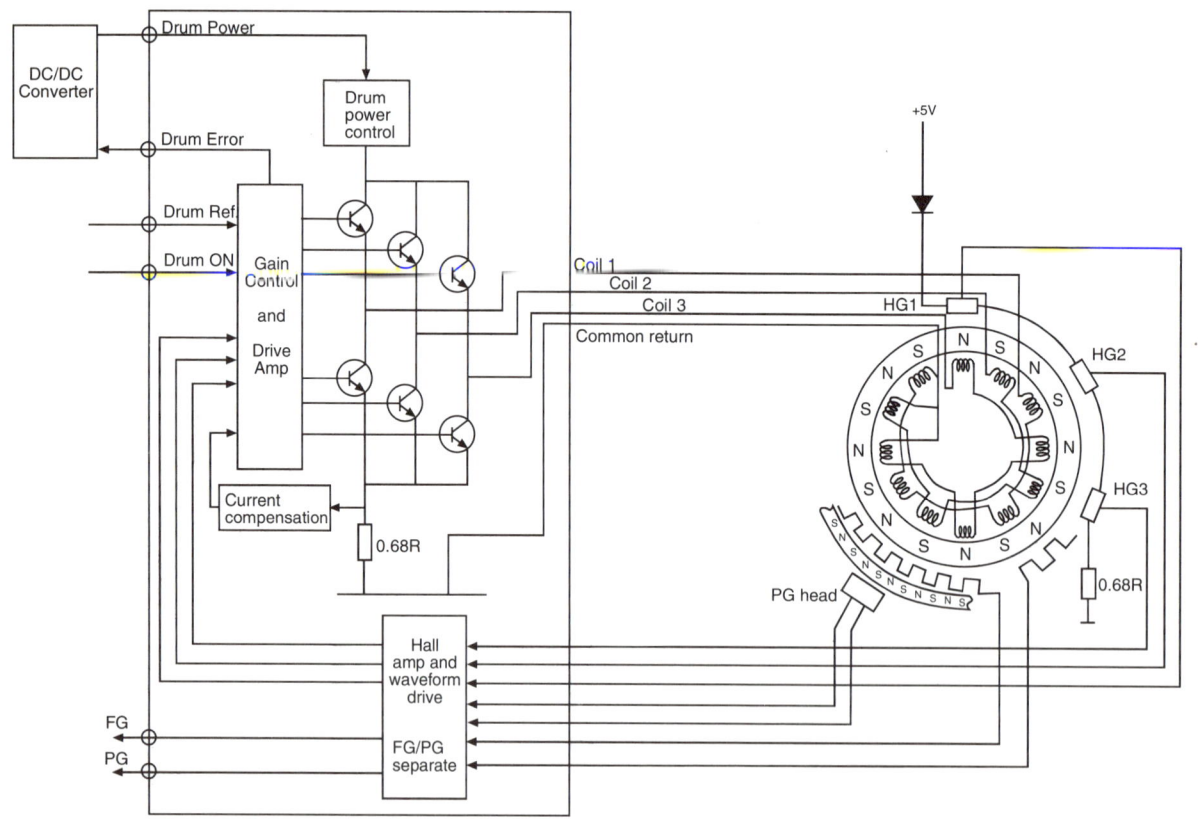

Figure 4.44 *Drum motor circuit diagram*

output.

A PG pulse is embedded into the FG pulse train. It cannot be easily seen on an oscilloscope, the pulse is isolated in an FG/PG separator within the motor drive IC and can be confirmed at the output.

In the event of a measured problem, or if the FG or PG pulse outputs are missing, then the motor should be replaced. If a motor drive fault does occur with a drum motor then rotational power is reduced and it may not start to rotate unless given a push, or it may just wobble a bit. In either case the replayed picture may have a severe side to side wobble.

Capstan motor circuit

Figure 4.45 is the equivalent capstan motor. It has a smaller number of drive coils and fewer N/S impregnations, and it rotates at a much slower speed than the drum motor. It still has 120° three-phase drive and a larger flywheel assembly. Speed is varied by controlling the power supply to the drive transistors in the same way as the drum motor. Either a biased magnetic resistor (MR) or an inbuilt PCB coil can be used to generate the FG signal from the magnetic segments on the flywheel rim.

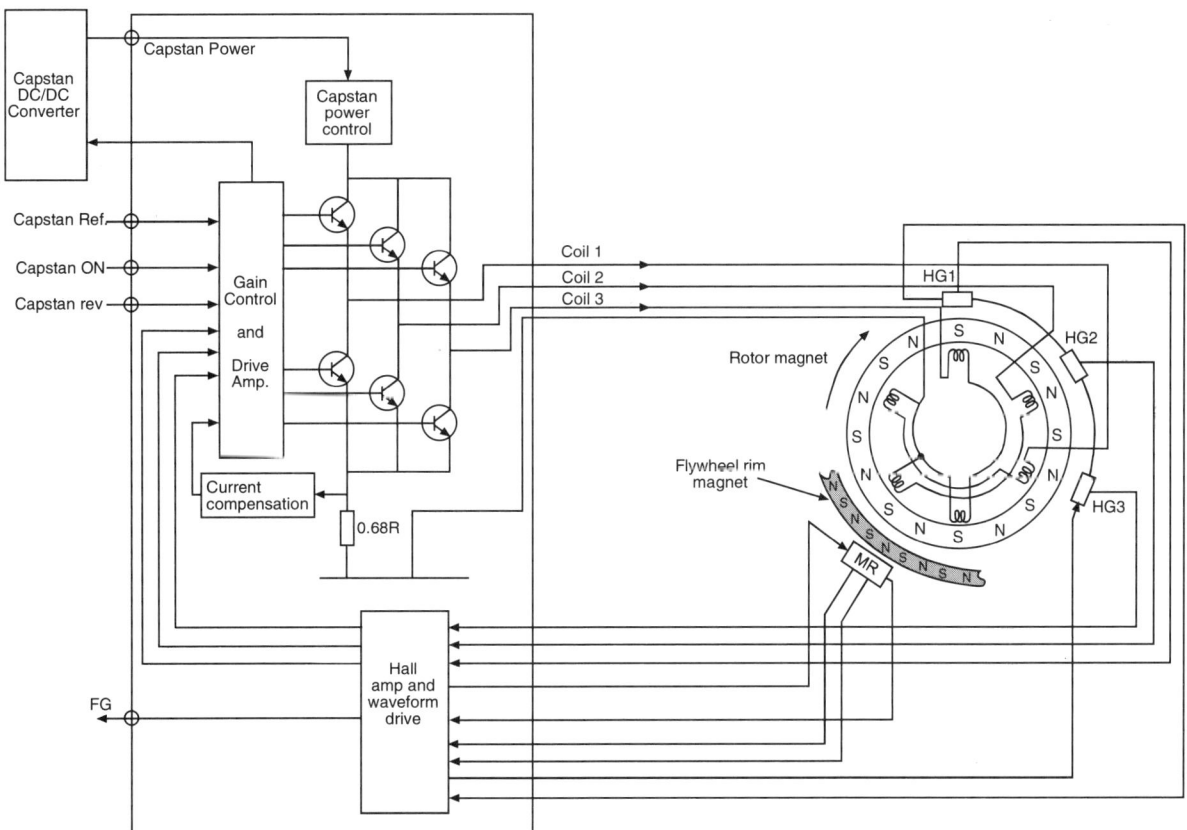

Figure 4.45 *Capstan motor circuit diagram*

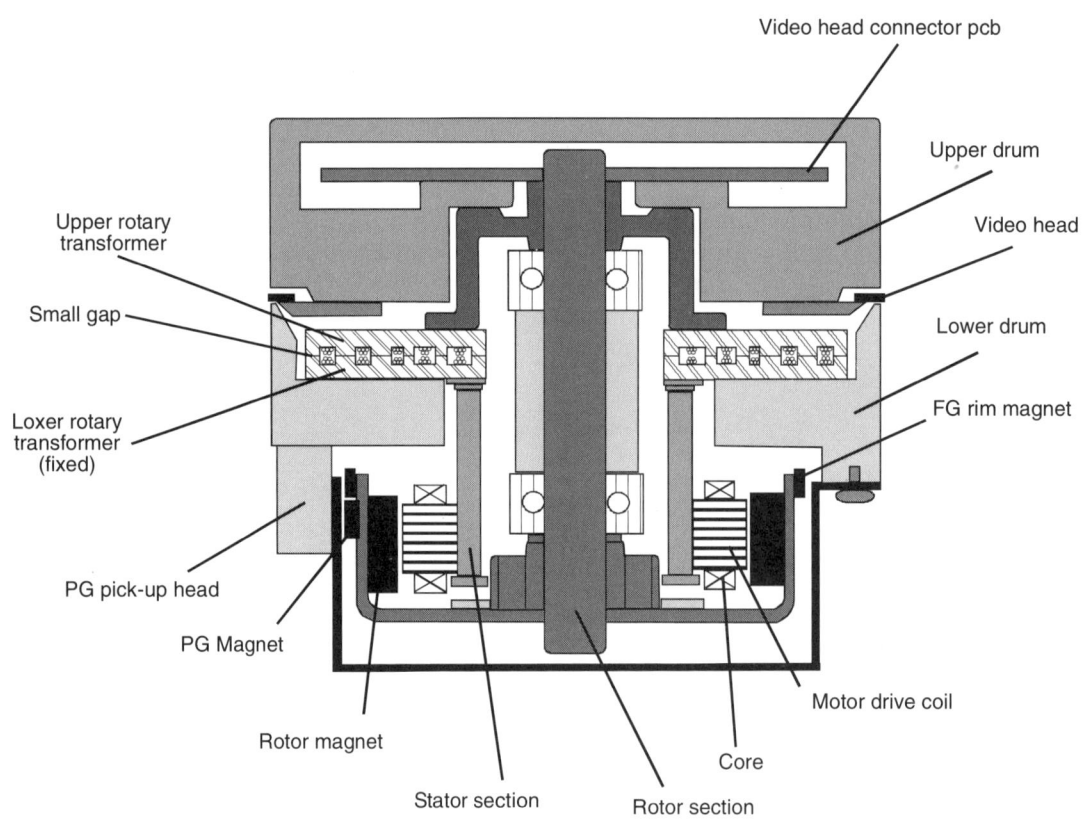

Figure 4.46 *Video head drum motor cross-section showing construction*

Drum motor

Figure 4.46 shows a section of a typical video head drum motor assembly. Parts of the lower drum and lower rotary transformer are static and do not move. Both the motor drive coils and magnetic core are fixed to the lower drum and set up the rotating magnetic field.

Rotary parts are the upper drum and the central shaft connecting it to the lower rotor section. The lower 'U'-shaped section is a flywheel and has the impregnated rotor magnet mounted to it, adding to the weight. Around the rim of this part is another circular impregnated magnet for generating the FG signal on PCB etched coil (not shown). A single PG magnet generates the PG pulse as it passes by the PG pick-up head.

Signals pass between the fixed lower section and the rotating upper section via the very small gap between them. Between the upper and lower rotary transformer sections is a very small gap. Magnetic transfer is between the upper and lower coils sections, each in its own slot and isolated from adjacent ones by the ferrite sections either side.

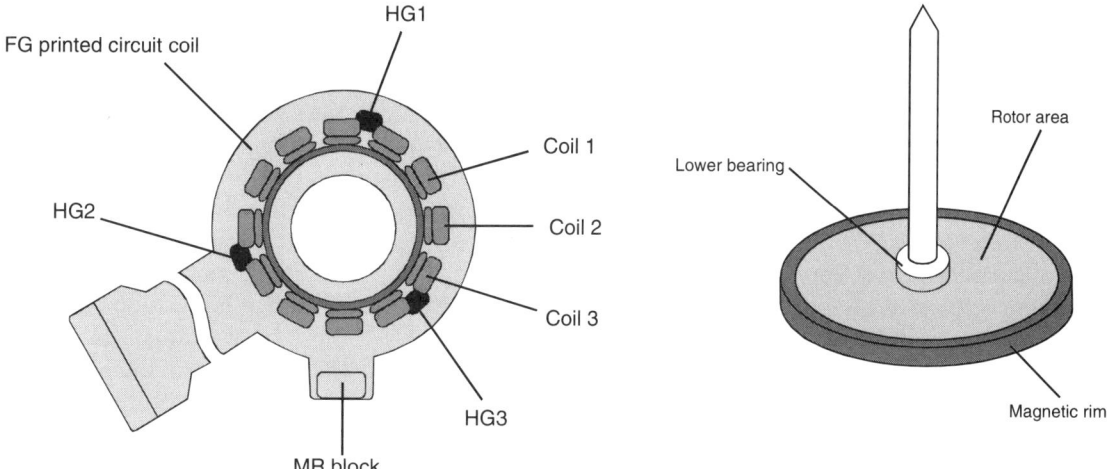

Figure 4.47 *Capstan motor stator and rotor*

Capstan motor

A capstan motor is shown in two parts in Figure 4.47, a stator and a rotor.

On the stator are shown the coils mounted 120° apart and the Hall effect devices, also mounted upon this assembly are either the PCB FG pick-up coil or the magnetic resistor for the FG output. The rotor is made up of the flywheel and shaft and a lower bearing. The centre part of the flywheel is heavily impregnated with N/S segments and sits over the coils. An impregnated magnetic rim of the flywheel FG is magnetically isolated from the more powerful centre section.

If it is fitted, a PCB coil is mounted over the flywheel and close to the magnetic rim. Rotation generates a sine wave which can be used as part of assembly edit functions for monitoring backspace reversing, particularly in camcorders.

Faults may be due to Hall effect devices or coils failing, although this is more likely due to solder joints. It is more common to lose the FG signal due to open circuit windings or joint problems on the MR pick-up coil.

Either part can be replaced separately, and rotor phosphor bronze bearings can be lubricated with a teflon-based oil.

5
Colour recording and playback

The first section in this chapter deals with VHS, the second section Betamax, and the third section describes 8mm. It is important to understand the way in which the colour is processed and I hope that it is presented in a way that is easily understood.

Betamax is no longer a current format but it is described here as a foundation and is complementary to VHS and 8mm. VHS has been developed and advanced to S-VHS and ET, and both have been joined by the 8mm and Hi8 systems. Digital colour recording and playback is unlike the colour under system, as the chrominance and luminance are inextricably locked together in the data stream and are not separated during the recording process, nor does digital colour recording and playback have any adjacent track crosstalk problems. Digital colour recording and playback is described in Chapter 11.

As is shown later, some of the techniques used in the Betamax system have been used in S-VHS. Also the 8mm system is similar to both VHS and Betamax and so there are no apologies for the in-depth treatment of the Betamax colour recording and replay techniques.

First though, let us consider the basic problem. The video tracks are laid down on the tape side-by-side, with no guard band between. If the heads do not maintain very accurate alignment with their appropriate tracks, there is the possibility of interference due to signals on adjacent tracks being picked up as crosstalk, from both luminance FM carrier and the chrominance signal.

The problem of crosstalk is not difficult to overcome for the luminance signal, which is frequency modulated onto a carrier in the higher part of the bandwidth. Head gaps are tilted with respect to one another (the slant azimuth technique); one will not significantly reproduce anything recorded by the other, or at least they will provide a high level of adjacent track rejection. The problem with the chrominance signal is that it is modulated onto a low-frequency carrier: 626.9 kHz for VHS, 685 kHz for Betamax and 732 kHz for 8mm. At the lower frequency spectrum crosstalk rejection by gap azimuth tilting is not as effective as it is for the much higher frequency luminance FM carrier. A technique has been devised that will give high colour crosstalk rejection. First though, we must give some consideration to the parameters that form the basis of a colour crosstalk rejection before it can be understood in detail in the record and replay processing.

The PAL 1½ line offset

The PAL TV system does not start in the same burst phase on any specific line after the field syncs. In fact it is varied over two (odd) and two (even) fields such that the complete repetition of the PAL signal is over four fields.

Figure 5.1 shows the four fields of PAL. If the starting point is taken as line 7 in the first even field, it can be compared to a line of similar phase in the second (odd) field. It is commonly known that the offset between two successive fields is half a line; that is, the interlace of two fields in a TV frame. It is less commonly known that there also exists a 1 line offset that is applied to the PAL colour phase so that the total colour offset between two fields is 1½ lines as shown in Figure 5.1. The comparison between the first field and the second field is from line 7 to line 321, from the second field to the third, line 321 to 10, from the third to the fourth field,* line 10 to line 324. From this comparison it can be deduced that the PAL phase is displaced by 1½ lines to the right in each

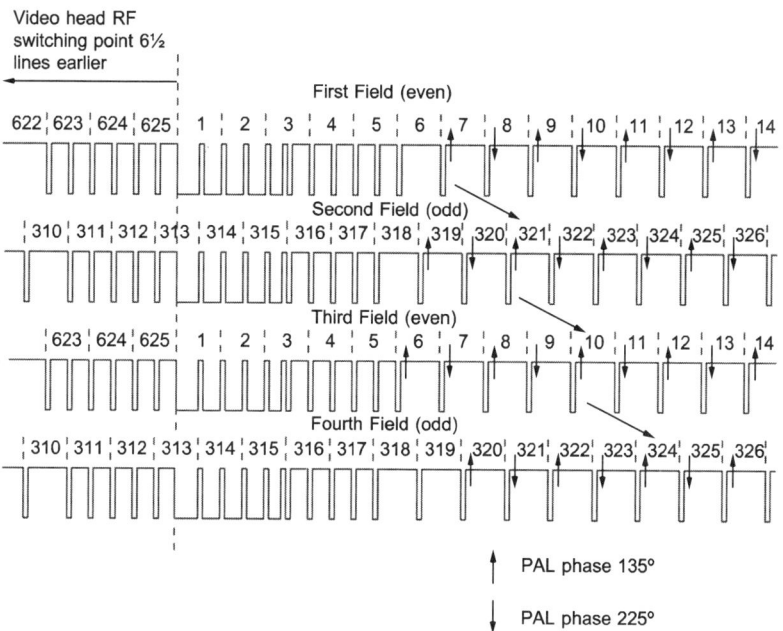

Figure 5.1 *PAL four field sequence showing 1½ line shift in PAL phase between fields*

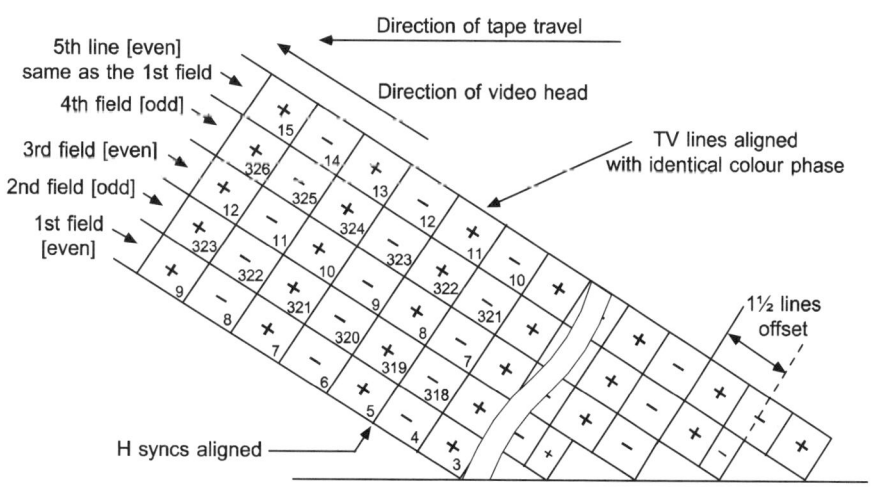

Figure 5.2 *1½ line offset of recorded tracks*

successive field, 1½ lines later.

During recording if each successive field were to start 1½ lines earlier, then the PAL phase and the line syncs would line up and this is achieved by the physical magnetic track pattern written by a video head.

By the specification of track angle and linear tape speed each successive video track of field starts 1½ lines before the previous one, as shown in Figure 5.2. This is an example of the magnetic tape track that is written (not withstanding the FM carrier upon which it is modulated).

Further up the track after field syncs have been recorded, the alignment of the numbered lines of Figure 5.1 is shown: line 7 in the first field aligns up to 321, 10, 324, 13 etc. in each subsequent field. The fifth field is the same as the first and any given line (for example line 10) is the same distance up each track from the start of the track at the edge of the tape (count the squares!). The method is called 'correlation'. Line correlation is achieved by aligning line sync pulses, shown as the edges of the squares, and colour correlation is achieved by aligning the same colour phase in each adjacent line in adjacent fields.

H sync and chroma correlation

In the early Philips N1700 system, manufactured in the late 1970's no special circuitry was used to overcome the problem of colour crosstalk. As the lines of adjacent fields were laid down next to each other, care was taken that the phase of the colour component of the signal was the same on adjacent lines. Then if one head picks up the other's colour information it will 'add to' rather than 'interfere with' the colour output. The eye does not notice this (the human eye is not very sensitive to minor colour inaccuracies). Also the colour content of any two interlaced lines that are next to each other, but come from different fields, does not differ much between these two fields. In practice, this worked out satisfactorily.

When we come to the VHS, Beta and 8-mm systems, however, we encounter even higher storage density on the tape, i.e. narrower track widths so both luminance and chrominance crosstalk increases. Additional signal processing is used within these systems to deal with the problem of colour interference (crosstalk) pick-up from adjacent field tracks.

PAL colour system

To start off, a brief summary of the PAL colour system vector diagrams will help to gain an understanding of vectors. Figure 5.3 shows the basic, well-known phasor diagram. The subcarrier frequency is 4.433619 MHz, with the (B-Y), or U chroma phase, at 0° and the phase selected for the (R-Y), or V signal, at 90°. The vector rotates anticlockwise, as showed by the little arrow, (w), one complete circle being 360° or one subcarrier cycle. Our vector is rotating at 4.433619 million times every second. In generating the PAL chrominance signal, we take two 4.433619 MHz subcarriers with a 90° phase difference. One is modulated in amplitude with the V (R-Y) video signal, with phase inversion on alternate lines, and the other is modulated with the U (B-Y) video signal. Then the two are added to form a single signal of two carriers, amplitude modulated, with one of them (V) shifted by 90° (quadrature modulation). The receiver separates the U and V components of the chrominance signal, using a comb filter (formed by the chroma delay line etc.), synchronous demodulators and, a phase-locked reference signal source.

The result of adding together the two quadrature-modulated subcarriers is that we get a single subcarrier that is modulated in phase and amplitude to encode the full range of colours and their saturation; see Figure 5.3(b). The maximum saturation of each colour is indicated by the dot at the end of the appropriate line, or phasor. In the PAL system, the V signal is inverted on alternate lines, which is phase shifting by 180°, so that the colours on these lines have the phases that are shown in Figure 5.3(c). Note that the burst is at 135° when V is at 90°, and 225° when V is at 270°. By drawing Figure 5.3(a) on a piece of paper and folding it along the U axis, it can be

Colour recording and playback

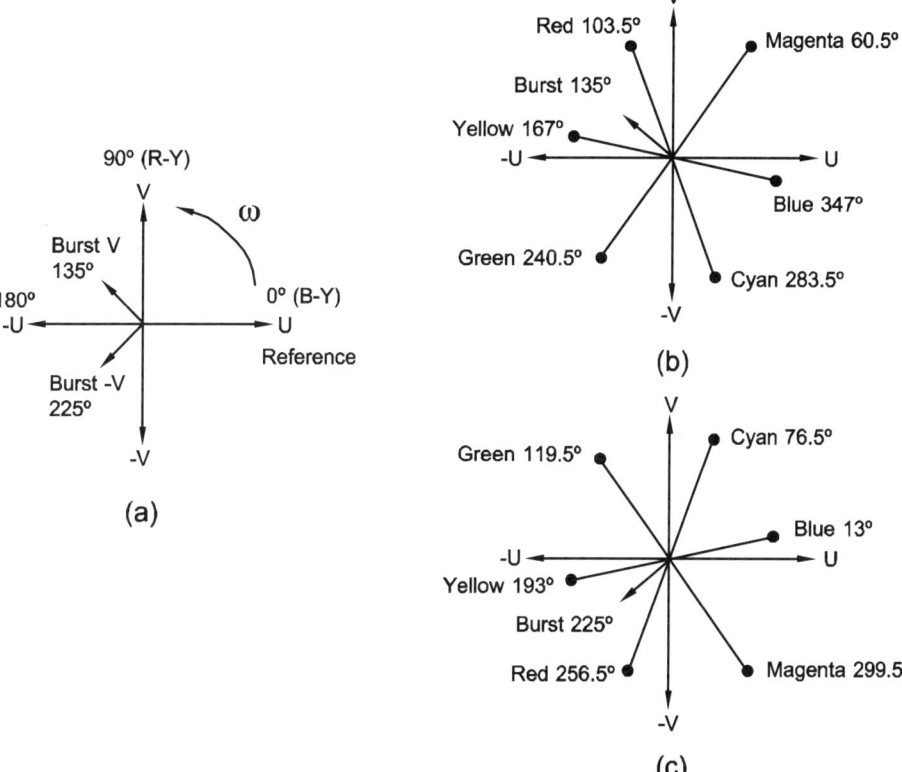

Figure 5.3 *PAL chroma vectors*

seen that the V and burst are mirror images. The same applies with Figure 5.3(b) that becomes Figure 5.3(c) when folded along the U axis. It is important to remember that the signal pattern recurs on every second line, i.e. V is at 90° on lines 1, 3, 5 etc. and at 270° on lines 2, 4, 6 etc. The idea is that by adding instantaneous signal voltages from two successive lines, the effect of transmission phase errors is largely overcome.

The VHS and 8mm record and replay systems both make use of this 2-line repetition sequence to cancel the effects of colour crosstalk in a 2-line comb filter.

TV comb filter

Staying with the basic PAL system for a moment let us recall how a delay-line PAL decoder operates (see Figure 5.4).

The delay line provides a signal delay of one line period, so that the output at any instant consists of the signal from the previous line. This is fed to an 'add' and 'subtract' network along with the direct, i.e. undelayed, signal so that the U and V signals can be recovered.

In detail, the real time signal entering the delay line is the V phase at 90° and the U signal at 0°

illustrated by the upright and right-hand vectors. Coming out of the delay line is the signal that went in one TV line earlier; this should have been the V signal at 270° and the U signal still at 0°. However, the delay line is an inverter and both vectors are shifted by 180° so that the V vector is at 90° and the U vector at 180°. So why is the delay line an inverter? It is due to the tailoring of the delay line to fit a specified number of cycles of 4.433619 MHz into it. If the delay line were exactly 64 microseconds long corresponding to a TV line, then there would be 283¾ cycles in it at any time. Therefore, the length is trimmed to 63.943 µs that correspond to 283½ cycles. This means that when a positive cycle is going in, a negative cycle is coming out, i.e. an inverter. If the number of cycles were whole (say 283) then the input and output would be in phase.

As in any standard colour decoder the direct and delayed signals are both added and subtracted. The sum network cancels the U signal leaving only the V (R-Y) signal. The subtraction network cancels the V signal leaving only the U (B-Y) signal. In video recorder colour processing circuits, a 2-line delay is used and as it can be seen later, has a delay of 127.88645 µs. There is a whole number of cycles during that delay time, numbering 567. This means that the video recorder delay line is not an inverter and that the delayed output is in phase with the direct input.

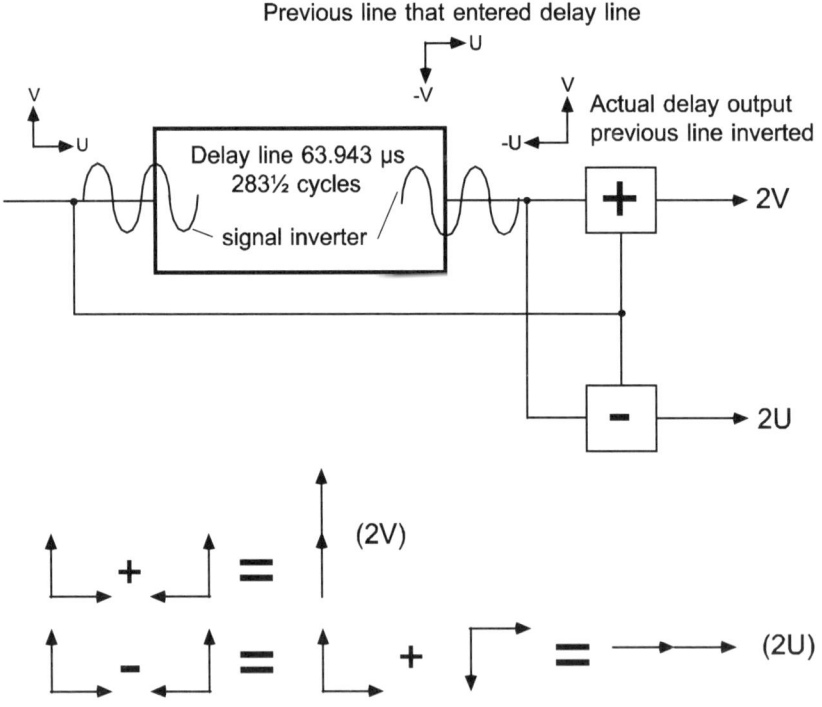

Figure 5.4 *U and V separation by a PAL delay line comb filter*

Colour recording and playback

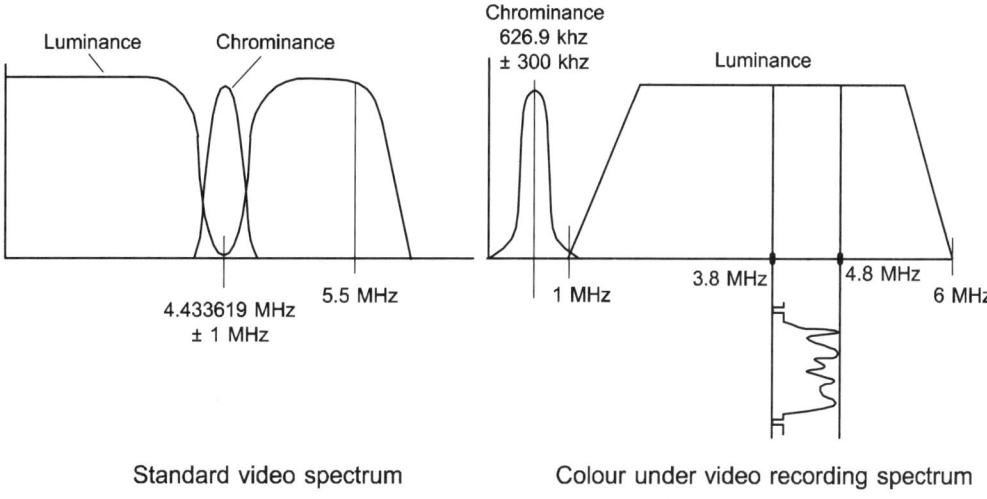

Figure 5.5 *Conversion from a composite video signal spectrum to a colour under recording spectrum*

The colour under system

This is the technique that is used by low band video recorders to get the colour signal onto the videotape. A composite video signal consists of both luminance and chrominance occurring in the same time domain as a mixed signal. For video recording the chrominance signal is down-converted to a lower carrier frequency that is situated below the luminance FM carrier lower sideband Hence the term colour under, as illustrated in Figure 5.5

It is the function of the colour processing circuits that the colour system not only down-converts the signal in record and up converts it in replay but it also provides for other correction techniques It cancels cross-talk between video heads and their tracks, it compensates for tape jitter affecting the colour quality, and corrects for phase errors in the replay signal, as we shall see in detail later in this chapter.

The VHS colour system

The VHS video recorders use two rotating heads to record and replay successive fields. One head records the colour subcarrier in the normal phase while the other records the carrier phase offset by 90° per TV line. This means that for the field tracks laid down upon the tape every other one has its colour phase modified by 90°/line throughout the field. The technique is illustrated in Figure 5.6.

The first row of the chart shows the burst phase recorded by head Ch.1 with the normal burst phase swing of 135° and 225° on successive TV lines (n) to (n+8). The signal recorded by head Ch.2 is a different story: it is being phase retarded by (-) 90° per line (retarded, the vector rotation is clockwise), not continuously but in steps of 0°, 90°, 180° and 270° on successive lines. This is compared to head Ch.1 phase in the row above. Compare head Ch.1 phase in line (n); there is no

111

Video and Camcorder Servicing and Technology

Line	n	n+1	n+2	n+3	n+4	n+5	n+6	n+7	n+8
Head Ch.1 record phase									
Head Ch.2 record phase delayed by 90°/line	0°	-90°	-180°	-270°	0°	-90°	-180°	-270°	0°
Head Ch.1 replay phase with head Ch.2 cross-talk									
Head Ch.2 replay phase advanced by 90°/line with head Ch.1 cross-talk									
2 line delay output	n-2	n-1	n	n+1	n+2	n+3	n+4	n+5	n+6
Delayed head Ch.1 with head Ch.2 cross-talk									
Delayed head Ch.2 with head Ch.1 cross-talk									

Figure 5.6 *Table of vector phases for head Ch.1 and head Ch.2 in the record and playback modes*

rotation for this line so the record phase for head Ch.2 is the same as head Ch.1. For line (n+1), head Ch.2 phase is retarded by -90° clockwise from the reference phase of head Ch.1 in the square above it. For line (n+2), head Ch.2 phase is retarded by -180° clockwise and so it is in the opposite phase. For line (n+3), head Ch.2 phase is -270° clockwise to head Ch.1 phase, and then the sequence is repeated. As a result of the combined effects of the PAL switching and the -90°/line phase delay, head Ch.2 records 2 lines in phase at 135° followed by 2 lines at 315°. Note the 2 line pattern that is set up!

The third line of the chart shows the head Ch.1 replay signal, which is obviously the same as the record signal being unchanged but with cross-talk from the adjacent head Ch.2 track. The smaller arrows indicates the phase of the cross-talk signal picked from the adjacent head Ch.2 tracks due to mis-tracking.

The fourth line shows the head Ch.2 replay signal, with a +90°/line phase advance (anticlockwise) replay correction so that it now corresponds to the head Ch.1 replay with the standard swinging burst. The small arrows are head Ch.1 cross-talk picked by the Ch.2 head. These are phase advanced by +90°/line as a side affect of head Ch.2 replay correction. Note again the line-paired pattern of this cross-talk: they are 135° for 2 lines and 315° for 2 lines. The last two sections of the chart show the relative outputs of the 2-line delay and are therefore delayed by 2 TV lines compared to those above, starting at (n-2).

To illustrate the action of the 2-line delay comb

Colour recording and playback

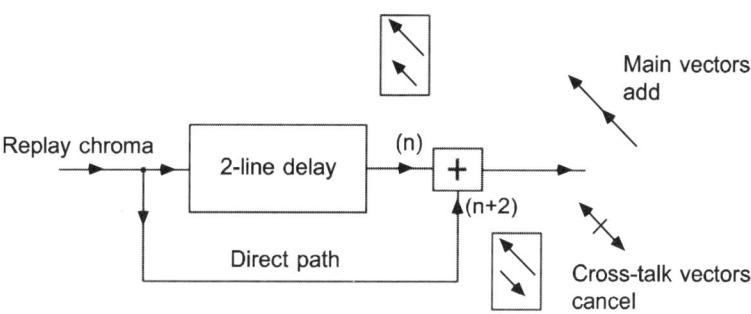

Figure 5.7 *Two line delay comb filter illustrating crosstalk cancellation between lines (n) and (n+2)*

filter, shown in Figure 5.7, four squares are highlighted. Two of them are head Ch.1 replay with cross-talk on line (n+2) in real time, below it is delayed head Ch.1 replay with cross-talk on line (n). This was replayed two lines earlier and is now coming out of the delay line.

The real time signal passes to the adder network via the direct path and is added to the delayed signal out of the delay line. Lines (n) and (n+2) are added; the main colour vectors are in phase and add, whereas the cross-talk vectors are out of phase and subtract. As an exercise try the head Ch.2 replay boxes which are highlighted for lines (n+6) and (n+4).

A study of the cross-talk shows the 2-line pair patterns. This means that at any time the direct and delayed vectors are always in antiphase and will cancel in a 2-line delay comb filter. This cross-talk cancellation is usually done after up-conversion at the chrominance frequency of 4.433 MHz so that standard glass delay lines can be used. In the digital domain the glass delay line is replaced by a CCD delay line or precision counter.

The VHS record system

The block diagram for the VHS chrominance record system is illustrated in Figure 5.8. Input video is passed to a 4.43 MHz filter to remove the luminance content and separate the colour signal. Simultaneous sync pulses from a sync separator stage, usually on the luminance/chrominance board, are fed to lock up a 2.5 MHz phase-locked loop. Line sync pulses, (fh), are split to a phase selector and a phase comparator. The phase comparator, which is part of the phase-locked loop, compares counted-down line pulses (f'h) to line pulses (fh). The main oscillator is a voltage-controlled oscillator running at a frequency of 2.5 MHz, which is 160 x fh. It is counted down by 4 in the first counter to 625 kHz. This counter also outputs 625 kHz in four phases: 0°, 90°, 180° and 270°. A further count by 40 completes the divide by 160 count and so the output from this counter is 15.625 kHz or line frequency (f'h). This is compared to incoming line syncs in phase and any error is sent to the voltage controlled oscillator

Video and Camcorder Servicing and Technology

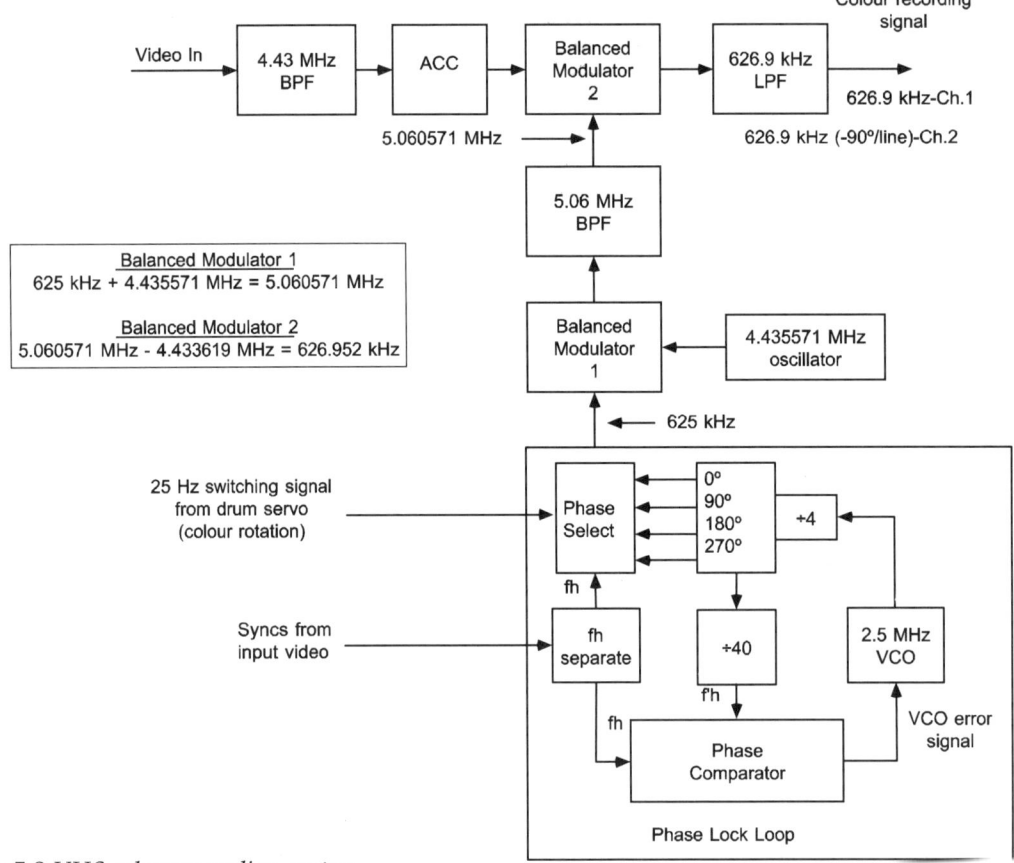

Figure 5.8 *VHS colour recording system*

completing the phase-locked loop.

The phase selector with its four inputs is responsible for the modification of -90°/line phase rotation of video head Ch.2. When the head switching flip-flop pulse is in head Ch.2 mode (low) the phase rotator runs, it is clocked by the line pulses (fh) from the line sync separator, and steps through the phases. When the head switching signal is high in head Ch.1 mode, the phase selector stops at 0°.

A 4.435571 MHz oscillator is fixed in record mode and its output is mixed with the 625 kHz output from the phase-locked loop (625 kHz - 90°/line for head Ch.2) in balanced modulator 1. Balanced modulator 1 adds the two inputs to give the sum of 5.060571 MHz (5.060571 MHz - 90°/line for head Ch.2). A tuned bandpass filter selects the sum output and rejects the difference component. In balanced modulator 2 the incoming colour signal is mixed with 5.06 MHz to give a sum output of 9.49 MHz, and a difference output of 626.952 kHz; the latter is selected by a low pass filter. This is the chroma record output to be mixed with the luminance FM carrier for recording. It is 626.952 kHz for head Ch.1 and 626.952 kHz (-90°/line) for head Ch.2.

The VHS replay system

Figure 5.9 shows the block diagram of the chroma replay system. The off tape replay signal is passed through a filter to select the 626.9 kHz component, or the –90/line phase shift in the head Ch.2 signal, from the luminance FM carrier. It can be seen that it has a similar structure as the record

Colour recording and playback

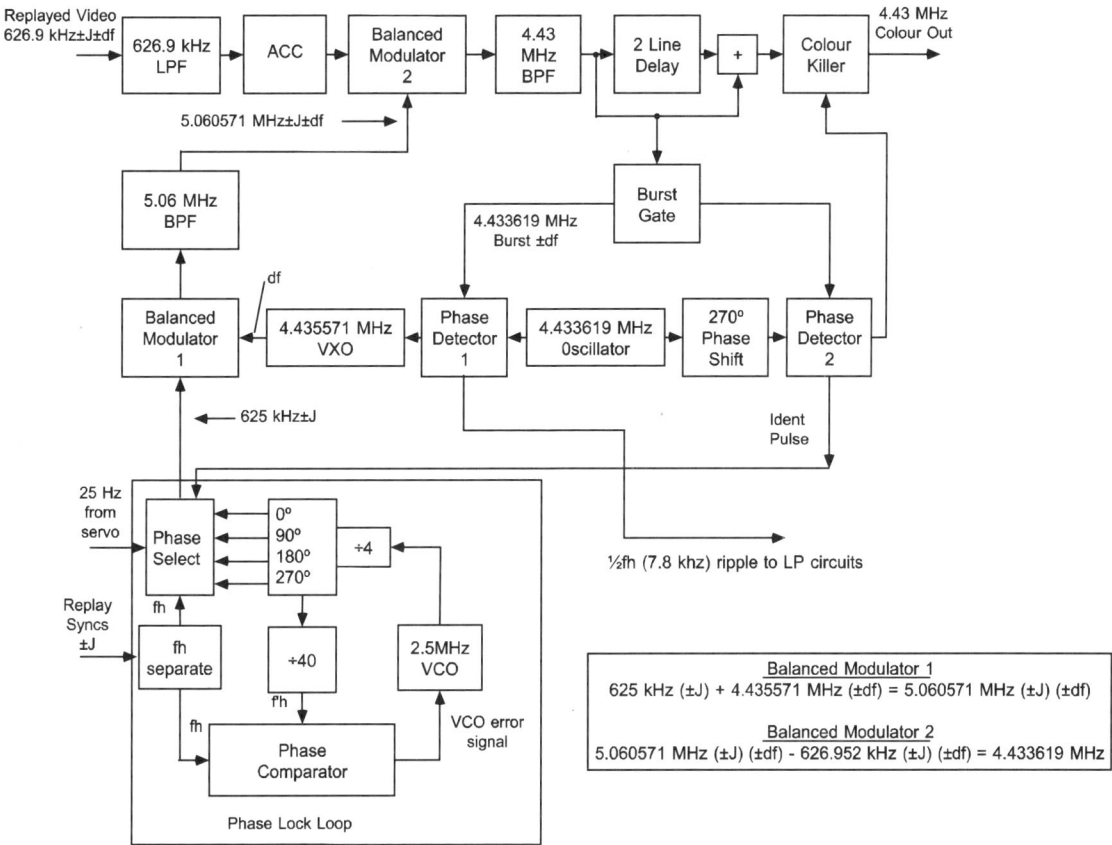

Figure 5.9 *VHS colour replay system*

block diagram with the addition of two phase detectors.

Starting at the bottom of the diagram, sync pulses obtained from the luminance replay signal are used to phase lock the 2.5 MHz oscillator, as in the record mode, producing a 625 kHz signal at the output. When head Ch.2 is replaying, the 25 Hz switching signal is low and line frequency pulses (fh) are used to step the phase selector. An intermediate carrier is obtained by adding 4.4335571 MHz to the 625 kHz to give 5.060571 MHz. In record mode 4.433619 MHz was subtracted from 5.060571 MHz to give 626.952 kHz and in the replay mode 626.952 kHz is subtracted from the 5.060571 MHz carrier in balanced modulator 2 to result in 4.433619 MHz output. It is in this balanced modulator that the head Ch.2 phase retard for recording phase shift is corrected in replay by reversing the –90°/line process.

The replay system has to be able to correct for frequency errors caused by small changes in the tape speed (jitter) and phase errors caused by tape tension fluctuations. Jitter causes small changes in tape speed that affects the head-to-tape relative speed; it is shown in the diagram as ±J.

VHS replay phase correction process

Replayed colour signal at 626.9 kHz contains the error ±J and so do the replayed sync pulses into the phase-locked loop. As the loop is phase locked to the replayed syncs then the loop will also be subject to the jitter error, and therefore the loop output at 625 kHz will also contain jitter. In the up-conversion of balanced modulator 1 the jitter

Video and Camcorder Servicing and Technology

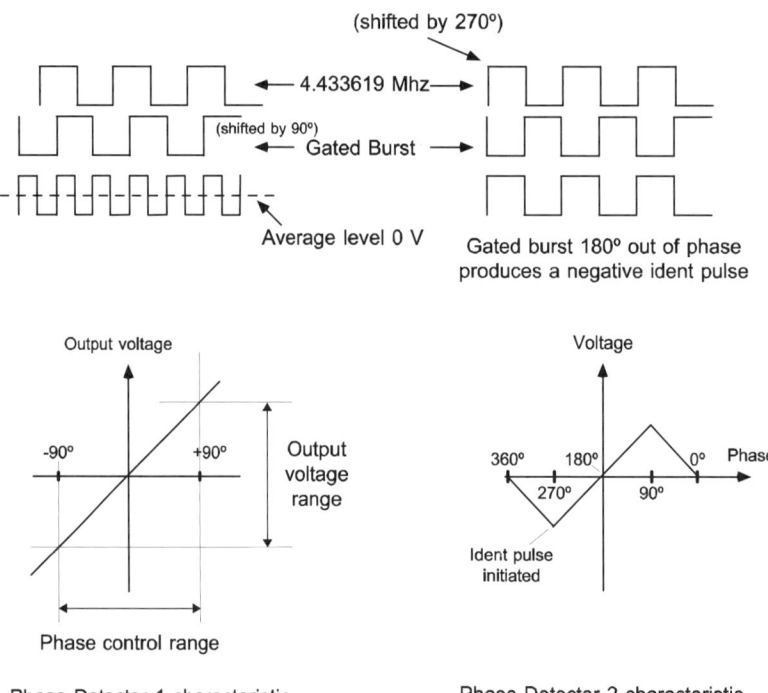

Figure 5.10(a) *Phase detector characteristics*

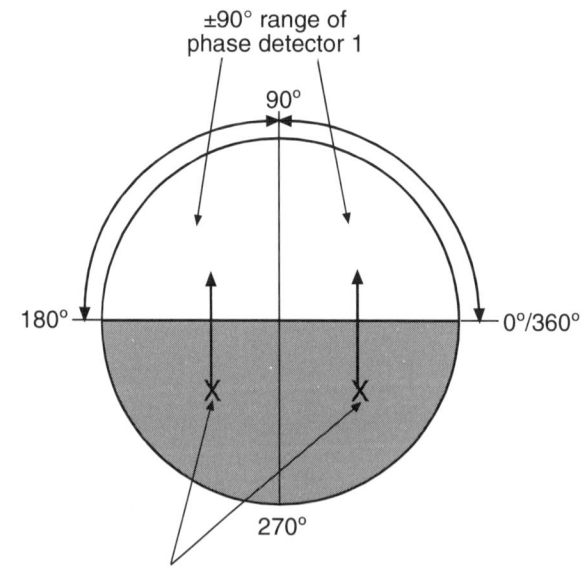

(b) *Phase correction range vector diagram*

component is carried through to balanced modulator 2 at 5.06 MHz. As balanced modulator 2 is a subtraction process then the jitter component is cancelled.

Phase errors are shown as (df) in Figure 5.9. If the 4.433619 MHz up-converted signal contains (df), the output from the burst gate will also contain (df). This signal goes to phase detectors 1 and 2. Detector 1 applies a corresponding error correction signal to the 4.435571-MHz variable oscillator, so that (df) is added to the up-converting path via balanced modulator 1 and appears within its output signal. Balanced modulator 2 thus receives two inputs, they are the replayed 626.9 kHz and the subtraction up converting carrier 5.06 MHz, both with the error signal (df). These inputs are subtracted in this balanced modulator, and phase error component, (df), is cancelled, leaving a fairly stable 4.433619-MHz chroma output.

VHS phase correction in detail

Phase detector 2 is concerned with phase errors much larger than (df). On replay there could be a bit of tape drop out or a sync pulse could go missing, and the phase error could be as large as 180°. In this event phase detector 2 sends an 'ident' pulse to the phase selector part of the phase-locked loop, shifting the 625-kHz signal by 180°. As a result, the coarse error is eliminated and fine control returns to phase detector 1. Phase detector 2 also provides a colour killer output to switch off the output amplifier. In consequence the detector doubles as a burst presence detector if the recording was in monochrome and stops spurious colours in the picture, or if the chroma errors cannot be corrected it acts as a colour killer.

The action of the phase detectors is shown in a little more detail in Figure 5.10. There are two inputs to phase detector 1: 4.433619 MHz from a reference crystal oscillator, and the gated bursts from the replay signal. When these two signals are 90° apart, the output waveform is symmetrical and averages out at 0 V. This becomes the stable condition with the replayed burst being shifted 90° to the crystal oscillator. If the replayed burst signal drifts by +90° to the left, then the output is not symmetrical and is high for a longer period than it is low. This causes the average integrated level to rise positively; a drift by −90° gives a negative output. This provides a positive and negative linear correction range with the burst centred on 90° with respect to the crystal oscillator. As the PAL burst is alternating line by line this output contains a ½fh (7.8 kHz) ripple that is used by the LP search correction circuits.

Phase detector 2 is concerned with ensuring that errors fall within the scope of phase detector 1. The 4.433619 MHz reference oscillator signal is phase shifted by 270° to be in phase with the gated burst. With correct conditions the output of phase detector 2 is positive, to the right of the correction range graph that is centred on 90°, similarly for phase detector 1. If the phase shift is greater than 90°, either way, then the output of phase detector 2 changes polarity and becomes negative; this is the 'ident' pulse.

Full correction is applied to the replayed colour signal in three stages.
1. Frequency compensation for jitter by replayed syncs into the phase locked loop.
2. Phase detector 2 to ensure that phase errors remain within the domain of phase detector 1.
3. Phase detector 1 maintaining correction within ±90° referred to a crystal reference oscillator.

Additional phase information

In any explanation of VHS colour recording and playback using the vector principle, it is reasonable to say that if head Ch.2 is subjected to a −90°/line phase retard then it must be corrected by +90°/line phase advance in replay. However, this is not strictly true as it intimates that the phase selector within the phase lock loop rotates in one direction in record and in the opposite direction in replay. Careful examination of the phase lock loop circuit does not show any record/

Video and Camcorder Servicing and Technology

replay switching, so there is nothing to cause reversal of the phase selector between record and replay. So what is the answer? In fact there are two possible explanations.

Simple balanced modulator

Figure 5.11 shows a basic up/down balanced modulator converter. For down-conversion the input at 4.433619 MHz (A) is subtracted from an intermediate carrier at 5.060571 MHz (B) to produce the colour recording signal at 626.9 kHz (C). In replay up-conversion, the replayed colour signal is still the same value (C). Therefore in order to obtain the correct colour carrier value of 4.433619 MHz as an output, it follows that the intermediate conversion carrier (B) must be the same in both record and playback! Following this back to the phase lock loop it is reasonable to deduce that the output value of the loop of 625 kHz must also be the same in record and playback.

-90°/line to a frequency

It can be argued that by retarding the phase of the signal by -90° per line it is the same as retarding by 360° or 1 cycle for every 4 lines (4 x 90°). The line frequency is 15.625 kHz or 15625 cycles per second so by dividing this by 4 lines we can obtain the equivalent number of cycles that are 'deducted' per second that is 3906 cycles or 3.906

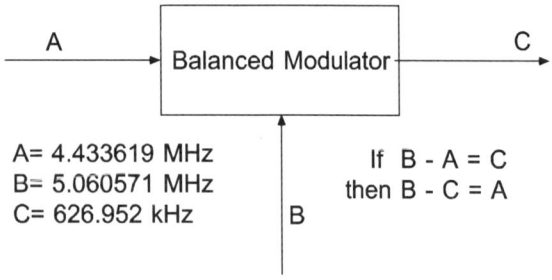

Figure 5.11 *Balanced modulator converter*

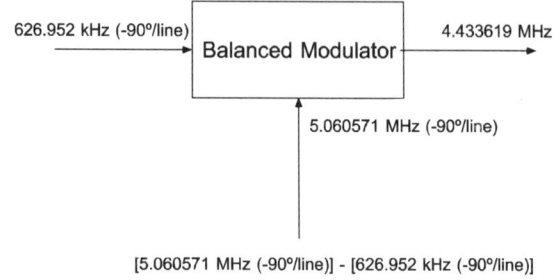

[5.060571 MHz (-90°/line)] - [626.952 kHz (-90°/line)]

= 5.060571 MHz (-90°/line) - 626.952 kHz (+90°/line)
= 5.060571 MHz - 626.952 kHz
= 4.433619 MHz

Figure 5.12 *-90°/line replay correction*

kHz. Conversely phase advancing by +90° per line is the equivalent of adding 3906 cycles per second.

Taking the 625 kHz output from the phase lock loop and retarding the phase by -90°/line is the same as subtracting 3.906 kHz to give a frequency of 621.094 kHz. A rotation of +90°/line is equivalent to a frequency of 628.906 kHz.

If the phase lock loop produces a frequency of 625 kHz for head Ch.1 and 621.094 kHz for head Ch.2 in record mode, it must also be the same frequency for each head in playback. Head Ch.1 PLL is 625 kHz and head Ch.2 is equivalent to a PLL frequency of 621.094 kHz. Head Ch.2 cannot be a PLL frequency of 621.094 kHz in record and then 628.906 kHz in playback, or the sum (B - C=A) in Figure 5.11 will not be correct. The colour signal (A) in playback would be 4.441430 MHz and incorrect. So +90°/line correction in playback cannot be achieved in this manner.

By accepting that the phase shift of 90°/line is the equivalent of a frequency shift of 3.906 kHz, the values of the recorded colour carriers for each head can be obtained. Head Ch.1 we know to be 626.952 kHz, so head Ch.2 is 3.906 kHz less or 623.045 kHz.

The shift of 3.906 kHz in the frequency of the recorded colour signals between adjacent video tracks is important for cross-talk and it will be shown to exist in both the old Betamax system and the 8mm system.

Colour recording and playback

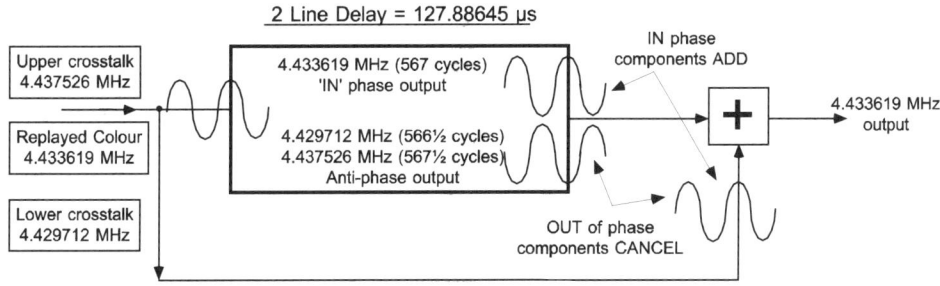

Figure 5.13 *2-line delay comb filter*

+90°/line replay correction

The second explanation is also based on the fact that the phase lock loop is the same frequency in both record and playback for head Ch.2. As the phase lock loop produces 625 kHz (-90°/line) with 4.435571 MHz added to it, then the down conversion intermediate frequency is 5.060571 MHz (-90°/line). Basic playback correction is shown in Figure 5.12 where the phase lock loop runs the same PLL frequency as in record, and where the up-conversion frequency was also 5.060571 MHz (-90°/line). In the subtraction process of the balanced modulator 2 we have:

(5.060571 MHz (-90°/line))
- (626.952 kHz (-90°/line))

By expanding the brackets we get:

5.060571 MHz (-90°/line)
- 626.952 kHz (+90°/line)

The phase rotation components cancel:

5.060571 MHz - 626.952 kHz = 4.433619 MHz

VHS frequency conversion explanation

So is there a frequency explanation rather than a vector explanation for the 2-line delay filter given in Figure 5.7 to cancel cross-talk? Yes, there is!

In the situation that head Ch.1 is scanning the tape in replay and its replayed colour signal is up-converted then the cross-talk that is picked up from adjacent Ch.2 tracks either side will also be up-converted but at a frequency that is 3.906 kHz lower. If head Ch.2 is scanning the tape and its replayed colour signal is up-converted then any crosstalk from head Ch.1 tracks either side will also be up-converted by the same process, but at a frequency that is 3.906 kHz higher.

Therefore two possible cross-talk frequencies can be obtained, which are 3.906 kHz higher or 3.906 kHz lower than the colour subcarrier frequency of 4.433619 MHz; these are 4.437526 MHz and 4.429712 MHz.

When either of the two cross-talk frequencies are applied to the 2-line delay comb filter network they cannot pass through as they do not match the comb filter response spectrum. The reason for this is shown in Figure 5.13.

With a delay line that has a delay time of 127.88645µs it can be easily calculated how many cycles exist within the delay line at a given instant; for 4.433619 MHz this is 567. By calculating the period of each cross-talk frequency and dividing the result into the delay time, we can confirm that there are 567½ cycles for 4.437526 MHz and 566½ for 4.429712 MHz. The presence of the odd half cycle at the delay line output proves that the two cross-talk signals undergo inversion through the delay line and will therefore cancel at the output sum network when added to the direct signal.

The Betamax system

Before carrying on to the present 8mm system it is worth a look at the way the colour was recorded and replayed in the Betamax system and how the cross-talk problem was approached. Although similar to VHS there are quite a few differences in the details. Instead of a 90°/line phase shift at a single carrier frequency, two carrier frequencies are employed. The frequency, for head Ch.1 is (44 - 1/8)fh, or 685.54688 kHz, while that for head Ch.2 is (44 + 1/8)fh or 689.45312 kHz. Two factors become clear if the following equation is solved:

$$\cos(w_c t + \emptyset) = \cos[(44 \pm 1/8)f_h t + \emptyset]$$

The first is that 685.5 kHz (for short) has a clockwise phase rotation of -45°/line with respect to the line frequency (fh), and the second factor is that 689.4 kHz has an anticlockwise phase rotation of +45° with respect to fh. This is from the ± 1/8 factor of the formula, cos (44fh ± 1/8fh)t + Ø, for Betamax where 1/8 of fh is equivalent to 45°. From

$$\cos(44f_h \pm 1/8f_h)t + \emptyset$$

Where 44fh = 687.5 kHz and 1/8fh = 1.953125kHz we get

44fh + 1/8fh = 687.5 kHz + 1.953125 kHz
= 689.45312 kHz

and

44fh - 1/8fh = 687.5 kHz - 1.953125 kHz
= 685.54688 kHz

This means that on every line the phase of the recorded chroma signal has been shifted by ± 45°, or 90°/line for one head with respect to the other. So instead of holding the signal of one video head constant while shifting the phase of the other line by line, as in the VHS system, Sony phase-shifted both signals with respect to each other in opposite directions. This results in similar recording vectors to those illustrated in Figure 5.6 for the VHS system. Thus applying replay correction for either head, the cross-talk picked up from the other head's track will be in opposite phase every 2 lines, so that it can be cancelled using a 2-line delay, as explained for VHS. As an exercise it is possible to work out the vector chart similar to Figure 5.6.

In explaining how the Sony system works, we will be considering two carrier frequencies rather than a single frequency with a stepped phase shift for one head.

Betamax chroma record system

The Sony Betamax chroma record system is shown in block diagram form in Figure 5.14. The 4.43 MHz chroma input signal is first filtered and gain controlled, then fed to the pilot burst adder. This inserts a pilot burst of 4.43-MHz carrier with a 90° phase shift in the horizontal sync pulse period, just before the normal burst; it is cleaned off the signal during the replay processing by a clamping pulse. The idea is to provide a replay reference for off-tape phase errors (df) which does not swinging about as the normal PAL burst is. (The phase errors we are considering here are those due to changes in tape tension fluctuations). It may be surprising to see the same technique adopted for S-VHS.

The input video via a sync separator provides the phase comparator with line sync pulses (fh) to phase lock the variable frequency 5.5 MHz

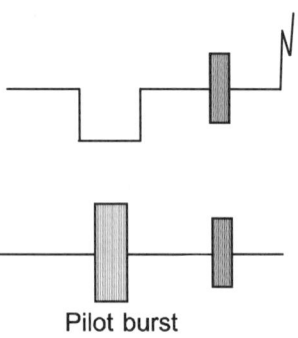

Figure 5.14 *Pilot burst position*

Colour recording and playback

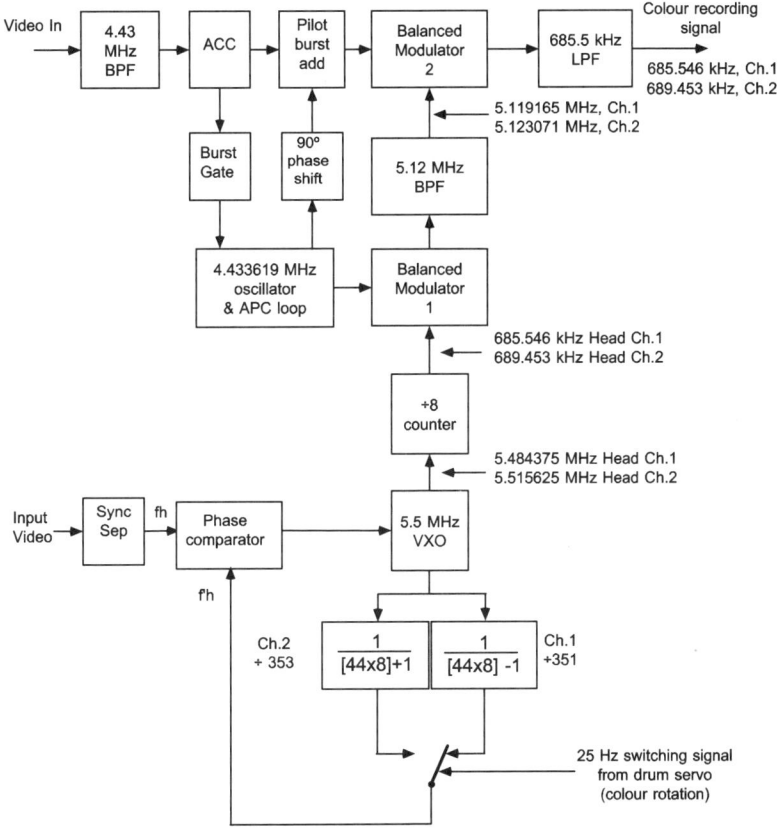

Figure 5.15 *Betamax colour recording system*

oscillator (VXO). There are two counters in the phase lock loop: divide by 351 (head Ch.1) and 353 (head Ch.2). A 25-Hz head select pulse from the drum servo controls the count selection. As a result, the VXO oscillates at 5.484375 MHz when head Ch.1 is recording and 5.515625 MHz when head Ch.2 is recording. The VXO output is then divided by 8 to give 685.546 kHz and 689.453 kHz, respectively. These two signals are then added to a 4.433619-MHz carrier, that is phase-locked to the input chroma signal, in balanced modulator 1. The intermediate carrier output from balanced modulator 1 is then mixed with the 4.43 MHz chroma signal in balanced modulator 2. After filtering, we get a 685.546 kHz recording output for head Ch.1 and a 689.453 kHz recording output for head Ch.2.

Betamax chroma replay system

The Betamax replay system is shown in Figure 5.16. As in the record mode, the divide-by-8 counter produces reference carriers at 685 kHz and 689 kHz (rounded). As previously shown with VHS it is the sync pulses containing jitter that are used to phase lock the 5.5 MHz VXO as they are sourced from the replayed luminance signal. Frequency/head selection is again provided by a 25-Hz PG signal from the head servo.

The replay chroma signal, at 685 kHz and 689 kHz, will contain both jitter and (df) phase errors introduced by the tape transport system. While the replay chroma carrier frequencies are restored to 4.43 MHz in balanced modulator 2 and the

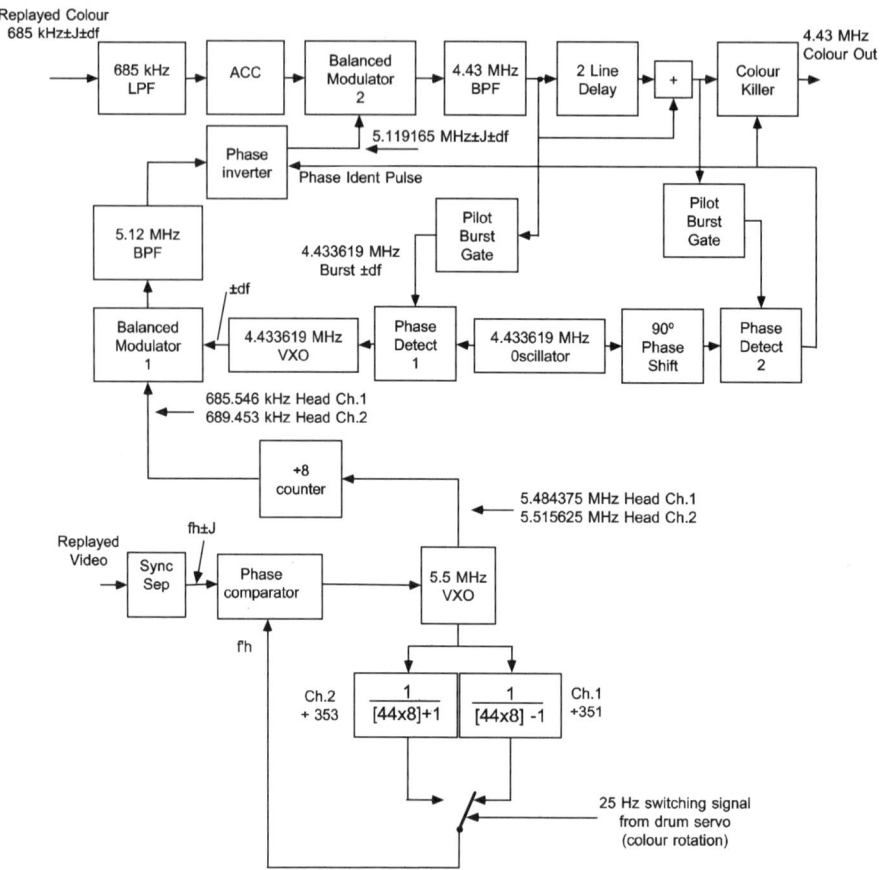

Figure 5.16 *Betamax colour replay system*

jitter component is cancelled, the output may still contain (df) errors. This signal is then passed through to a band pass filter, after which the pilot burst is gated out (with df) and fed to a phase detector 1. This compares the pilot burst with the output from a free-running 4.433619 MHz reference oscillator. The resulting error signal, containing (df), is modulated onto the output from the 4.433619 MHz VXO, hence the 5.1 MHz intermediate carrier that is fed to balanced modulator 2 also contains (df); in the subtraction mixing process, then, (df) is cancelled.

A second pilot burst gate feeds the ident detector in phase detector 2. This compares the pilot burst with the signal from the free-running 4.43-MHz reference oscillator, phase shifted by 90°. If the pilot burst is correct, no problem. If it is 180° out of phase, however, the ident signal is sent to a phase inverter circuit and inverts the 5.119 MHz (or 5.123 MHz) intermediate carriers.

These corrections are similar to those carried out in the VHS system where phase detector 2 shifts the output of the phase lock loop by 180°. It is possible to do this by inverting the 5-MHz carrier signals, as it is the same as a 180° phase shift. Both phase detectors carry out corrections over 360° between them, and their operational areas are the same as given for VHS in Figure 5.10.

Frequency correction or jitter is compensated for in the same way as for VHS. As the replayed colour signal is subjected to ±J so are the replayed syncs into the phase comparator. Each head output from the VXO also contains ±J and this is carried up the conversion route to balanced modulator 2 in which the ±J component is cancelled along with (df).

Colour recording and playback

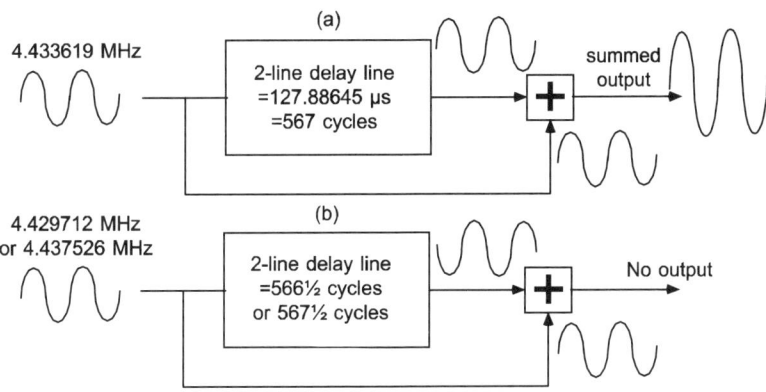

Figure 5.17 *Betamax 2-line delay comb filter*

Action of the Betamax delay line

The 2-line delay system works as follows. Let us consider the case of a 4.433619 MHz replay signal from head Ch.1. The replayed colour signal is 685.546 kHz and it is subtracted from the 5.119165 MHz conversion carrier in the main balanced modulator, the result is a correct 4.433619 MHz. The direct and delayed carriers are in phase and thus add; see Figure 5.17(a).

Cross-talk picked up from an adjacent head Ch.2 track will also be replayed. If the 689.453 kHz cross-talk is subtracted from the 5.119165 MHz carrier the result is a 4.429712-MHz signal. Delaying this signal by 2 lines and adding it to the direct signal produces complete cancellation; see Figure 17(b). Let us consider this in a little more detail.

As mentioned earlier for VHS a standard delay line introduces a 180° phase shift. This is because its delay time is not exactly 64 μs, i.e. one line, but 63.943226 μs. The duration of one 4.433619 MHz cycle is 0.2255493 μs. If we divide 64 μs by 0.2255493 μs, the result is 283¾ cycles. For the adding and subtracting business to work, the odd quarter cycle must be avoided. So the delay line is given a duration of 63.943226 μs, i.e. 283½ subcarrier cycles. If we now take a 2-line delay, we must double this figure to give 127.88645 μs; 567 cycles of subcarrier are now passing through the delay line, which means that the output and input are in phase for the correct frequency.

In the case of a 4.429712 MHz cross-talk signal, 566½ cycles pass through the delay line. The output is in opposite phase to the input and cancellation takes place.

With a head Ch.2 replay signal the intermediate carrier is 5.123072 MHz and the head Ch.1 cross-talk 5.123072 − 0.685546 = 4.37526 MHz. Each cycle of this takes 0.2253507 μs, so there will be 567½ cycles passing through the delay line. Again the output is in opposite phase to the input and cancellation occurs. This proves that the ±45° phase shift between head Ch.1 and head Ch.2 in the Betamax system produces identical cross-talk frequencies as VHS does with its 90°/line phase rotation. The 2-line delay comb filter cancels the cross-talk frequencies that are produced ± 3.906 kHz either side of the 4.433619 MHz colour carrier. In this respect it proves that ±45°/line gives the same results as 90°/line, in so much as the cross-talk frequencies produced are identical.

The 8mm colour system

Due to the influences of both VHS and Betamax the 8mm system has technical aspects that contain the best attributes of the previous technologies. In particular the colour cross-talk has a 90°/line phase rotation as this has proved to be more stable than the continuous frequency change in Betamax. It is a similar method to that used in VHS insomuch as it is a phase rotation system applied only to one video track. The phase of the colour signal is shifted by 90°/line and stays in that phase for the duration of the line period. In the Betamax system with two separate frequencies the phase shift does change gradually throughout the line period and replay phase correction may be less effective than on a line-by-line basis.

As the 8mm video system has a tape width of 0.25" and a video track width of 34 µm the constraints of signal-to-noise and stability are more demanding. The circuit techniques are similar to VHS and Betamax but they are much tighter in their operation to achieve the improvements.

Colour is down-converted as in the other systems by the use of the two balanced modulators that are referred to in 8mm text as frequency converters. However, to maintain continuity I shall refer to them as balancedmodulators 1 and 2. The colour under frequency is given as 732 kHz; the exact frequency is derived below, it is determined from the formula $(46\ 7/8 \times fh)$.

(Read as 46 and seven eighths, times line frequency).

This can be expanded:

$$(46\tfrac{7}{8} \times fh) = \frac{368 + 7 \times fh}{8}$$

$$= \frac{375\ fh}{8}$$

$$= \frac{375 \times 15625}{8}$$

$$= 732.421875\ \text{kHz}$$

The cross-talk separation phase shift is +90°/line and applied to head Ch.1; the positive indicator means that the phase rotation is anti-clockwise. The shift of the colour phase in record is therefore similar to that of VHS and can be proved to set up the recording phase as inverted line pairs in the same way that VHS does. Replay correction is also identical and as an exercise you could draw up the phase diagram for 8mm in the same way as Figure 5.6 for VHS.

The 8mm colour recording system

The block diagram for 8mm is shown in Figure 5.18. The voltage controlled oscillator (VCO) and phase lock loop is shown in the bottom section. It runs at 375 times the line frequency (375 fh) and so the counter in the phase locked loop is a divide by 375 to obtain H sync frequency f'h.

The output of the VCO is divided by 8 and the resulting 732 kHz is added to 4.433619-MHz carrier frequency in balanced modulator 1. This 4.433619-MHz carrier is phase-locked to the incoming burst in an APC loop to maintain close phase control.

The intermediate carrier frequency is therefore phase-locked to the colour signal burst, as both components to balanced modulator 1 are phase-locked to the video signal. A band pass filter selects the sum output of 5.166041 MHz and it is passed to balanced modulator 2.

Input chroma is level controlled in the ACC amplifier then it is pre-emphasised to maintain signal-to-noise and frequency response. The burst is additionally emphasised to increase its level by 6 dB as it is used in replay as the phase correction reference. This further improvement in burst signal-to-noise ratio is desirable due to the increase in levels of replay noise and cross-talk. Incoming colour at 4.433619 MHz is subtracted from the 5.166041 MHz in the main balanced modulator. The difference output is selected as the down-converted colour signal in a low pass filter and is then passed for recording with the FM luminance carrier signal.

Colour recording and playback

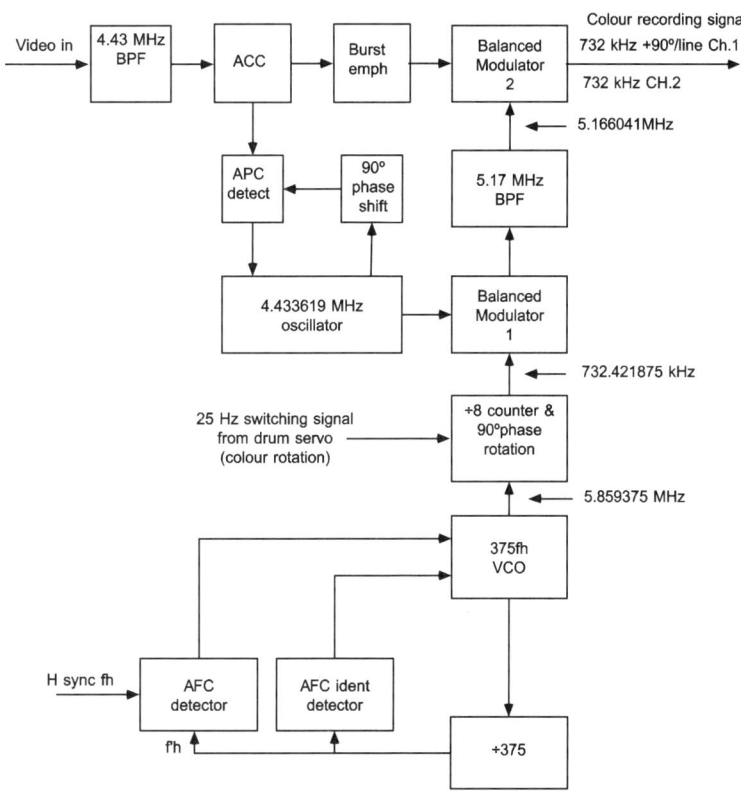

Figure 5.18 *8mm colour recording system*

8mm phase control

Two control systems are responsible for the frequency and phase control of the phase lock loop. An AFC detector for phase control and an AFC ident for frequency control may be considered as fine and coarse frequency control, fine control being approximated to phase lock. This type of control is suited to use in the digital domain as the operation depends on precision counters.

The VCO is oscillating at 375 × fh (equivalent to 5.859375 MHz) and then is divided by 8 at its output. This results in the input frequency into balanced modulator 1 which is 732.421875 kHz. As I have previously shown in Betamax, by adding this input frequency of 732 kHz, to a frequency that is the same as the colour carrier at 4.43 MHz, the result is a 5-MHz intermediate carrier. For stability this 4.433619-MHz oscillator is phase-locked to the recording signal. It follows that in the subtraction process of balanced modulator 2, where the incoming colour is then subtracted from the 5-MHz carrier, the result is 732 kHz. This is the colour recording signal and it is the same frequency as the signal derived out of the phase-locked loop.

I have already shown that the phase rotation is

in effect a frequency shift of 3.906 kHz. Head Ch.1 is effectively 3.906 kHz higher in frequency than head Ch.2 which is similar to VHS where the head Ch.1 is 3.906 kHz higher than head Ch.2. You can work out for yourself what frequency the head Ch.1 colour recording frequency is!

The counter and phase-lock loop works differently between record and replay so I will discuss the operation in detail and point out the subtle differences.

The counter as a divide-by-375 works as a preset counter, see Figure 5.19. It counts by 8 in the first counter section and by 47 in the second counter section, a total count of 376. However, at a count of 46 in the second section the logic selects a divide-by-7 counter instead of a divide-by-8 in the first section. The counter then effectively counts 46 counts of 8 and one count of 7. That is (46 x 8) or 368, plus 7, which equals a total count of 375.

The 47 count periods are specifically labelled within the second section of the divide by 375 counter, as M1T – M47T, with the specific count of 7 being during period M2T.

As is normal colour conversion practice the VCO is phase-locked to incoming syncs from the input video signal, and phase locking also operates on a counting principle. Within limits this has an operating 'window' wide enough for some degree of frequency control. Therefore the loop is able to lock up to frequency and phase with a single detector.

AFC detection takes place within periods M19T –M28T of the 47 periods of the counter, there are 10 count periods centred on period M24T as shown in Figure 5.19. Within this 'window' area the horizontal sync pulse maintains phase lock with the 375fh VCO. Outside the 'window' area there is no phase lock and the AFC ident circuit provides an error signal for a coarse frequency correction.

The 8mn colour replay system

The replay block diagram is shown in Figure 5.20. Again there is a phase lock loop around the 375fh voltage controlled oscillator with a wide control range. Replay up-conversion is fairly standard; the output of the phase lock loop is divided by 8 to 732 kHz and added to 4.43 MHz in balanced

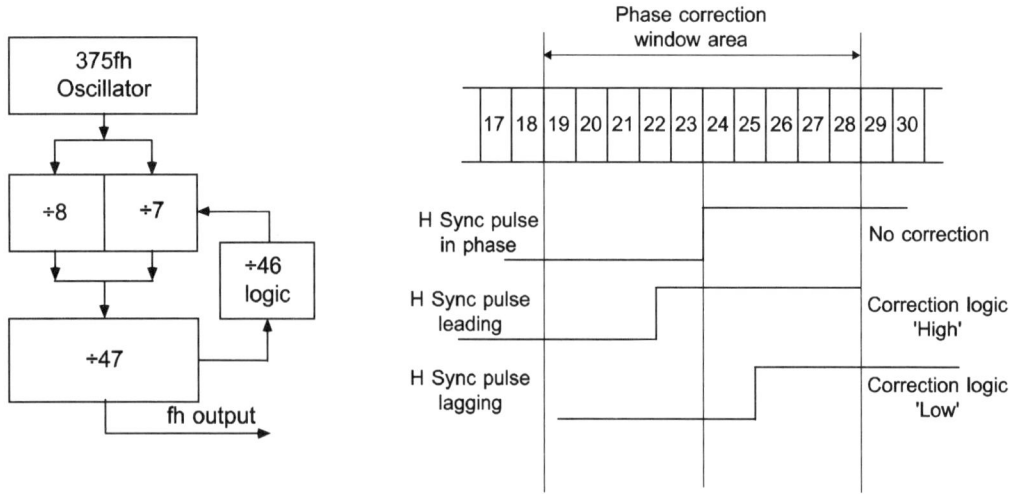

Figure 5.19 *AFC phase locked loop*

Colour recording and playback

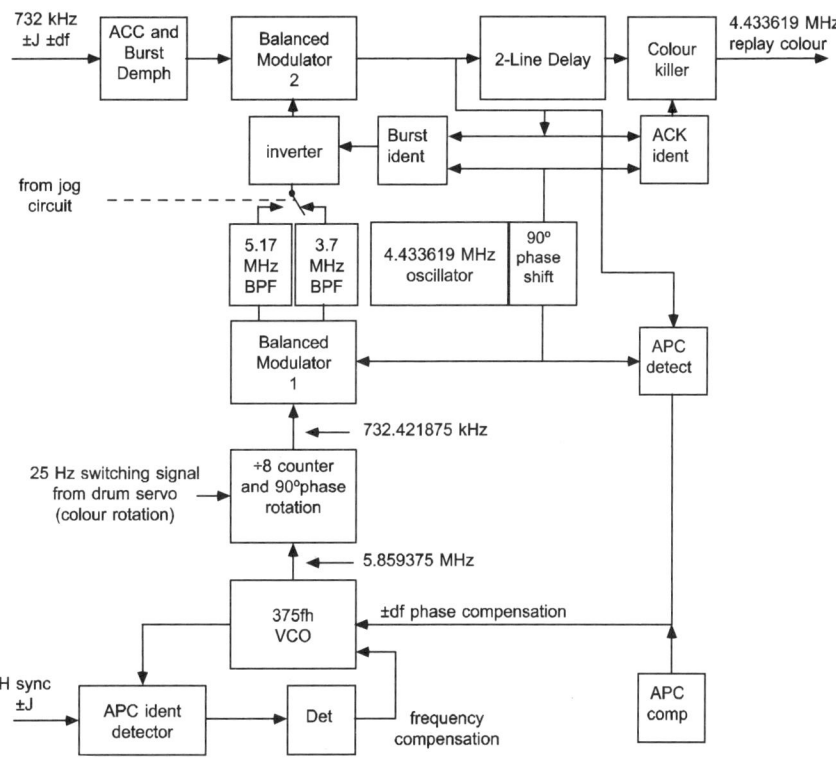

Figure 5.20 *8mm colour replay system*

modulator 1. There is a slight change in the output of balanced modulator 1 as a switch has been added and another filter, but for normal replay purposes the switch selects the 5.17-MHz filter. A carrier inverter has been added in the 5-MHz path its function is described later as part of the APC loop. In balanced modulator 2 the replayed 732-kHz colour signal is subtracted from 5.17 MHz and results in 4.433619-MHz colour carrier, which passes through the 2-line delay and out to be mixed with replay luminance through a colour killer.

Although 8mm text refers to the loop as an automatic phase control (APC) loop it also operates as an automatic frequency control (AFC) loop, similar to its use in the record mode, and it is part of the jitter compensation as described for VHS and Betamax. It operates differently in playback as shown in Figure 5.21.

The operation 'window' is over five periods (5H) of the replayed horizontal sync. From the 375fh counter a count over five sync periods is 1875 and this forms the clock period. If 5fh is clocked as 1875 ±4 clock pulses the phase lock is achieved, as in record. If the count is outside the window then coarse drive is applied to bring the VCO frequency back into the APC control range and the count back to 5fh = 1875 ±4. It can be calculated that the window range of 4 cycles is equivalent to 4.433619 MHz ±9 kHz, which is within colour specifications of ±50 Hz.

Four VCO clock periods are: -

127

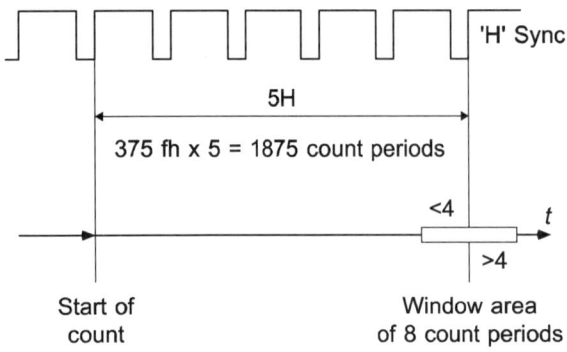

Figure 5.21 *APC loop timing*

$$\frac{1 \times 4}{375 fh} \ \mu s = 0.68 \ \mu s$$

And 5fh = 320 μs

The range is: $\frac{0.68 \times 4.433619 \text{ MHz}}{320} = 9 \text{ kHz}$

By following the 5fh period in this way the VCO will correct frequency or large phase shifts that will be ±J and these errors will be passed up the conversion process to cancel in balanced modulator 2.

Replay APC is achieved in two sections in the same way as VHS and Betamax. The block labelled burst ident is in fact similar to phase detector 2, if the replayed burst is more than 180° out of phase then 180° phase shift is applied by inverting the up conversion carrier.

The 8mm APC loop

There is a slight difference in the fine phase control at ±90° in the APC block and this is shown in more detail in Figure 5.22. The APC or phase detector is supplied with the replayed gated burst and a 4.43-MHz reference, and outputs an error signal. This error signal in the 8mm system is sent to the 375fh VCO phase lock loop and so the APC loop control is applied to the phase lock loop as an additional signal rather than via the up-conversion carrier in balanced modulator 1. It provides for a wider range of control. Refer back to VHS and Betamax for clarification on this point.

The 4.43-MHz reference signal, via a 90° phase shift, supplies both the burst ident detector and the APC detector; it also supplies balanced modulator 1 for up-conversion and the phase detector circuit within the automatic colour killer (ACK).

Due to the relatively long time constant and the

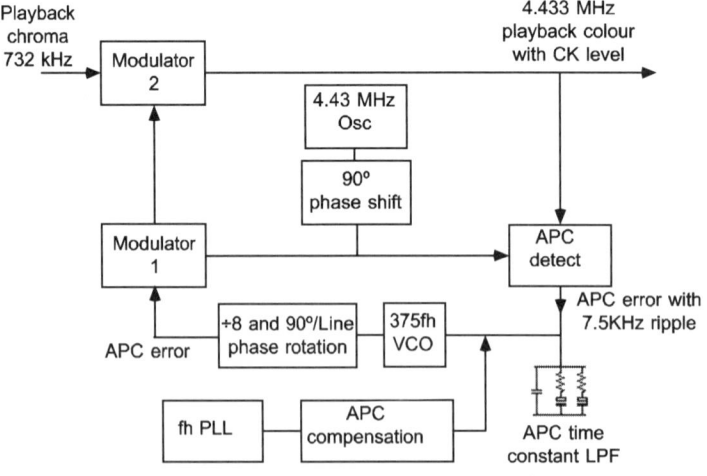

Figure 5.22 *APC loop*

Colour recording and playback

The 8mm comb filter

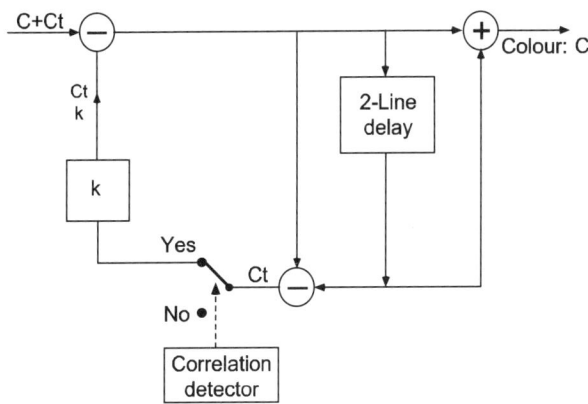

Figure 5.23 *Comb filter sharpener and correlation detector*

Further to this APC compensation there is a refinement of correlated cross-talk cancellation in the 2-line delay comb filter as illustrated in Figures 5.23 and 5.24. The loop first establishes that correlation exists in the replay signal. This means that the signal is replaying on a regular sync and swinging burst basis.

Correlation does not exist for instance in picture search when many video tracks are crossed over and the replayed burst phase is irregular.

It has been shown that in the sum output of the delay line network the cross-talk cancels leaving only the wanted signal. It is true to say that if a subtraction network were in place then the wanted signal would cancel leaving only the cross-talk component (Ct). This is part of the larger picture of the comb filter as a portion of Ct is then passed back to the input signal to subtract some of the cross-talk before the delay line. If you consider this as a sort of negative feedback the result is to sharpen up the response of the comb filter.

In Figure 5.24 the spectral response of the comb filter is shown. The 'period' of each arch is line frequency and so each arch is a harmonic of fh. In the top waveform the darker signal is the

fact that the burst is swinging 90° on each line, then the APC error signal would have a 7.5 kHz ripple on it. In VHS this is averaged out, or smoothed, whereas Betamax had the constant phase pilot burst. 8mm is more precise in its design and employs a further (fh) phase lock loop and amplifier to produce an equal and opposite ripple which cancels that caused by the swinging burst; this is APC compensation.

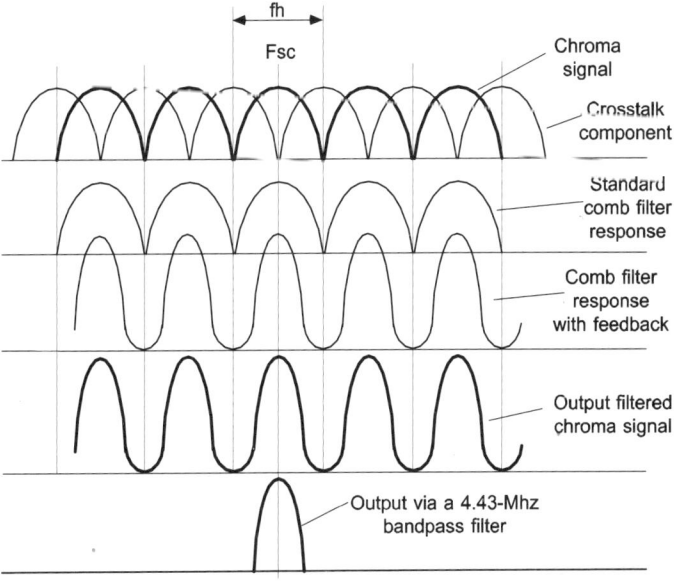

Figure 5.24 *2-line delay comb filter spectrum response*

replayed chroma, and within it but 180° out of phase is the cross-talk. Below that is the standard response curve of a comb filter as arches, and below that the almost sinusoidal signal is the modified sharper response of the 8mm comb filter. The bottom signal is the chroma output showing that the cross-talk has been removed and then by adding a single bandpass filter in series with a comb filter only the main frequency passes through, and a single response at 4.433619 MHz is shown.

When any video recorder is in the picture search mode the video heads crossover many tracks, and colour replay is subject to random rather that sequential phase changes. Normal phase control is too slow and a solution in 8mm is to switch in a 3.7 MHz filter.

Whereas in normal replay the 732-kHz signal is subtracted from 5.17 MHz in balanced modulator 2 to result in the 4.43-MHz colour signal, in picture search it is added to the 3.7- MHz signal. This has the effect of inverting the colour signal.

In picture search the Jog circuits detect instantaneous phase changes of more than 180°, and can rapidly switch the filter from 5.17 MHz to 3.7 MHz and achieve a rapid phase inversion by changing the filters over.

Hi 8mm

It may come as a surprise to some but there are no changes in the colour systems between standard 8mm and high band 8mm.

S VHS

The main change in S VHS is the use of a pilot burst in the same way as Betamax. It is added to the record colour signal in a position equivalent to line sync and it is cleaned off in replay. There are no other changes to the colour record and replay systems.

Fault finding in the colour circuits

In order to fault find in the colour circuits the service engineer must be equipped with a minimum of test items: an oscilloscope, frequency counter, colour bar generator and if possible a vector scope. A videocassette containing a known good recording of colour bars is also essential as a reference tape in order to fault find in replay.

The main objectives in the first instance are to carry out basic checks to locate the fault area by reducing the options and the amount of circuitry that has to be checked.

In order to do this the service engineer has to determine if the problem is either in record or playback, or both record and playback. The owner can sometimes assist with this.

Playback

A good start is to playback the known good tape of colour bars; check the replay for colour, no colour, or unlocked colour patterning. If there is a good colour replay then the problem may be a recording fault whereas unlocked colours or a monochrome replay confirms a replay fault and fault finding in playback can commence from this point.

Recording

Record a section of colour bars and play that, recording back on another known good, control, video recorder and note if there is colour or not.

Both

If there is a recording and playback colour fault, confirmed by a monochrome or unlocked colour playback of known good tape and monochrome or unlocked colour playback on control video recorder, this indicates that the problem lies in a section that is common to both recording and playback.

The best way to work on this type of fault is in playback.

The first step is to check the phase lock loop and prove that both line syncs and drum flip-flop are present at the loop inputs. Whilst the drum flip-flop, or RF switching signal, is the same in both recording and playback, the source of line syncs differs. In recording they are derived from the input video and in playback from the replayed video signal. The sync separator may be common to both modes.

It is difficult to measure directly the phase lock loop output frequency in the area of 600–800 kHz as the frequency counter is subject to errors, caused by the alternating frequency shifts of the phase rotation system. However, some video recorder designs allow the engineer to add resistors around the first balanced modulator and measure these frequencies at its output by disabling the summation system. This is usually carried out in the E/E monitoring mode.

Where it is not possible to carry out any measurements then a vector scope can be used to check if the phase lock loop is locking, to line syncs, by using syncs as a reference to lock the vector scope and then by looking at the input or output of the first balanced modulator. It can sometimes be judged if the loop is locking. This may not be a clear indication to an inexperienced user and some experience of what to look for is required. The display is a bold circle with faint sub-ellipses around its perimeter, often much larger than the circle. If the loop is locked these faint ellipses may be seen to be stable. If the loop is not locking then the ellipses will spin and are difficult to see.

Without the availability of a vector scope the next best method is to check the input and output of the 5 MHz band pass filter between the two balanced modulators. The presence of a signal at the output at least indicates that the frequency is probably correct and usually that the phase lock loop is also functioning normally. No output indicates that the frequency is not correct and that the fault lies within the phase locked loop, or the possibility that the filter is open circuit.

Assuming that the signal path is fully operational so far, the next step is to check the main balanced modulator (2). A low frequency colour playback signal (about 700 kHz) should be present on one input and the 4.433619-MHz conversion carrier on the other. There will be an output, unless the modulator is faulty, but the frequency may not be correct. The frequency can be confirmed to be correct by an output from the following 4.43-MHz band pass filter. An input to the comb filter may be at 4.43 MHz but it may not contain the head switching components. If there is no subsequent output from the comb filter the head switching may be incorrect. One possibility is that a previous repairer may have assembled the video head drum the wrong way around such that a head Ch.1 replay is up-converted in head Ch.2 mode. This can happen if the drum flip-flop is out of phase, but this is rare. A further option is that the phase rotation for head Ch.2 in the phase lock loop is not functioning, in which case the output from the comb filter will be alternate fields only.

If an output colour signal can be seen on the oscilloscope but the TV monitor is in monochrome then the frequency of that colour signal may be only very slightly incorrect. Check the reference oscillators into balanced modulator 1 with the frequency counter. Check also the colour output signal paths and filters for signal continuity up to and including the Y/C mixing prior to the video output. If the replay colour signal is off frequency it may not pass through the comb filter. Also a faulty delay line can cause hanover bars of alternating saturation at line rate, just as in a TV.

Colour recording fault, playback OK

As for playback it is worth checking that the phase lock loop has its line syncs and drum flip-flop signals, although the problem may not be in this area. Check the frequency of the 4-MHz reference into the first balanced modulator (1) with a counter; a wide tolerance is not good and the frequency should be within 50 Hz. Check the presence of the 5-MHz conversion signal out of it and through the 5-MHz band pass filter to confirm the correct frequency range. Confirm the 4.43-MHz colour signal into balanced modulator 2 and the low frequency colour out of it. Then check the low band colour signal following it through the recording circuits and filters up to the point where it is mixed with the recording luminance carrier.

Burst level is used for playback phase control and automatic colour level in both the VCR and the TV. The picture may appear to be over saturated if the burst is incorrectly low, or desaturated if it is too high.

With Betamax and S-VHS it is important to keep an eye on the pilot burst. Pilot burst level is important and so is its phase, which is dependent upon an input phase locked loop. If this loop is not locked then the record colour killer may operate, or even cause coloured bars in the subsequent playback.

As the colour system contains some varied types of filters, e.g. low pass, high pass and band pass, these are often multi-section for frequency and phase/delay response and are prone to failure or partial failure. The signal may pass through the filter and come out low in level and distorted; this may be due to failure of one section of the filter. Patterning and ghosting due to multi-reflections can result from a partial failure; for instance if the filter loses a common ground connection .

If the record colour level is too low then excessive noise and a 'grassy' effect results. On the other hand if the record level is too high then saturated colours, such as bars, have a diamond type of pattern within the colour, accompanied by over saturation of yellows and patterning within the cyan bar.

Colour flicker can be due to one video head being inefficient, although this is much more likely to affect the luminance signal. Some video recorders use two replay colour level AGC control systems, one for each video head, and they are switched by the 25 Hz RF flip-flop signal. This can be a source of colour flicker.

A colour fault that has nothing to do with the colour circuits is evident on replay where faint coloured patterns like butterflies flicker away in the background. A test for this is to compare a recording on virgin tape with a recording over previously recorded colour bars. If the flicker is evident on the over-recording only, check that the full erase head is functioning, this may be the cause as previous colour recordings are not being fully erased. Luminance is not affected by the erase head fault due to full erasure by the recording FM carrier.

Digital chrominance circuits

Chrominance down conversion, recording and playback can be carried out in the digital domain with the advantages of greater stability and the elimination of adjustment procedures. Large-scale integration allows the whole process to mounted within a single IC.

While the method of down-conversion for recording and up-conversion in playback differs from that shown earlier in this chapter, the result is a more stable replay with reduced phase errors as time base correction is incorporated.

Figure 5.25 shows the basic recording system where the input chrominance signal is converted to a digital signal. The chrominance signal in its analogue form is a 4.433619-MHz phase-modulated carrier and it is sampled by a phase locked 14.3-Mhz clock. In its digital format the chrominance signal is demodulated by a subtraction process by subtracting a 4.433 MHz carrier leaving the base band burst and chrominance components (R-Y) and (B-Y). A burst detector gates and samples the burst component and feeds it to a phase detector. An output error signal controls the 14.3-MHz clock oscillator which samples the input chrominance. Sampling is therefore phase-locked to the chrominance signal.

After filtering, gain control is applied to the base band chrominance and a pilot burst is added in the S-VHS recording mode. Then the chrominance signal is modulated onto a 627 kHz carrier, phase-locked to video line syncs, and the frequency is varied between 626.952 and 623.045 kHz for alternate heads to provide phase rotation for cross-talk cancellation. Before recording the digital signal is converted back to

Colour recording and playback

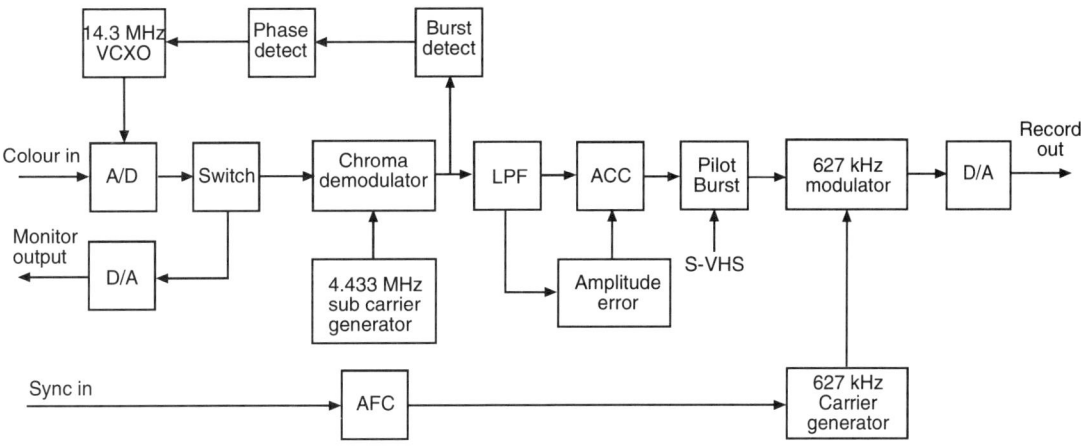

Figure 5.25 *Digital colour recording system*

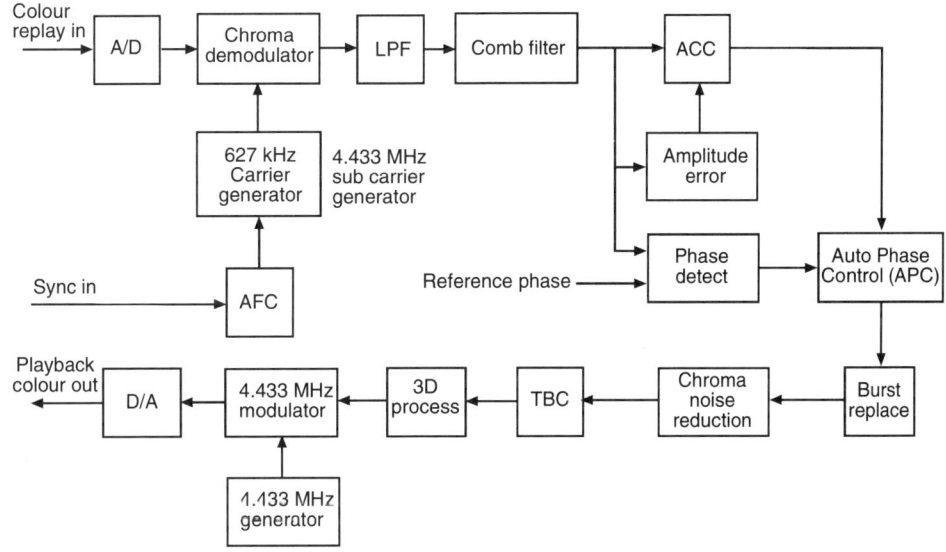

Figure 5.26 *Digital colour replay system*

analogue format for compatibility with standard VHS recordings.

In replay the chrominance signal is separated from the luminance FM carrier and digitised. The process is reversed for up-conversion and phase error correction, as shown in Figure 5.25. First the low frequency chrominance carrier signals are demodulated by a 627-kHz carrier generator that is locked to replay line syncs for AFC correction. Then the base band signal passes through a 2-line delay comb filter to cancel adjacent track crosstalk in a similar way to that of the analogue chrominance circuits described earlier in this chapter.

After amplitude control the first part of the replay phase correction takes place. A reference phase signal is used in a phase detector with the replayed burst to develop an error signal that is then used to correct the replay signal phase. This application is that of automatic phase control (APC). In the following sections the burst is replaced to maintain its form and shape, and noise reduction is applied by the addition of sequential lines to reduce the noise by √2 (see Figure 3.36, p.47).

Still in the base band form, the chrominance signal is processed by the digital time base correction circuits in parallel with the luminance signal as shown on p.53.

Finally a three-dimensional (3D) chrominance noise reduction is employed to clean up the signal according to the picture content and playback mode.

The 3D chrominance noise reduction system is shown in Figure 5.27 where the input signal is split into two paths to subtractor 1 and subtractor 2; the output of subtractor 1 is stored in a frame store. Correlated noise is obtained by subtracting two lines in subtractor 1 from different frames: line (n) and line (nf) from a previous frame. As the difference signal contains both noise and picture components a limiter is used to reduce the picture components to a level that will not affect the resolution of the picture. This is a similar technique to that of luminance cross-talk cancellation on pp 41 and 42.

In subtractor 2 noise is subtracted from the chrominance signal, and a motion detector is utilised to decrease the subtraction noise components where movement occurs due to a high content of picture components as picture degradation would occur. Over large areas of colour the subtraction noise signal level is increased to reduce the noise by a greater amount; there are no significant picture components in such areas.

Where there are amplitude differences between two video heads this can cause a flicker effect in the replayed colour that cannot be easily removed by the ACC control because it has a longer time constant than frame rate. Two lines from sequential frames are subtracted in subtractor 3. The resultant error signal contains flicker and is processed in the flicker compensator to a frame rate amplitude compensation signal. This is then applied as a correction to the chrominance signal in subtractor 4.

After the CNR circuit the base band chrominance signal is modulated onto a 4.433-MHz carrier before being converted back to an analogue format and added to the replayed luminance signal as a composite video output.

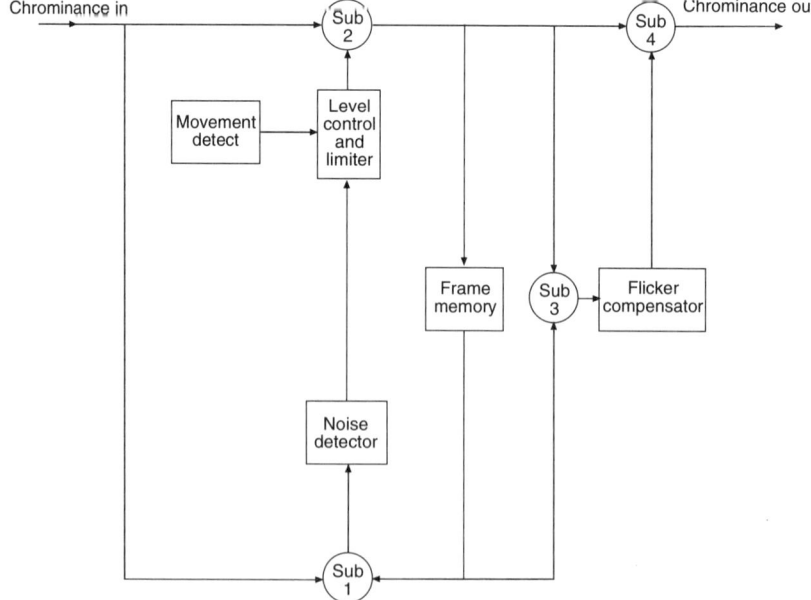

Figure 5.27 *3D chrominance noise reduction system*

6

Systems control

The function of the systems control circuits within a video recorder is to organise the handling of the videotape in order to minimise damage. This is achieved by monitoring the operating modes and controlling the operation of the mechanical section in an orderly manner. In early models, with mechanical keys, this was partly done with mechanical interlocking levers operated by solenoids. In later models control was under a microprocessor programme; operating solenoids, levers and interlocks were eventually replaced by operational motors. Basic models had logic ICs and transistor interfaces to drive the solenoids. Expansion of the systems capabilities were brought about by the appearance of large scale integrated circuits that contained the logic ICs in one package and direct drive motors.

The advent of front loading machines with touch button controls meant that more sophistication was required and the microprocessor took over control functions. The systems control is abbreviated by some manufacturers to 'Syscon', whilst others abbreviated the mechanism control to 'Mechacon'.

General operation

Until a cassette is present, the systems control ensures that all operations are inhibited In the case of VHS this is also the situation if the lamp in the cassette compartment fails.

The end of tape sensors are optical and consequently the cassette would be damaged if the lamp failed and tape transport was not disabled. If the lamp fails during play, the control system enters 'stop' and then 'unthreading', and finally into 'stop' once more. The tape is also unthreaded if the end of tape sensor is activated in either play or record.

End of tape detection in fast forward, rewind, cue or review will result in stop mode being entered. All forward functions are inhibited when the end of the tape is reached and the end sensor is activated. Only rewind can be selected, but in more automated recorders the rewind function is automatic after completing play or record. At the completion of rewind mode the tape start sensor is activated, this initiates stop and further rewind is inhibited, and only forward functions can be selected.

Counter Search operation will cause the systems control to enter stop when the counter passes 0000.

The system control resets all operations whenever power is applied, although if power is removed during play the tape remains threaded. Unthreading takes place when power is re-applied and the system resets to the stop mode.

Tape protection control

The systems control circuits monitor the tape transport mechanics and serve to protect the tape, as far as possible, from damage. Take-up spool rotation is monitored and if it ceases to rotate during any transport mode then unthreading takes place and stop is entered. There is a short time delay of 4–5 s to prevent erratic operation due to a momentary pause during normal working rotation. In basic models only the take-up spool is monitored; in the sophisticated models the supply spool is also monitored. If the tape does not travel after an operational function is selected within 5s again monitored by the take-up and supply spools, stop mode is entered.

The video head drum is also monitored, using the derived flip-flop signal, and if rotation ceases in play or record then unthreading takes place

Video and Camcorder Servicing and Technology

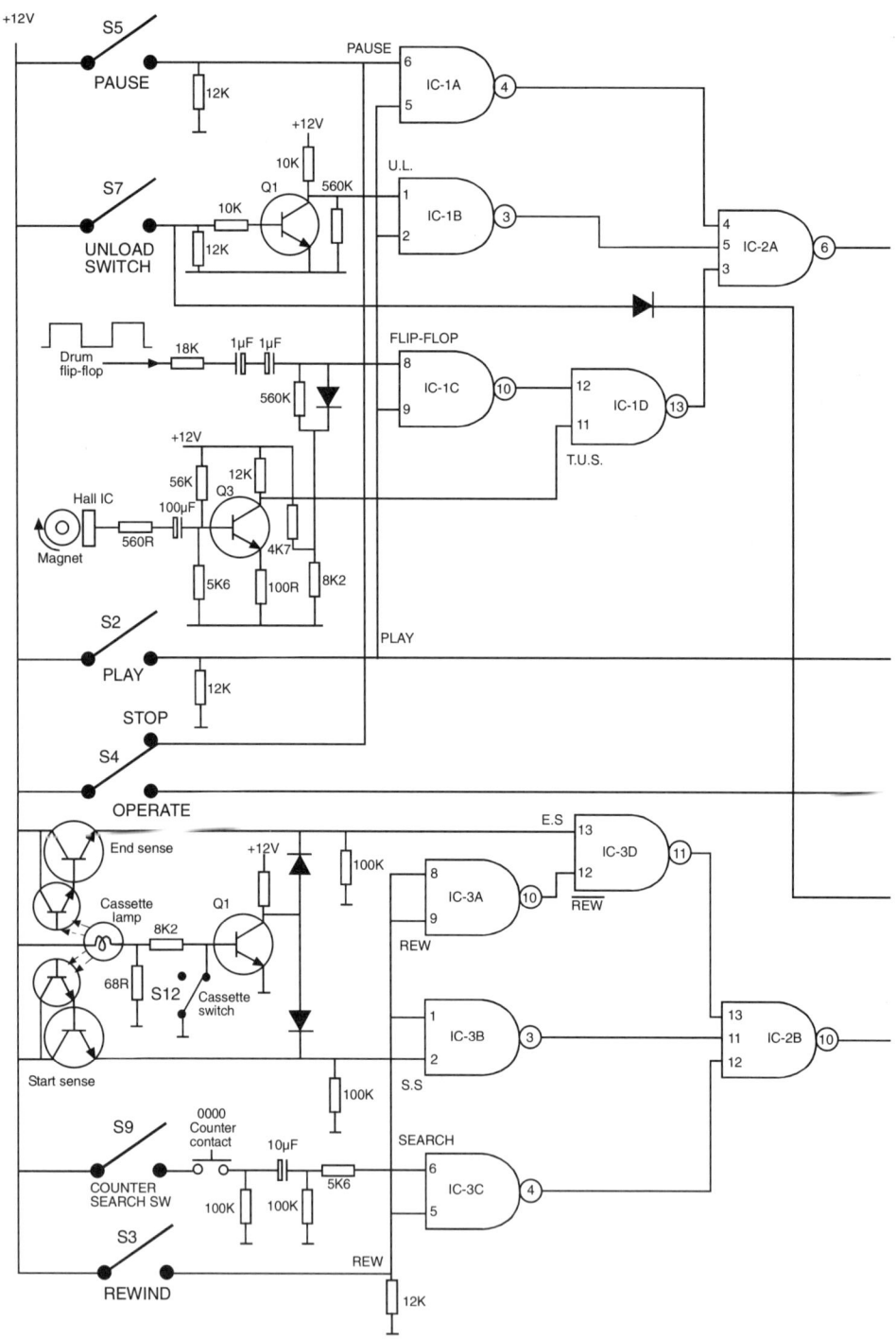

Figure 6.1 *Basic systems control circuit diagram*

Systems control

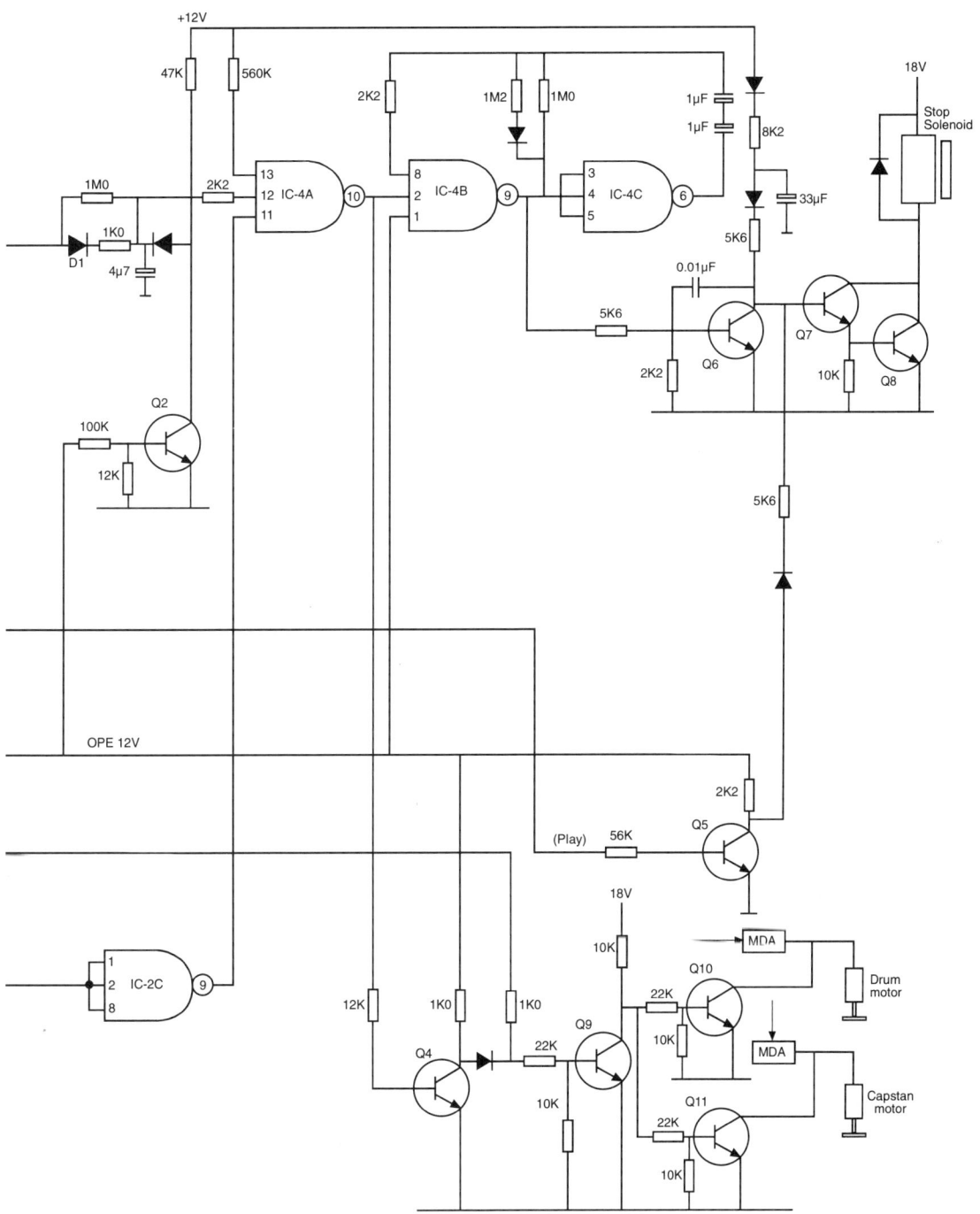

after 5–6 s.

Some video recorders have a slow motion or still frame facility, which requires that the supply and take-up spools stop rotating in a play mode but without stop or unthreading being initiated.

This is the 'Play Pause' or even 'Record Pause' mode of operation. The system control circuits inhibit unthreading during still frame and also start a time elapse counter to prevent tape damage by limiting the length of time that the stationary tape can be wrapped around the rotating video head drum; about 6 minutes.

Basic systems control

As most modern control systems are totally integrated it is worthwhile to first take a look at a basic systems control and how the mechanism control and tape protection were achieved with just discrete transistors and logic gates.

This example is from a very early video recorder with mechanical keys that interlock when selected. The stop key releases the mechanical interlock, as does a stop solenoid driven by the systems control logic. Cassette loading is mechanical from the top, the cassette being inserted and then pushed down. Tape loading drive is accomplished by the capstan motor via a belt drive with a mechanical lock at the fully loaded position; there is no reverse search facility.

Figure 6.1 is such a basic systems control circuit diagram. The logic gates used in this circuit are NAND gates. The operation of a NAND gate is as follows: the output of a NAND gate is normally high when it is in the 'off' condition. It requires both inputs to go 'high' before the gate switches on and the output changes state to a 'low' state.

In the stop mode the following conditions are present: play switch S2 and OPE switch S4 are in the open position, there is no OPE 12V therefore transistor Q9 is off, and therefore Q10 and Q11 are on. Q10 and Q11 are MDA control transistors; they shunt the motor drive voltages to ground, so the motors will not rotate.

With a cassette inserted, switch S12 is opened, and Ql in the end sensor circuit is switched on via the 8K2 resistor, the cassette lamp and reverse biases diodes D3 and D4. When the play key is pressed S2 and S4 close providing OPE 12 V and Play 12 V. The gates IC3A, B and C are all off because the rewind key is not pressed and so the REW line is low, held down by a 12 K resistor; the outputs of these gates are therefore high. Gate IC3D is also off as the End sensor is off due to a cassette being present and cutting off the lamp source. Pin 13 of IC3D is low so the output of IC3D is also high. All three inputs to gate IC2B are high, being driven from 1C3, so this gate is on and its output is low. IC2C is used as an inverter, which is off and so its output is high.

Switch S4 provides the OPE 12 V power supply. Transistor Q2 is turned on and reverse biases diode D2 that has previously charged Cl to 12 V; Cl cannot discharge quickly. IC4A is on as all three inputs are high, and its output is low so Q4 is held off enabling the 12 V OPE supply to turn on Q9. Q9 switches off Q10 and Q11 and the motors rotate driven by their respective motor drive amplifiers.

When threading takes place, switch S7 is in the open position so the transistor Ql is off and with the Play switch pressed S2 is in the closed position. Gate IC1B is turned on with both inputs high, the low output holds IC2A off so IC2A output is high, which via Dl and R9 maintain Cl charged during threading. At the end of threading S7 changes to the closed position IC1B turns off and IC2a is turned on, and its output goes low.

In order to maintain rotation of the motors after loading Cl must be kept charged to a high level to prevent IC4A from turning off. This is achieved by combining the flip-flop signal at 25 Hz, which is derived from the rotating video head drum in the audio/servo PCB, with a 1-Hz pulse derived from the rotating take-up spool. The two rotational signals are combined in IC1D and the signal on pin 11 is the combined waveform as in shown in Figure 6.2. The signal is then inverted through IC2A. The high portions of the signal top up the charge on Cl. In the low periods Cl discharges at a very slow rate, but not enough to switch off gate IC4A. This is due to the discharge path being only through a 1 M resistor to ground via the low output of IC2A.

Systems control

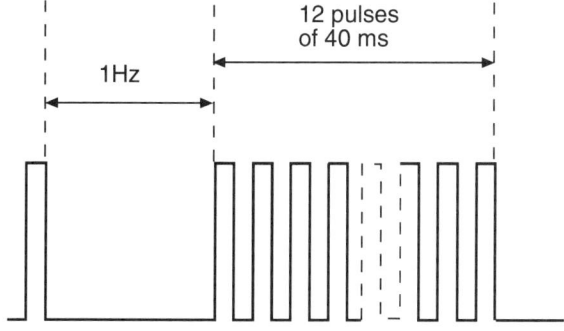

Figure 6.2 *IC1D output monitoring tape motion*

Safety function

If the video head drum ceases to rotate, for whatever reason, the flip-flop signal is no longer present. In this case pin 8 of IC1C remains high as pin 9 is biased via 560 K and 4K7 to 12 V so IC1C switches on, and its output goes low. IC1D then switches off, cutting off any signals to IC2A, which switches on due to all three inputs being high. C1 discharges below the 6 V input threshold of IC4A pin 12 and IC4A switches off; pin 10 goes high and Q4 switches on to cut off motor power. IC4A operates the stop solenoid via IC4B, turning off Q6, which releases all of the function keys. A similar train of events occurs if the take-up spool stops rotating and the 1-Hz switching signal is no longer present.

In this case transistor Q3 remains turned on and the low voltage present on its collector also turns off IC1D causing the stop solenoid to operate and the motors to switch off. At this point in time, S2 and S4 are reset to the open position but the tape is still in the threaded up position.

The unloading switch S7 remains closed and 12 V travel via diode D7 to transistor Q9 to hold it on, maintaining rotation of the motors whilst unthreading takes place. When unthreading is completed, S7 returns to the open position, Q10 and Q11 then switch on to stop the motors.

If the end sensor is activated when the videocassette has played its full length, the end sensor turns on and the ES line takes IC3D, pin13 goes high, IC3D turns on as both inputs are then high, and its low output turns off IC2B which turns on IC2C. The resultant low output turns off IC4A, which as previously described, turns the motors off and operates the stop solenoid.

In this condition, no other function will operate other than rewind. The rewind key will turn on IC3A that in turn will turn off IC3D; this will allow IC2B to turn on again, turning IC2C off and allowing IC4A to turn on again and also the motors. If the start sensor is activated, or the counter search switch operates, IC2B will be turned off again causing IC4A to turn off and hence stop the motors. Note that if the start sensor is activated, then IC3B, which turns off IC2B, will prevent the rewind operation.

The tape rewinds until: the start sensor is activated and IC3B turns on, IC2B turns off, IC2C turns on and pin 9 goes low again , and IC4A turns off stopping the motors operating the stop solenoid.

Rewind is allowed if the counter search operates as this is in pulse form to operate the stop solenoid via IC3C, IC2B or C and then IC4A, B and C.

Pause mode

In the pause mode it is necessary to maintain the motors in the operate condition whilst the take-up spool has ceased rotation, by inhibiting the safety circuit. The main pause operation is a mechanical leverage system, but our systems control has to take care of the logic as a precedent for future microcontrolled systems.

The pause switch S5 is put into the closed position, turning on IC1A. Pin 4 of IC1A goes low and turns off IC2A, the output of which goes high. The high output of IC2A holds C1 charged and IC4A on, preventing C1 from discharging in the absence of any of the pulse waveforms, and of

Video and Camcorder Servicing and Technology

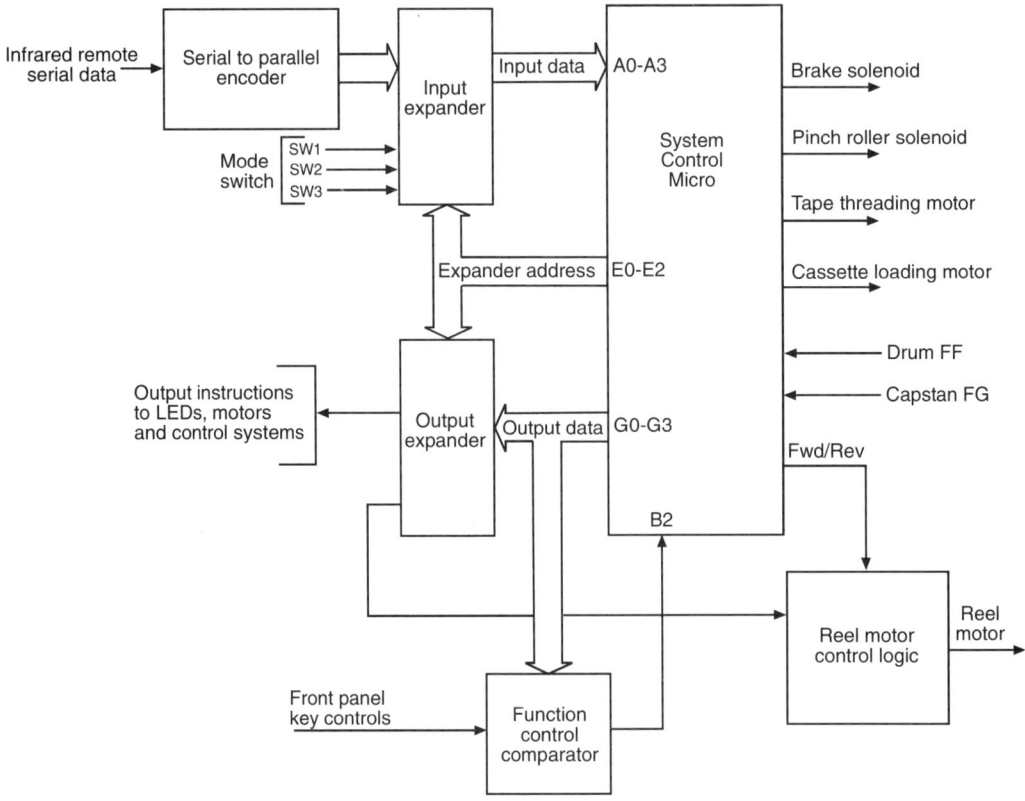

Figure 6.3 *Typical advanced microcomputer video recorder system control*

course the motors still rotate. The mechanical section of the pause key lifts off the pinch wheel drive and brakes the tape spools.

More sophisticated systems control can be fitted into video recorders and in general the first ones to emerge were based on 4-bit microprocessors. In these initial micro based systems control circuits the micros were standard 4-bit ICs that were mask programmed during manufacture.

In the mechanical control or systems control of the early video recorders a large number of interface ICs and transistors can be found to decode the microprocessor outputs and arrange them into a drive more suitable for the solenoids and motors to be driven from the micro. Inputs in the form of data had also to be arranged into a form suitable for the micro by the use of data selectors. Data selectors are a method of strobing input data, up to 32 different inputs into only eight input ports. Later versions of data selectors have emerged as input expanders.

Later generations of systems control microprocessors are much more 'custom designed' for the job that they have to do. In this respect less peripheral discrete components are required to match the micro in/out ports to the pushbuttons and tape deck mechanical components. Only interfaces to motors are required to perform the functions of being able to reverse the motors and run them at various required speeds. The more functions that a video recorder has within its facilities then the more complex the systems control circuit is. Facilities have increased from still picture, then cue and review, then the addition of backspace editing and slow motion with noise free replay. The curent peak of video recorder capabilities is insert editing with clean-in switching and clean-out switching.

The next step in systems control development

An outline of a more advanced 'mechacon' (mechanical control) system is shown in Figure 6.3.

Functions from the infra-red remote control or from the front panel pushbuttons are fed to a series-to-parallel converter. These are the Stop, Play, FF, Rew, Pause, Record etc., and are usually called the mechacon functions. Each individual function is assigned a 7-bit binary code, preceded by a 3-bit 'key code'; in binary it is 100. This key code is to tell the video recorder that it is a specific manufacturer's remote control that is generating the binary information and not another. It is an identification code to prevent cross interference between different brands.

A 10-bit binary word is issued by either the IR remote control or by the front panel function pushbuttons, which are matrixeed as a pulse encoder to produce the required serial binary word to be transmitted over the infra-red link.

Serial-to-parallel converter

Series binary information is a train of sequential pulses, but the microprocessor does not read a series of pulses, it wants them all simultaneously in parallel. Consequently we have a serial-to-parallel converter, that is really a shift register where each 8-bit byte is fed in as a serial word and then out as a parallel 8-bit word.

There are eight outputs from the serial-to-parallel convertor as seen in Figure 6.4. Seven of the outputs are the function binary information and are labelled D0–D6; the eighth is an interrupt output which goes high whenever function data is being sent to the microcomputer. This eighth bit is used to tell the microcomputer that it is receiving valid function control binary data.

The data output from the serial-to-parallel converter goes to the microprocessor via an input expander that converts the 7-bit data to 4 bits, D0–D3 plus the eighth as an interupt and then D3–D6 plus other inputs, in a 7 x 4 matrix.

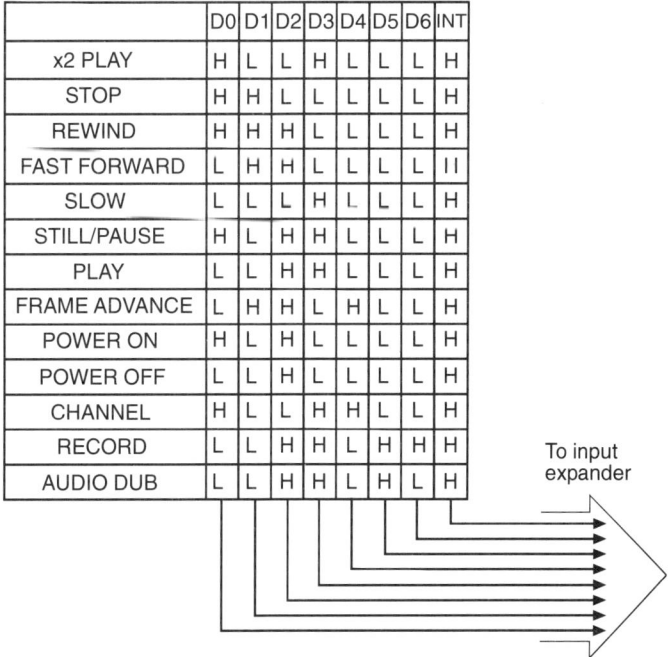

	D0	D1	D2	D3	D4	D5	D6	INT
x2 PLAY	H	L	L	H	L	L	L	H
STOP	H	H	L	L	L	L	L	H
REWIND	H	H	H	L	L	L	L	H
FAST FORWARD	L	H	H	L	L	L	L	H
SLOW	L	L	L	H	L	L	L	H
STILL/PAUSE	H	L	H	H	L	L	L	H
PLAY	L	L	H	H	L	L	L	H
FRAME ADVANCE	L	H	H	L	H	L	L	H
POWER ON	H	L	H	L	L	L	L	H
POWER OFF	L	L	H	L	L	L	L	H
CHANNEL	H	L	L	H	H	L	L	H
RECORD	L	L	H	H	L	H	H	H
AUDIO DUB	L	L	H	H	L	H	L	H

Figure 6.4 *Input serial-to-parallel converter*

Video and Camcorder Servicing and Technology

Input expander

Input expanders are used to route information into the microprocessor under the programmed control. In this case the micro has only four input ports and there are 28 information data lines. Data lines are therefore split into four sets of seven. Each set of seven is allocated to one micro input port (H1–H4) and each data line is switched into the micro port in turn, see Figure 6.5.

Each of the four sections of the expander is similar to a mechanical seven-way wafer switch. Hence; the whole expander is similar to a seven-way switch with four wafers on the shaft giving a four-pole seven-way switch with all of the four sections working as seven-way selectors in parallel.

However, in a mechanical switch you are restricted to a sequence of 1, 2, 3, 4, 5, 6, and 7. In this equivalent electronic four-pole seven-way switch no such restriction exists and the selectors, in parallel, can go from 1 to 4 to 7 to 3 to 2 or whatever is required. This control of the switching is done by the microprocessor via three output ports E0, E1 and E2.

Three-bit binary ports can produce 2 x 2 x 2, or eight possible binary combinations. However the expander only requires 7, so '000' (or LLL) is not used.

In this way the micro can issue a 3-bit binary word for instance 110 (H, H, L), then the start sensor is switched to port A0, the end sensor is switched to port A1, cassette lamp to port A2, and the record safety switch to port A3. In the next instant the micro may produce 010 (L, H, L) and we will have D5 to A0, D4 to A1, D3 to A2 and D6 to A3. Consequently, acting under the control of its programme the microprocessor can select four inputs out of 28, selecting any four particular inputs at any instant in time.

Function data, D0–D6 and the Interrupt are selected in two stages feeding D3, D4, D5, D6 (010) and D0, D1, D2, Int (011). The micro feeds this information into a temporary store in order to read the information fully.

Let us look at the information that is available for selection. We know that data D0–D6 and Int

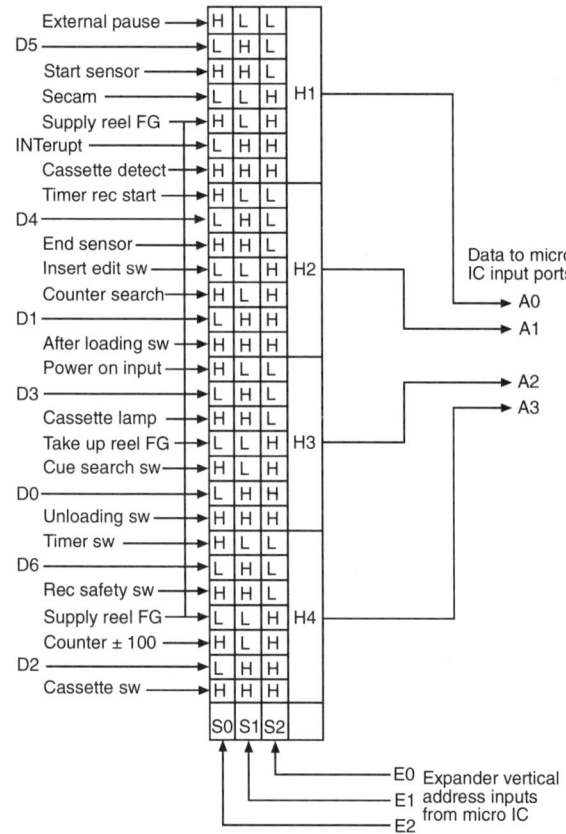

Figure 6.5 *Input expander*

is one of the 13 possible mechanical functions.

Camera pause is for an external camera or editing control that will also put the recorder into backspace editing (see Editing).

Start sensor, is an optical sensor that will be used to stop the tape at the completion of rewind.

Supply reel FG pulses, are generated by the supply reel rotating, and are integrated to maintain a low level; this goes high when stopped.

Cassette detect is an optical sensor or switch that is used to determine that a cassette is inserted into the cassette compartment, it is usually low with a cassette in.

Timer record start, is a signal from the timer circuit to put the system into record for a timer recording.

End sensor is an optical sensor to stop the tape

at the end of the cassette after completing play, record or fast forward.

Insert edit switch puts system control into insert edit mode (see Editing).

Counter search switch ensures that the recorder stops at 0000 or the tape counter in fast forward or rewind.

After Loading Switch closes when threading is completed and may be part of the mode cam switch.

Power on is a power-up reset to prevent random selection of any functions when the recorder is first switched on.

Cassette lamp detect shuts down all functions if the cassette compartment lamp fails.

Take-up reel FG low when take-up reel is rotating. If this goes high during play or any other mode then all functions stop, to prevent tape damage.

Cue search switch allows the recorder to stop in fast forward or rewind if a cue pulse is picked up by the audio control track head, an early version of indexing. Cue pulses are recorded each time the machine enters record mode.

Unloading switch, which is closed throughout the unthreading process, is used to rotate the supply spool in the reverse direction to wind the tape back into the cassette, and may also be part of the mode cam switch in later video recorders.

Timer switch, puts the recorder into timer recording mode.

Record safety switch prevents any record mode, including edits, if the cassette record prevention tab is removed.

Counter ± 100 used in the counter search mode is 'high' for a tape counter reading lower than 9900 and higher than 0100, it also goes 'high' at 0000. The effect is to slow down fast forward and rewind speed within the boundaries of ± 100 either side of 0000 to ensure that the tape stops accurately at 0000 without overshooting due to high tape speed.

Cassette switch, closes when the cassette compartment has lowered the cassette within the machine. Some video recorders utilise the start and end sensors for this function.

This is not all of the information that is required by the microprocessor; the drum servo head switching and capstan motor FG signals are fed directly to two other ports for the protection of the tape. If either of these two signals are absent then all other functions are inhibited.

Where fitted, a mode cam switch can be used into the input expander in place of other mechanical detection switches.

Operation controls

There are two ways to obtain operation control, such as play, record, fast-forward, search etc. One is a remote control that sends serial data directly into the microcomputer, the other is the use of front panel function keys.

Within the remote control a large number of paths are used to scan the keys in a matrix. It is possible to use this method for the on-board function keys but it involves an encoder IC. A more common method is that of a resistor ladder network and an A/D output from the microcomputer and a comparator. Advantages of such a method are that the keyboard only requires two wires and there are no key pulse paths with high frequency pulses flying about causing interference; such a circuit is shown in Figure 6.6.

A pulse count from the micro output data ports, G0, G1, G2 and G3, is buffered and mixed in a digital to analogue decoder. As the four ports give a binary count, a resistive network of 10 K and 20K resistors is able to provide a staircase voltage scale by adding the 4-bit binary pulses. A 4-bit binary word is capable of providing 2^2 or 16 steps when added in a resistive adder network. The staircase steps represent voltage levels in 16 values from 0 to 10 V, as shown in Figure 6.6. These values are inputted to the negative input of an op-amp, used as a comparator. The positive input to the comparator is determined by the fixed resistors connected to ground via various function switches and a 10 K 'pull-up' resistor to 10 V.

With no control functions selected the 10 K pull-up resistor holds the output of the comparator to a high level and the output of the comparator will be fixed irrespective of the D/A

Video and Camcorder Servicing and Technology

positive input, so the output will change from high to low. The change on the input port B2 of the microprocessor will signal function control selection. The micro will then look at the binary word on ports G0–G3 and recognise from its programming a pause/still command. Still Frame mode will then be entered (see Still Frame and Slow Motion).

Output ports G0–G3 also double up as an output data to the output expander. In this respect the output binary from G0–G3 will not necessarily be a straight count.

In practice a nice neat staircase signal will not be presented to the comparator, and inspection of the waveform will reveal seemingly random voltage levels. This does not matter too much as any binary word on G0–G3 will at any time produce a voltage reference to the comparator. Should any instantaneous reference voltage change the output of the comparator, then the binary word for that voltage will be recognised as a function control command; the micro will then confirm the status by producing a ramp count.

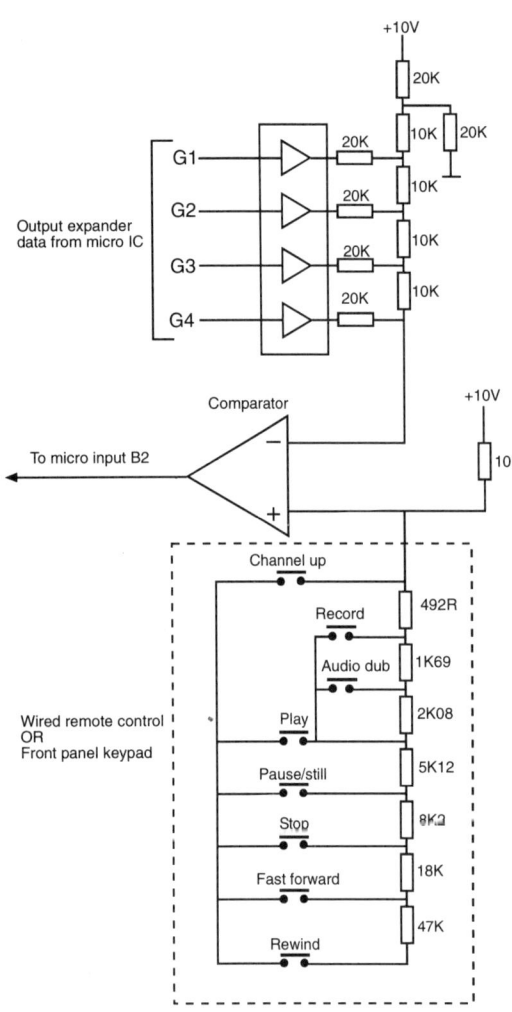

Figure 6.6 *Resistive ladder operations selector*

input. When a function is selected, for example if pause/still were selected in play mode, the voltage to the positive input to the comparator would be 4.84 V. The output of the comparator would remain high for the first seven steps of the staircase applied to the negative input, that is up to the 4.53 V step. When the binary count on G0–G3 steps onto the next level, the voltage level will change to 5.15 V. This will then make the negative input to the op-amp comparator higher than the

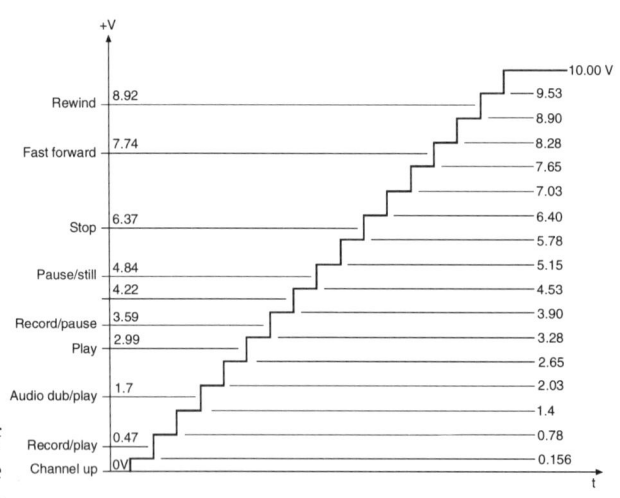

Figure 6.7 *Comparator voltages on the operations resistive ladder.*

144

Microprocessor outputs

Direct output controls are available from a custom micro for the brake solenoid, pinch roller solenoid, threading motor, cassette motor and reel motor, where such functions are fitted (refer to Figure 6.3).

These are termed direct outputs as they are coded for more than one operation and do not output through the output expander. Solenoid drives have two parts, one to provide a high current 'operate' pulse and the second to provide a holding current via higher power drive buffer transistors. The threading motor (or tape loading motor) is driven from three ports, F0, F1 and F2, and a bridge network of transistors. This provides for threading, unthreading and motor braking.

The cassette motor, the one that transports the cassette compartment up and down, is also driven by a bridge network of transistors to enable the motor to reverse drive. Four ports are used for this function; D0 and D1 to load, D2 to unload and D3 for motor braking. Two output ports are utilised for the reel motor that is fitted on this sample video recorder, which drives the spool carriers. This motor also drives forward and reverse, and ports C0 and C1 control the motor direction. Other reel motor functions come from the output expander.

Output expander

The output expander (Figure 6.8) works in the reverse manner to the input expander, as it is similar to a four-pole seven-way switch again. The output ports G0–G3 are each switched to one of the seven outputs in its own quadrant. The microprocessor address ports, S0, S1 and S2, select one of the seven outputs of each of the four sections in parallel.

For example, if S0–S2 were 100 (HLL) then G0 would output to reel idler control, G1 outputs to reel motor fast forward or rewind, G2 outputs to ±100 (fast forward or rewind), and G3 outputs to reel motor unloading. All four of these are reel motor control signals, but the actual output does not appear until the fourth address signal, when a strobe latching pulse has been inputted to the expander to hold the signal output until the next change is required.

If the reel motor is to be in the fast forward or rewind mode then G1 will be high when S0–S2 is 100. The reel motor fast forward output will be locked high by the next strobe pulse for the duration of the function. A subsequent strobe pulse for the next function used will reset the reel motor fast forward output to low again.

As S0–S2 and G0–G3 are continuously changing it is necessary to use the strobe pulses to set and hold the output from the expander for the duration of the selected function. It is possible to set more than one output high, as it may be necessary to perform more than one output function at any time; for example, whilst holding the reel motor in fast forward, the FF LED will have to remain illuminated also.

Let us have a look at some of the outputs from the expander.

Reel idler control, the reel motor idler pulley normally sits in contact with the reel motor shaft but not with either spool carrier turntable. An idler function is a short high level pulse to spin the reel motor and hence the idler pulley in order to 'throw' the pulley over to one of the spool carriers so that it may then be driven by the motor.

LED outputs are self-explanatory, i.e. to illuminate an LED.

Muting, this function is Audio muting in fast forward, rewind, cue and review.

Search fast forward (Cue), search rewind (Review), cue and review signals to capstan servo microprocessor.

CH UP advances the tuner channel selector in up count at a rate of 0.5 Hz.

Reel motor fast forward or rewind runs reel motor at high speed, approximately 45 times normal play. Fast forward or rewind is selected by micro ports C0 or C1.

E/E control audio and video switches video output and VHF output between E/E (stop or record) and playback signals.

Video and Camcorder Servicing and Technology

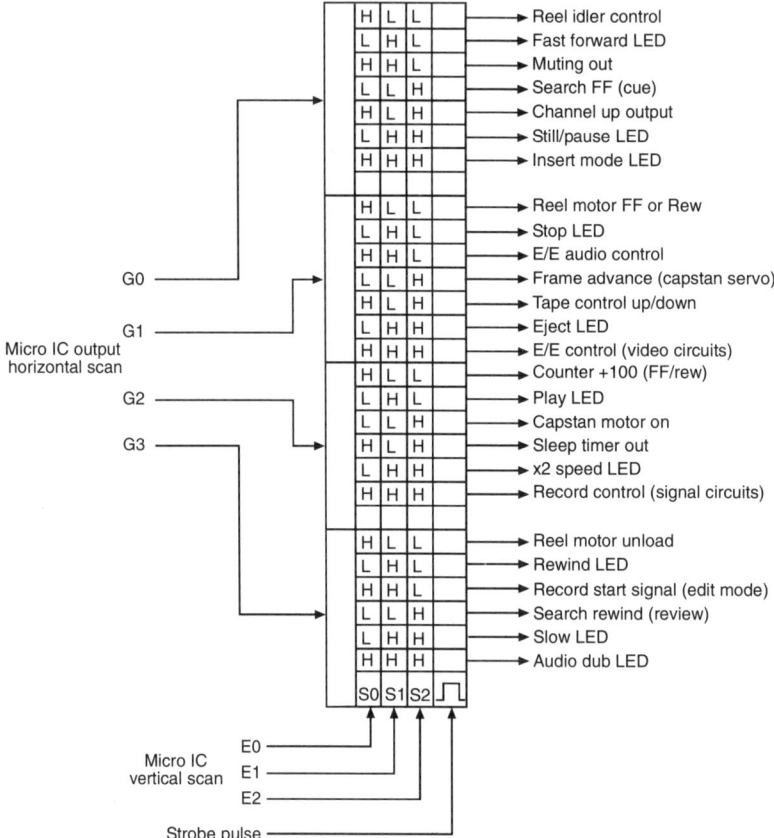

Figure 6.8 *Output expander*

Frame advance, puts the capstan servo control micro into frame advance stepping mode. See Chapter 9 for details on this function.

Tape counter up/down ensures that the digital tape counter counts up in all forward modes and down in all reverse modes.

±100 (fast forward and rewind): in fast-forward or rewind this signal appears when the tape counter is ± 100 either side of 0000, it slows the reel motor down.

Sleep timer out instructs tuner/timer micro which times down from 60 minutes or a preset time and then stops tape. It is used when 'Timer' is selected whilst the recorder is in Record Mode.

Record control switches audio and video circuits between play and record.

Reel motor unloading used during unthreading to drive supply spool in reverse to take up tape slack.

Record start is used to inhibit recording until the correct instant required, and is used in assembly and insert edit modes.

Mode cam switch

Original versions of video recorders until the mid 1980s used single switches for the mechanism, one for the final tape threaded position (loading) and the other for the unloaded position (unloading). As mechanisms developed to accommodate indexing in fast-forward and rewind, a half loading position was used. Additional mechanism positions required that a more comprehensive mechanism control be developed. To this end a cam is fitted to the mechanism drive gearing that will provide

Systems control

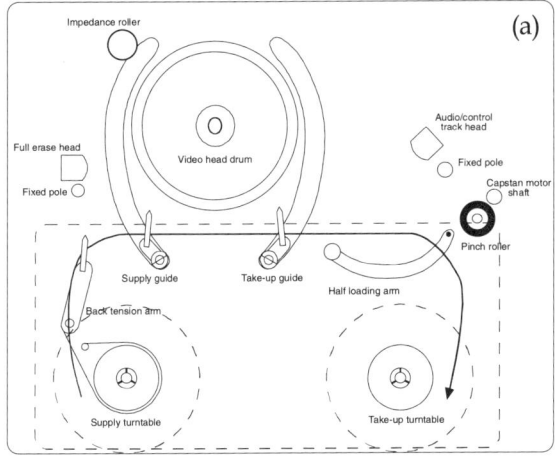

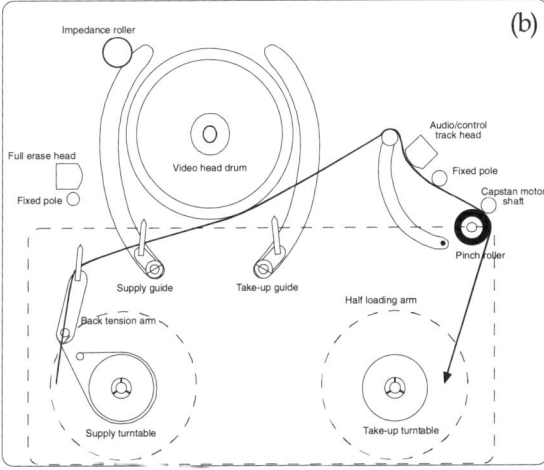

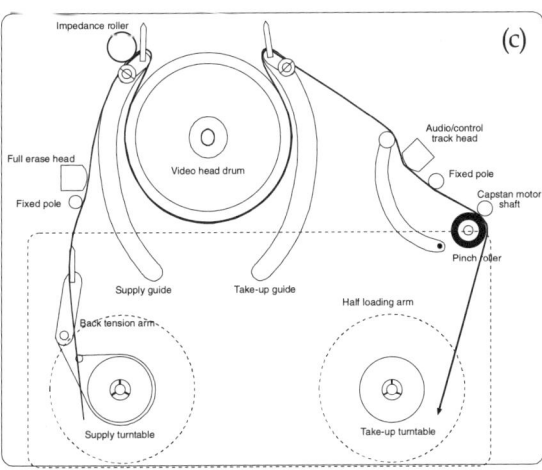

Figure 6.9 *Mechanism positions*

multiple positional information. This is the mode or cam switch consisting of a set of wipers connecting several concentric tracks.

Figure 6.9 illustrates three typical mechanism positions. Part (a) shows the standby position when a cassette is loaded and the video recorder is powered off.

Part (b) shows the half loading position where the tape is drawn out of the cassette up to and around the audio/control track head; this is referred to as the stop position. Both fast-forward and rewind are carried out in this position which allows for high-speed reading of the control track for rapid index search.

Part (c) shows the fully loaded position. Play and reverse play are carried out in this position, and forward search can take place. In some models fast-forward and rewind take place in the fully loaded position but there is a relaxation of the guides and pinch roller for this to take place.

The most forward position of the mechanism is that of reverse search. As both guides are fully engaged the result is to clamp the tension arm and release back tension braking to prevent interference with the supply spool during this function.

Figure 6.10 shows the details of a cam, or mode switch, at the heart of an electro/mechanical system. Cams A, B and C are three tracks found within the cam switch, each track is connected by wiper contacts to ground, the centre track. Each track is connected to electrical outputs SW1–SW3 to provide mechanism positional information to the microcomputer.

$0°$ is used as the reference position, standby; this is when the cassette has been loaded into the mechanism. It may be a temporary position on the cam switch as the tape is subsequently drawn out as the mechanism runs up to the half loading position where the cam switch is at $60°$. Fast-forward and rewind are carried out in the half loading position allowing CTL head to read the control track for indexing and other data information.

When either play or record is selected then full loading takes place, and the supply and take-up guides draw out the tape to the fully loaded

Video and Camcorder Servicing and Technology

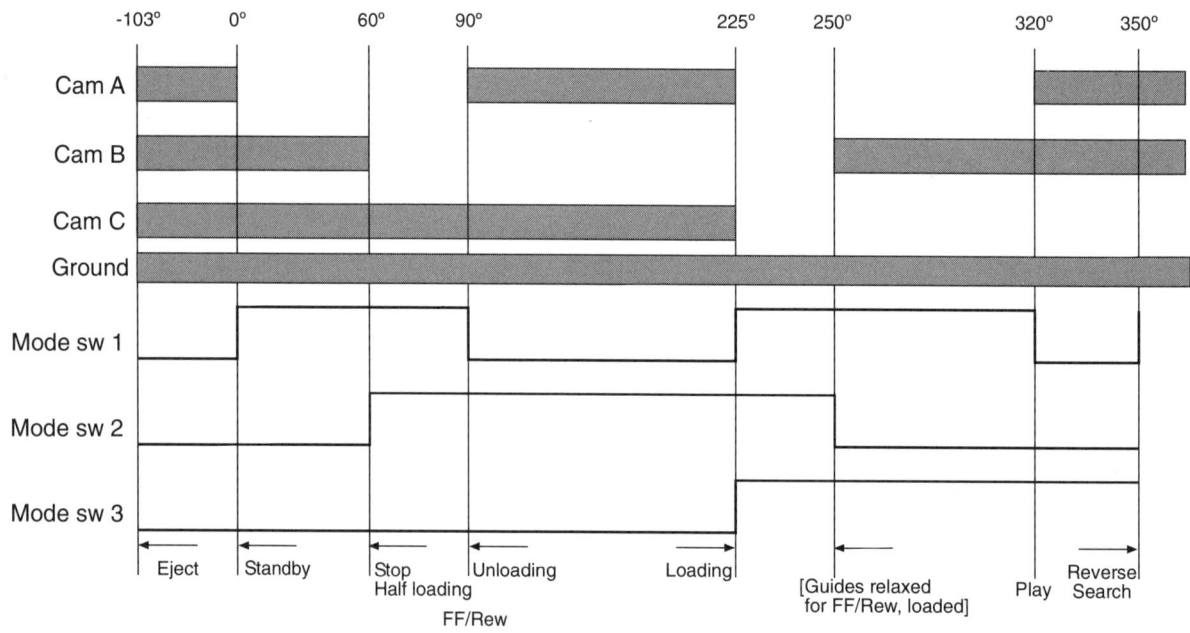

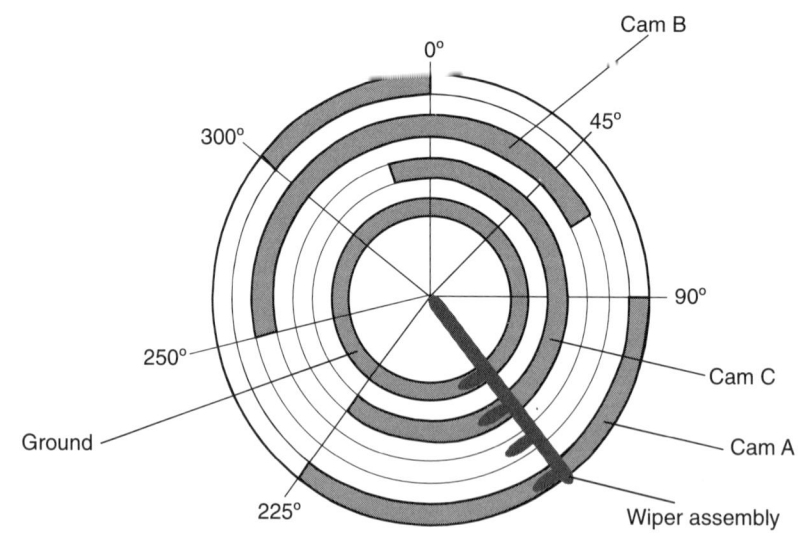

Figure 6.10 *Mechanism control cam switch and contact details*

position at 225°.

For forward search (cue) the mechanism stays in the play position as drive is supplied by the capstan shaft and pinch roller.

In reverse search the mechanism moves forward to a new position at 350°. This is to release the tension arm and fix it rigid otherwise the reverse tape drive on the supply spool would be adversely affected by the tension brake band.

Another position at 250° is where the tape guides drop back a fraction to reduce tape tension around the drum. It is a position used on video recorders with fast-forward and rewind in the fully loaded position for a 'quick play' function. Quick play reduces the time between fast-forward (or rewind) and subsequent play, and gives the user a quick response between these modes.

Three cam tracks can give up to eight mechanism positions including eject, more frequently found as a mechanism position in camcorders.

This is only an overall view of a sophisticated mechanisms and systems control. The circuit details are much more complex and of course the manufacturer's service manual will have to be referred to for these details. It has been used as an example to illustrate the types of switching signals involved and how these signals are matrixed in and out of the control microprocessor.

As video recorders and camcorders develop, more and more of these peripheral integrated circuits become an integral part of the microcomputer until there is only one IC... the systems control microcomputer.

7
Long play

Long Play (LP) was introduced into VHS video recorders in 1983. Playing time is doubled by running the tape at half speed (11.7 mm/s) and this created problems that had to be overcome. Due to the reduction in linear speed, video track width was reduced by half, from 49 μm to 25 μm and picture search presented the designers with both sync and colour stability problems as the video heads traversed many irregular tracks. Long play tracks are not recorded onto tape with the same regular pattern as those of standard play.

The form of the LP video track is determined by the slower tape speed. Usually there is an extra pair of LP video heads fitted to the head drum. In the first long play machines these addition, heads were spaced 70° round the drum from the standard play (SP) heads. Current practice is to mount LP and SP heads on common ferrite chips with opposing azimuth tilt. A cheaper option is to use the same heads for standard play and long play; however, the signal-to-noise ratio in standard play is reduced. The advantage of having extra LP

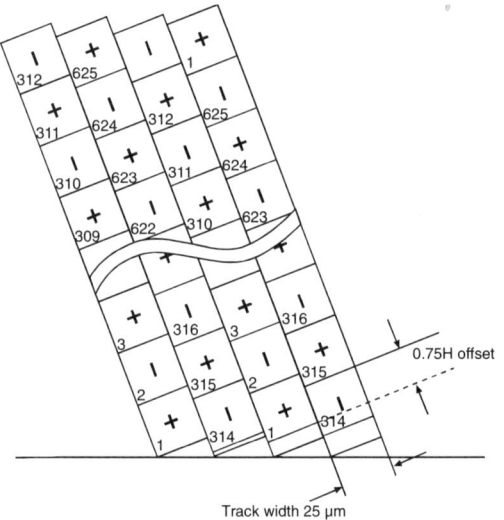

Figure 7.2 *Long play track layout*

heads is shown for still pictures in Chapter 9.

Video tracks recorded in the standard play mode are shown in Figure 7.1. Video tracks for long play are shown in Figure 7.2.

Standard play tracks have a 1.5 line offset between adjacent tracks and allow for sync and colour phase alignment as discussed in Chapter 5. With a reduction in tape speed the track angle becomes steeper and the track width is reduced. The offset between TV lines in adjacent tracks is 0.75H, half that of standard play and no line sync correlation as shown in Figure 7.2.

The main problems occur during picture search when due to the increase in linear tape speed a video head will crossover a number (usually about 5) of its own video tracks as it traverses the tape ribbon. In normal standard play these track crossings create no problem since video track line correlation ensures (given drum speed correction) that continuity of replayed line sync pulses is maintained at 64 μs intervals. In LP search mode,

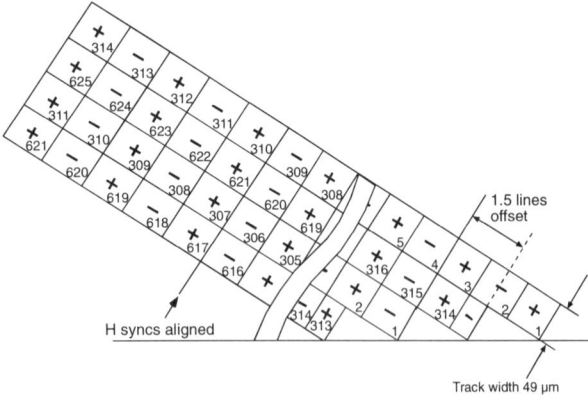

Figure 7.1 *Standard track layout*

however, each track crossing will give rise to a half-line jump in replayed sync. If this goes uncorrected the result will be severe picture skew (line pulling) on the monitor TV and colour phase is not contiguous, as two lines with the same phase can be replayed.

Advanced noise correlation and carrier interleaving techniques are less effective in long play as the TV line tracks are not aligned to give line correlation. Colour cross-talk cancellation is the same as standard play and is less affected by the lack of line correlation.

This severe break up of the video signal in long play visual search can be corrected by the use of two 'jump' circuits. One is a 1H jump to correct for colour phase inversion, the other is a 0.5H jump to correct for the 0.5H sync pulse errors.

Jump circuits

Extensive use is made of the 'D' type J-K flip-flop, or bistable, in the jump circuits. Two inputs are used in this type of bistable: a clock input and a 'D' input. When the logic input D is 0 then a clock transition from high to low sets the Q output to 0. When logic D is 1 then a clock pulse sets Q to 1.

Refer to the circuit diagram in Figure 7.3 and the waveforms in Figure 7.4. In the top part of Figure 7.3 the video head, shown as a broken line, passes over a number of tracks in search mode; it is the same azimuth as the dark shaded tracks and only reads these. The resultant playback video signal has a half line about one-third of the way through (white) and another at about two-thirds (dark).

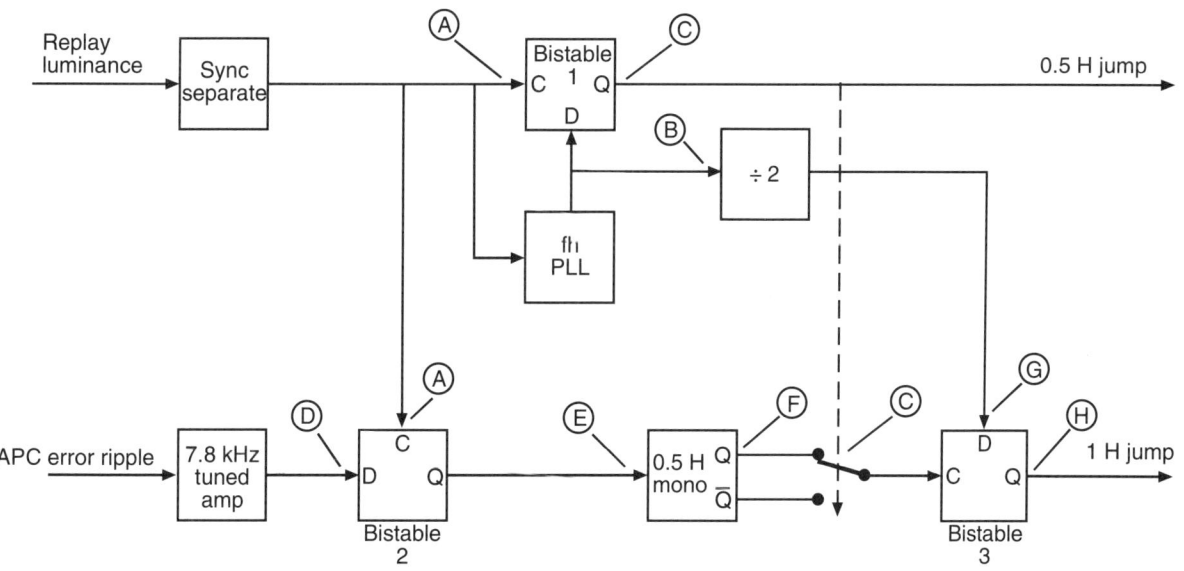

Figure 7.3 *Jump switching pulse circuit*

Video and Camcorder Servicing and Technology

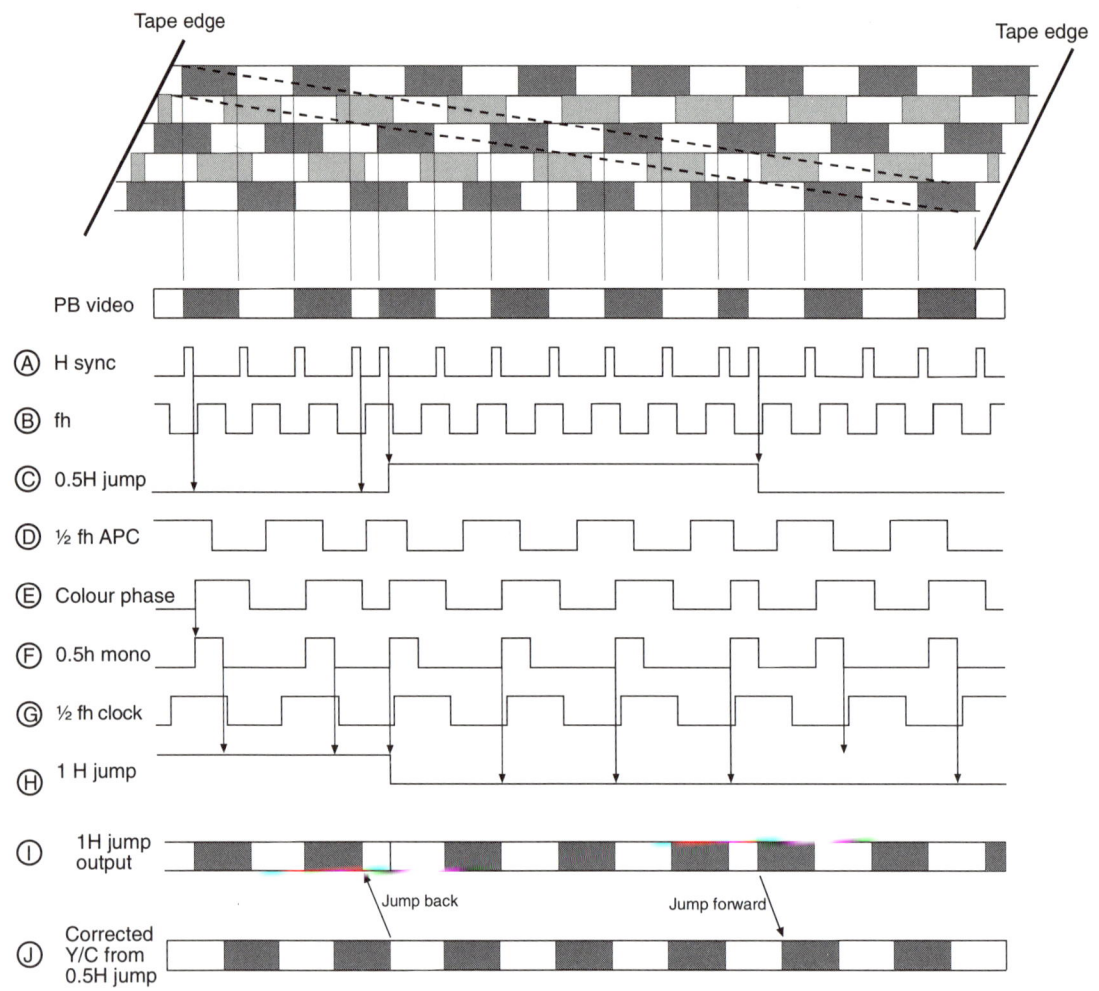

Figure 7.4 *Jump circuit waveforms*

Replayed video luminance is fed to a line sync separator, waveform (A) is the output and it has three paths. One is to the clock input of bistable 1, another is the clock input to a line sync phase locked loop, and the third is the clock input of bistable 2.

In bistable 1 the clock input of line syncs (A) sets the logic level of the output Q according to the logic level of D input, which is a square wave (B) from the fh phase lock loop. The characteristics of the phase lock loop are such that it is unaffected by the odd half line pulse occurring. The coincidence of the clock pulse (A) is when (B) is low and (C) remains low. An erroneous 0.5H pulse occurs when (B) is high and so (C) is set high until a flowing 0.5H clock pulse restores the original sequence (C) then returns back to the low condition. This is the derivation of the 0.5H jump output to Figure 7.5.

A second input in Figure 7.3 is a 7.8 kHz ripple

from the colour replay phase correction circuits to a tuned amplifier to give waveform (D) a ½ fh waveform to the D input of bistable 2. Line syncs (A) are the clock input, and the output waveform is (E), representing the colour phase. This signal is used to trigger a 0.5H delay monostable on the rising edge to give waveform (F). The phase of waveform (F) is selected by the 0.5H jump waveform (C) as the clock input to bistable 3. Waveform (G) is the D input to bistable 3 and is formed by dividing the fh PLL output (B) by 2.

When the 0.5 jump (C) is low, the falling edge of (F) sets the output (H) high when (G) is high and low when (G) is low. However, when the 0.5H jump (C) is high then the rising edge of (F) sets (H) according to logic (G), since the clock input is the inverted output of the monostable. Waveform (H) is the 1H jump output to Figure 7.5.

It can be seen from Figure 7.5 that the 1H jump signal inverts the replay colour phase before it is mixed with the replayed luminance component. After mixing the composite video signal is subject to a 0.5H jump shift, or not, according to the 0.5H jump switch.

Comparing the colour phase of the playback video signal in Figure 7.4 and the modified video (I), it can be seen that the colour phase of (I) is inverted when the 1H jump goes low. In the 0.5H jump section the extra half line (white) is eliminated as the switch changes from delayed to direct, and the single half line is extended when the 0.5H jump switches back from direct to delayed. This leaves the composite video signal corrected in line sync timing and colour phase.

Video recorders that have a monochrome long play search do not have the 1H jump circuit fitted; they do have a 0.5H jump circuit to maintain picture stability.

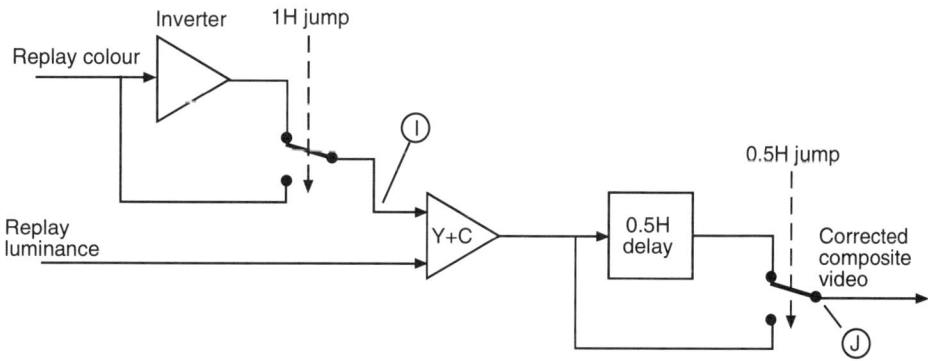

Figure 7.5 *Video signal path with jump switching*

8
Hi-fi audio

For a long time audio recording in domestic video recorders followed the standard sound recording/replay techniques that are used in any common cassette or reel-to-reel audio recorders. Audio frequency response in conventional video recorders is limited by the linear tape speed to around 8 or 10 kHz. Stereo video recorders simply split the 1 mm audio track on the upper edge of the tape into two tracks (each 0.35 mm wide) with a 0.3 mm guard band between them. The right-hand channel is written along the very top edge of the tape, and any distortion of the tape ribbon edge – caused by dirt, misaligned or worn tape guides or stretching – severely affects reproduction in the right-hand channel. Audio level fluctuation, drop out and phase distortion of the right-hand channel become evident as the wear and tear in a well-used tape takes its toll.

A world market was envisaged for a video recorder with a substantial improvement in audio performance, brought about by the public's growing appreciation of good sound quality, and by the increasing availability of NICAM television receivers with built-in stereo channels using hi-fi amplifiers and respectable speaker systems. A high quality audio system in the video recorder would also make it worthwhile to directly link video machines to the separate domestic surround-sound system.

The technical challenge faced by the designers was to improve the audio quality to a level that could be termed high fidelity and be comparable to current state-of-the-art stereo systems and digital audio players. This had to be achieved whilst retaining system compatibility between the hi-fi machine and the linear audio tracks of earlier models, without affecting video signal quality. A system was developed whereby the audio signals, as FM carriers, could be added to the video signal prior to recording. Although this technique worked with the NTSC system, and found some application in the USA, it was not applicable to the PAL system without severe mutual sound/vision interference.

Research into interference-free audio carrier techniques brought about the method of recording the FM audio tracks by means of separate audio heads mounted on the video head drum. This confers the obvious advantage of high head to tape speed, an increase from 2.339 cm/s to 485 cm/s. The video information is then recorded over the audio tracks, partially erasing the audio signal track patterns, but leaving enough for a good replay. Mutual cross-talk is reduced to an acceptable level by careful selection of the audio

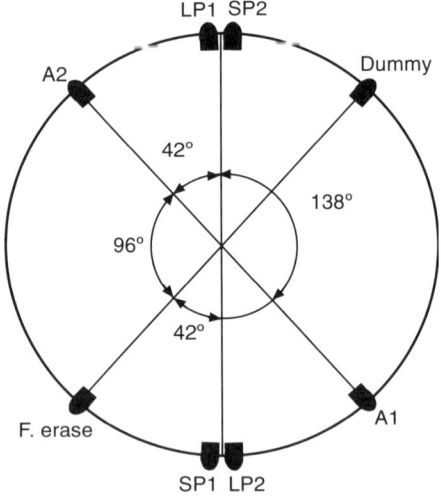

Figure 8.1 *Video drum arrangement for Hi fi audio heads*

carrier frequencies and the use of a large azimuth tilt on the audio heads.

The audio signals are frequency modulated onto RF carriers, one for the left and one for the right-hand channel. The reduction in level of 12 db brought about by over-recording by the video heads does not therefore affect the play back sound quality as the FM signal is unaffected by amplitude variations. A further reduction in crosstalk between the FM carriers of video and audio is achieved by the physical difference of azimuth tilt; co-axial tracks have opposing azimuth. With FM carriers, replay level drops off rapidly with increasing azimuth offset; whereas the video head gaps are set at ±6° the audio head gaps are cut at angles of ±30°. Consequently each audio and video head produces an output only when it traces its own recorded track and due to the high signal-to-carrier frequency ratio, audio track cross-talk is not significant.

Head drum construction

A sample VHS video head drum assembly is shown in Figure 8.1. Four video heads are mounted in pairs upon the drum, with SP1 and LP2 together and SP2 and LP1 together. The reason for the reversal of the position of these heads is that in still-picture and slow-motion modes the reproduced picture is a still picture, built up by the alternate replays of SP2 head and LP1 head; see Chapter 9 for details.

It can be seen that the audio heads are placed 48° ahead of the video heads and that the flying erase head is a further 90° advanced making it 138° in advance of LP1/SP2 video heads. Now we have to consider that the tape is moving laterally as the heads scan it at an angle. To this end the head that is in advance will have to be mounted above the deck reference plane. This is because the tape will move laterally in the period that the audio head passes over a datum point and when the following video head reaches that point (see Figure 8.2). Therefore the advanced audio head has to be on a higher plane from deck reference level by a value of 58.7 μm to accommodate the tape travel, such that the path of the audio head lies along the centre of the video track. In order that this tape travel is taken into account for all heads on the drum in both standard play and long play, each head is mounted at differing levels with respect to the deck reference plane; this arrangement is shown in Figure 8.3.

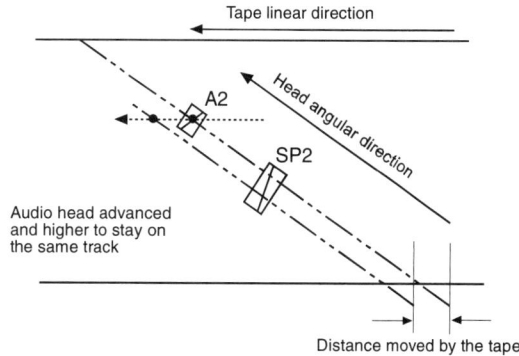

Figure 8.2 *Relationship beteen head height and tape movement*

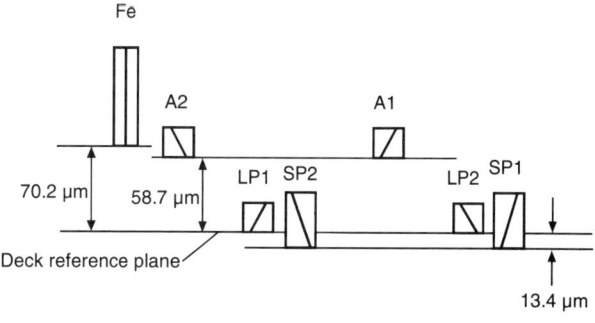

Figure 8.3 *Head height alignment about a reference plane*

155

Video and Camcorder Servicing and Technology

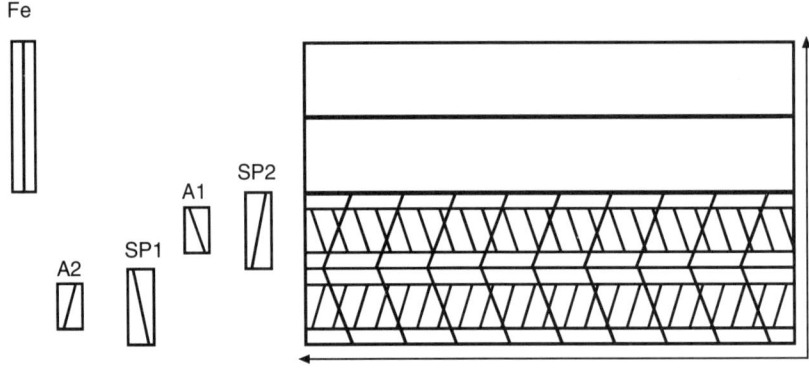

Figure 8.4 *Standard play speed video and audio track azimith and alignment*

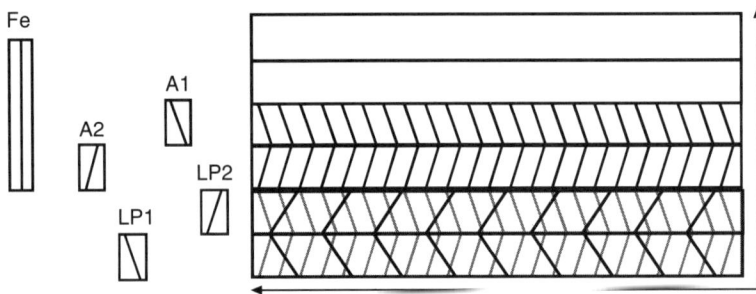

Figure 8.5 *Long play video and audio track azimuth and alignment*

In Figure 8.4 the relationship of the heads along the helical tracks brought about by the angular and reference height shift can be seen for standard play. In editing modes the signal heads are preceded by a flying erase head. Head A1 then records the audio track followed by the SP2 video head and A2 records the next audio track followed by video head SP2.

In long play the arrangement is slightly different (see Figure 8.5) as the track width is half that of standard play being, 25 μm, and this creates a 2-line shift between the audio heads and their corresponding video heads. After the flying erase head, each audio head A1 and A2 record their audio tracks and then the two video heads follow up two tracks later: video LP2 for audio A1, and video LP1 for audio A2.

Figure 8.6 shows the head scanning sequence for standard play, SP, taking into account both the angular displacement and height. It has been shown that due to its height the flying erase head is two tracks ahead of the video heads. What is not so obvious is that the audio heads are one track ahead. Audio head A2 follows the flying erase head by 96° and records CH1 audio on track 1, then audio head A1 records CH2 audio on track 2. Subsequently video head SP1 records over the audio that has been laid down by A2 along track 1. This is followed by video head SP2 recording video on track 2. From Figure 8.6 it can now be seen that the A2 head is in advance of SP1 head by 222° and is aligned with the same track as SP1 because of the height difference of 72 μm (58.7 + 13.4 in Figure 8.3).

The LP recording sequence is shown in Figure 8.7. In this example the flying erase head covers 4 tracks followed by audio heads A2 and A1, then LP1 and LP2. LP1 is 1 revolution and 222° (total 402°) behind A2. This puts audio head A2 almost two tracks in advance of LP1.

The sequences for SP and LP head recording is the same for playback, and the timing for signal switching for each head has to be taken into account when editing.

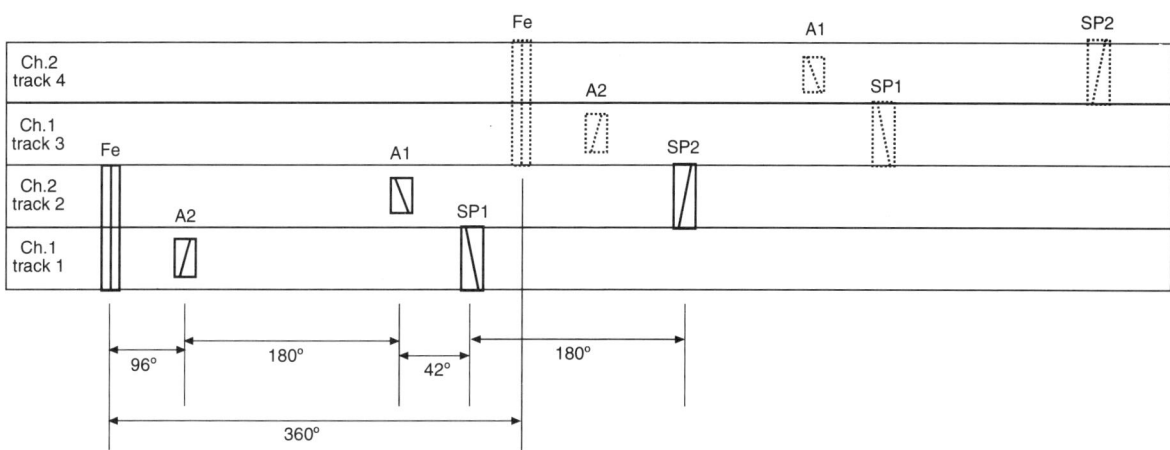

Figure 8.6 *SP head record/replay scanning sequence*

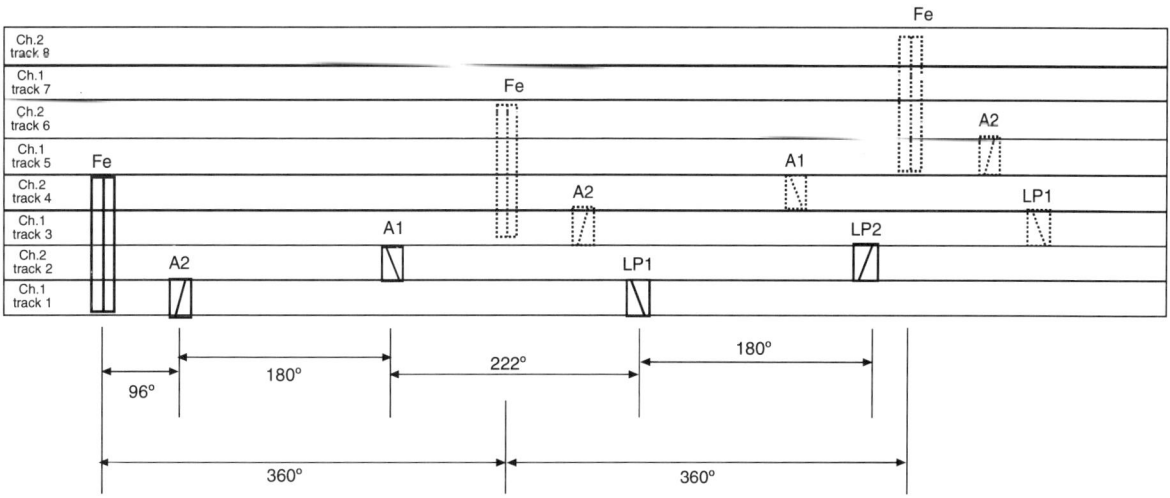

Figure 8.7 *LP head record/replay scanning sequence*

Depth multiplex recording

Figures 8.8 illustrate the depth multiplex recording principle with two heads recording a section of track; the audio head is in advance of the video head as mentioned previously. As the audio head has a wider gap and lower frequency signal than that of the video head its magnetic field penetrates deeper into the oxide layer on the tape. The surface layer of the audio magnetic pattern is subsequently erased by the following video head recording low frequency chrominance and luminance FM carrier. The replayed audio signal is attenuated by approximately 12 dB by the video layer recording, depending upon the energy level of the tape in use. Separation of the audio and video carrier signals during replay relies on the azimuth difference between the audio and video heads and the head response due to the head gap.

Although some of the first hi-fi video recorder models had stereo capability, it was only on the hi-fi tracks, with only a mono longitudinal audio track. Later models offered stereo in both hi-fi and longitudinal modes for complete compatibility with all tapes. This was later dropped in favour of a mono track again due to manufacturing costs.

Hi-fi recording and playback

The left-hand and right-hand audio signals are frequency modulated onto a 1.4-MHz and a 1.8-MHz carrier, respectively, with a deviation of ±150 kHz. The two carriers are added, and then recorded by the audio heads. It is important to appreciate that each audio head caters for both sound carriers during its 20 ms sweep of the tape. The hi-fi signal spectrum, relative to that of the video signals is shown in Figure 8.9. Although

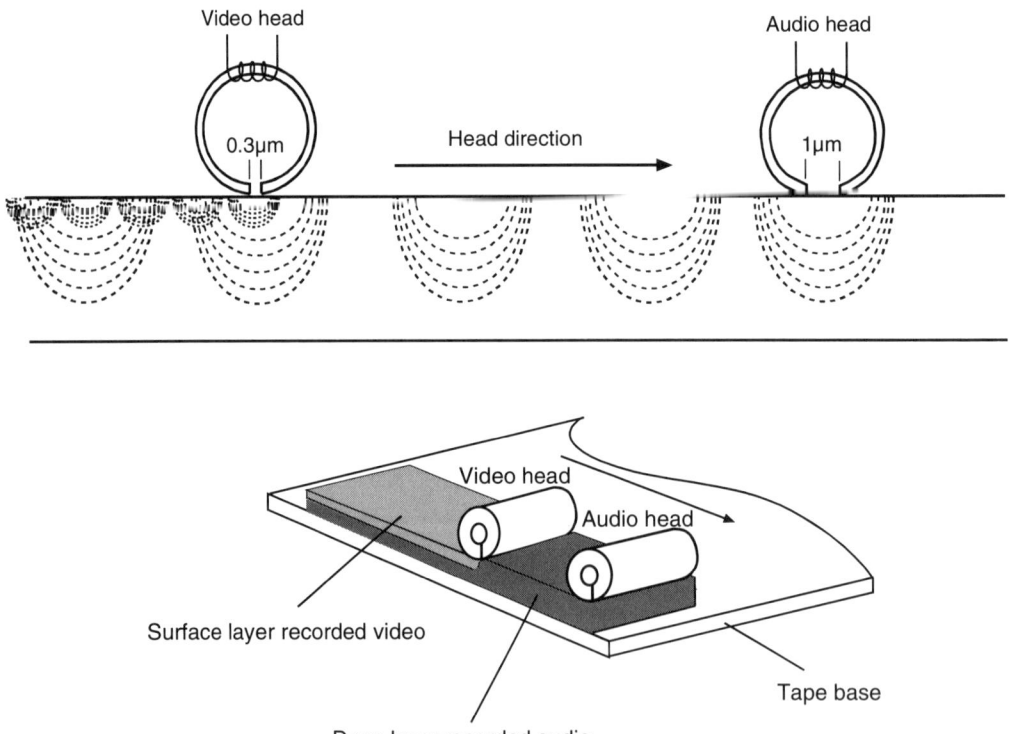

Figure 8.8 *Depth multiplex recording - illustrating video and audio recording layers*

Hi-fi audio

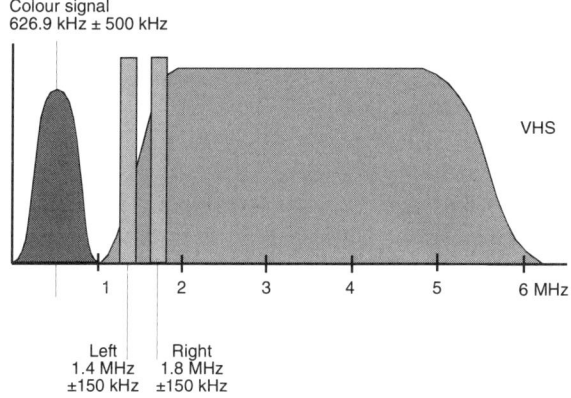

Figure 8.9 *Video and audio recording spectrum*

both carriers are shown as part of the video frequency spectrum they do not interfere as they are recorded deeper in the tape layer and are separated.

A block diagram of the basic record/replay system is given in Figure 8.10. The required programme is routed through the input selector, typically taking the sound channels of a simulcast (simultaneous radio and TV broadcast) from an FM radio tuner or NICAM stereo broadcasts. Recording sound level can be chosen to be either AGC or manual control, with levels indicated on LED ladder arrays or meters. Muting takes place during 'power up' and 'power down', and tuning and channel selection, to eliminate spurious noises before the signal enters the compressor. The function of the compressor is to reduce dynamic range, permitting a wide range of subsequent frequency modification and considerably improving overall signal-to-noise ratio.

After undergoing reduction of dynamic range and frequency equalisation, the audio signal passes through a pre-emphasis circuit and a level limiter on its way to the FM modulator. The limiter prevents any risk of driving the modulator to a deviation greater than 150 kHz. The modulated FM carrier for the left (Ch.1) signal at 1.4 MHz is added to the right (Ch.2) carrier at 1.8 MHz before being routed to the audio recording heads via a drive amplifier.

During replay the left–hand and right–hand signal carriers are individually tuned by band pass filters for routing to their respective demodulators and drop out compensators. Drop out compensation is necessary to mask interference arising from tape drop out and audio head switching transients. Demodulation is performed by using the recording modulator oscillator as a VXO in a phase lock loop so that the error output of the loop forms the recovered audio signal. De-emphasis followed by expansion corrects for the signal modifications carried out during record, reducing noise and restoring the dynamic range. Power muting eliminates spurious noise in the final output signal during servo lock up.

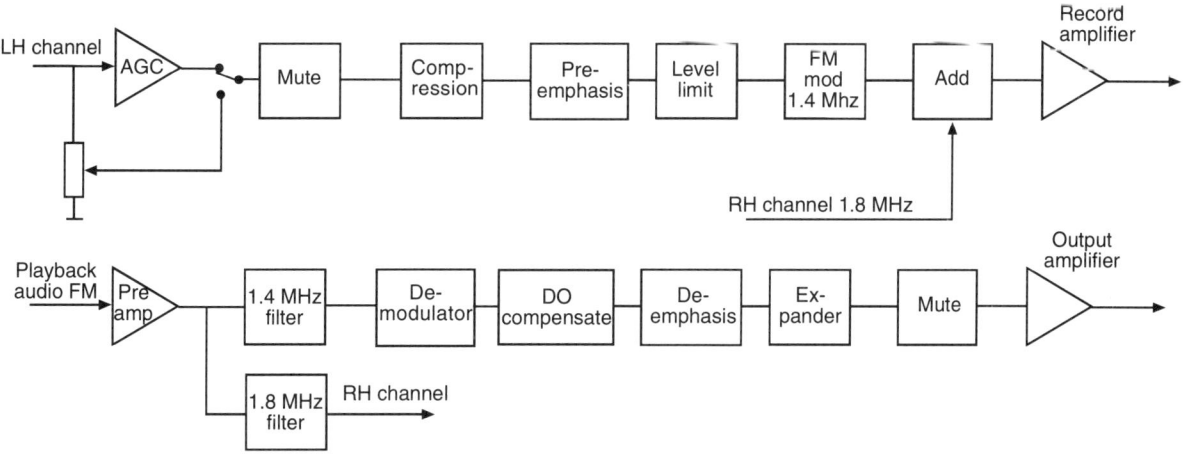

Figure 8.10 *Audio Hi Fi recording and replay diagrams*

Compander

In order to maintain a high dynamic range (and hence signal-to-noise ratio) of 80 dB in the overall record-to-replay signal, a compander (compressor–expander) system is used, see Figure 8.11. During record the dynamic level and frequency response ranges are compressed and modified. Correction is applied during replay to restore the signal to its original form in terms of bandwidth and dynamic range. As we shall see, the compression and expansion systems follow a logarithmic law and are both carried out within a specially developed integrated circuit.

In a Dolby system only the high frequency components of the signal are modified. Low level high frequency signals are amplified prior to being recorded, and then attenuated during replay to restore them; consequently off tape noise is also attenuated.

The control range of the logarithmic compressor is shown in Figure 8.12. At the 0 dB point it is neither an amplifier nor an attenuator. If the input signal increases by 120 mV the output becomes attenuated by 20 dB, whereas a decrease in input of 120 mV will increase the gain by 20 dB. As is well known the dB scale is a logarithmic one, and the compressor follows this law. For every 60 mV that the input signal falls below 0 dB, 10 dB of gain is applied, and for every 60 mV that the input signal rises above the 0 dB level, 10 dB of attenuation is introduced – compression indeed!

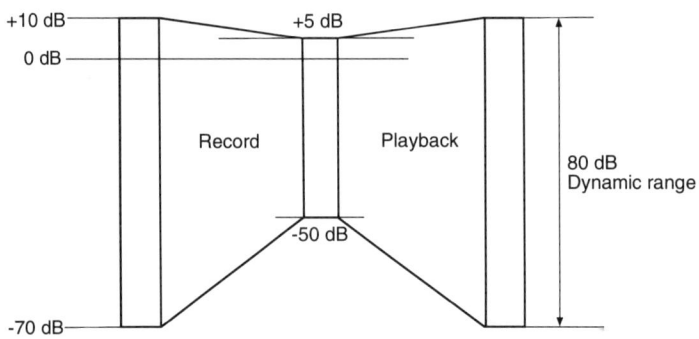

Figure 8.11 *Compander dynamic range*

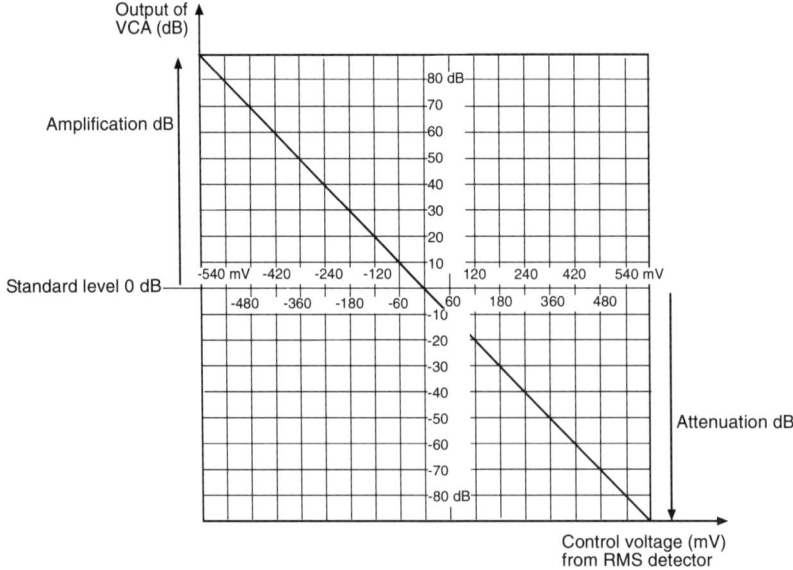

Figure 8.12 *Compander VCA charactaristic*

Hi-fi audio

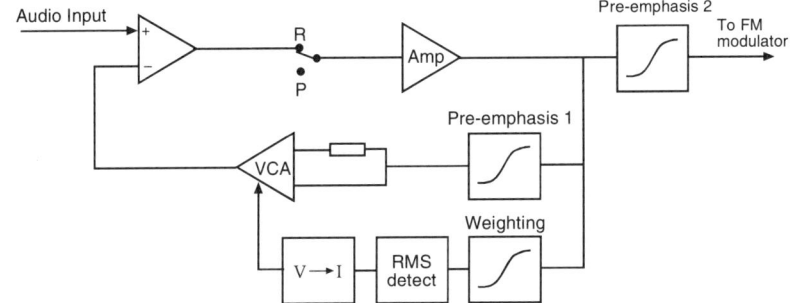

Figure 8.13 *Record compressor*

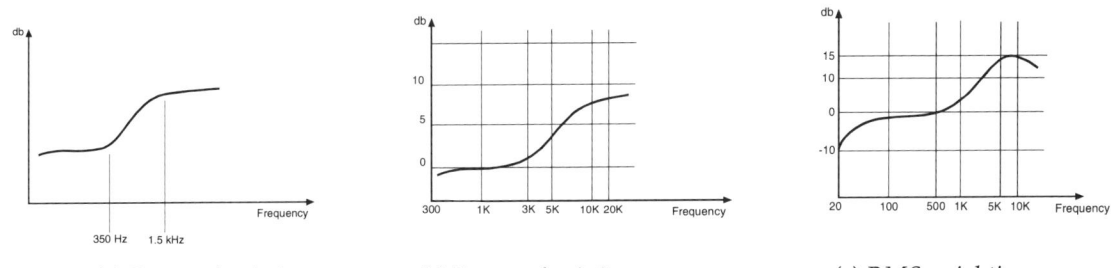

Figure 8.14(a) *Pre-emphasis 1*   (b) *Pre-emphasis 2*   (c) *RMS weighting*

If the compressor worked in this fashion over the whole audio frequency range, the necessary pre-emphasis process that follows may cause over-deviation of the FM modulator stage in the frequency range been 2 and 10 kHz, which contains the most energy in speech and music programmes. Extra compression within this range permits more pre-emphasis without the risk of exceeding deviation limits. The introduction of a filter within the compander to perform this function is called 'weighting' the frequency response of the compressor.

A block diagram of the recording compressor is shown in Figure 8.13. The input signal is applied to an operational amplifier then via a buffer amplifier to the recording pre-emphasis 2 circuit, the frequency response of which is shown in Figure 8.14(b). Compression is performed in the operational amplifier according to the feedback signal applied to its inverting input. A signal sample is fed to a voltage controlled attenuator/amplifier (VCA) through a filter, pre-emphasis 1, as Figure 8.14(a) shows. This boosts all frequencies above 1 kHz to improve signal-to-noise ratio. A weighting filter curve, Figure 8.14(c), progressively increases the signal level for increasing frequencies in the range 1-10 kHz, beyond which a slight roll-off takes place. This carefully filtered signal is presented to the control input of the VCA, where as a result, higher input frequencies boosted by

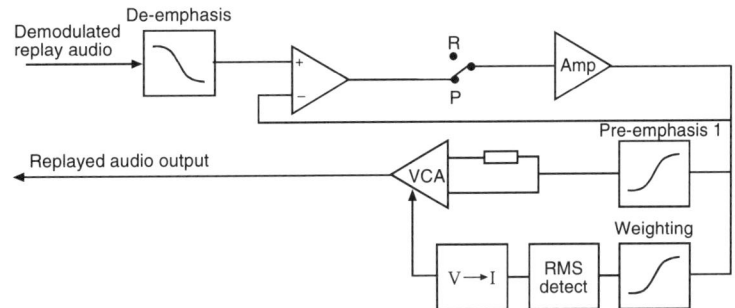

Figure 8.15 *Replay expander*

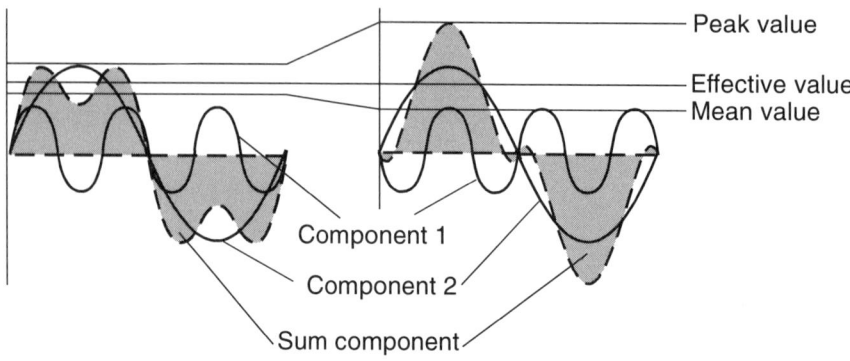

Figure 8.16 *Root mean square of a complex waveform*

pre-emphasis 1, will undergo more compression than lower frequencies. The extra compression in the treble range will compensate for the effects of pre-emphasis 2 in the recording circuits to ensure that FM deviation limits are not exceeded.

In playback mode the frequency-tailored compressor is inserted not in the feedback path of the input amplifier, but in the direct path, as shown in Figure 8.15. The effect of this is to reverse the operation of the circuit, which now becomes an expander whose amplitude/frequency response mirrors that introduced during record. All frequencies that were compressed and pre-emphasised are now de-emphasised and expanded with the result that tape noise is also attenuated and over a much wider frequency range than can be normally achieved by direct audio recording.

input signal. Conventional average or mean level detectors are not good enough for this application where a more accurate representation of the average level of a complex signal is required. A peak detector is not sufficiently accurate either. Consider the two signal forms shown in Figure 8.16 the complex waveform (shaded) is the sum of two components. It can be seen that neither the peak values nor the mean values are the same for the component waveforms 1 and 2, although the energy content (the shaded area within the waveform envelope) is identical in both cases. This can only be proved by taking the peak value, squaring it, and then taking the square root of the mean value of all the 'squared peaks' to produce the effective value:

Effective value = $\sqrt{\text{mean value of (peak values)}^2}$

RMS detector

An RMS (root mean square) detector is used in the compander to drive the voltage controlled attenuator from the 'average' level of the audio

Audio head switching

During replay it is necessary to switch between the audio heads, selecting each alternately as it scans the 180° tape wrap, in just the same way as

Hi-fi audio

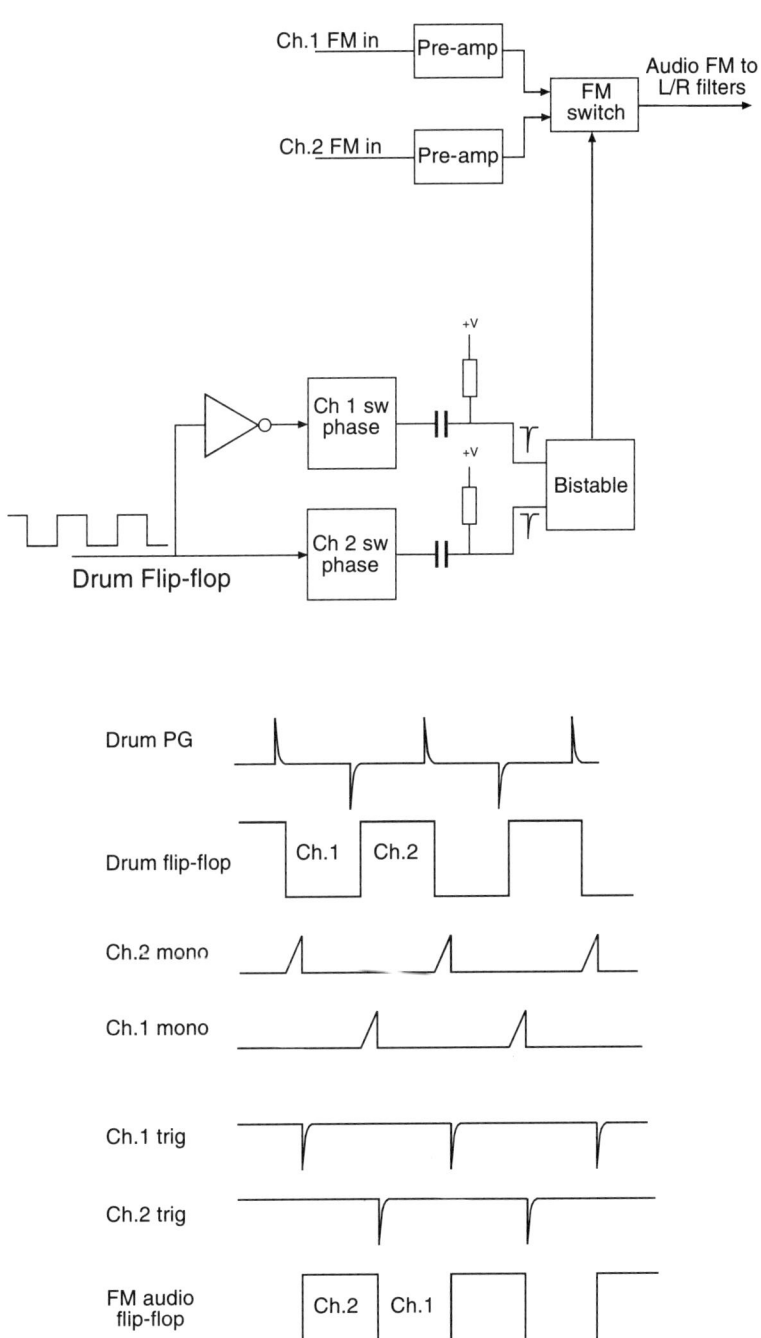

Figure 8.17 *Derivation of the audio head switching waveform*

for video heads. The same RF switching (flip-flop) square wave is used, but because the audio heads are lagging 42° behind the video heads their switching signal must be delayed by 42°, which corresponds to 4.7 ms.

The basic diagram and timing waveforms are given in Figure 8.17. The drum flip-flop signal is sent to two 4.7 ms delay monostables in 1C1 via direct and inverting paths respectively, so that they are alternately triggered by positive and negative edges of the square wave. The reset edges of the monostable outputs are sharpened into trigger pulses in a differentiating network then used to set and reset a bistable. Its output forms the audio head switching waveform, now delayed by 4.7 ms from the video head switch points, and is able to gate the alternating outputs of the FM audio heads into a continuous FM carrier.

Drop out compensation

A momentary loss in FM audio carrier is masked by applying a short 'hold' pulse to the demodulated audio signal to prevent it from falling to zero. This hold level is generated within the IC, and fixed at about 75% peak audio level. It is also inserted at the instant of audio head changeover where the hold pulse is derived from the audio head-switching flip-flop.

When a drop out occurs the result is a reduction in the level of the replay audio FM carrier, possibly to the point where noise or distortion will degrade the demodulated audio signal. At the moment of switching between audio heads there is further risk of introducing noise and distortion. In both cases a masking technique must be used to preserve good sound reproduction. For drop out the 'hold' period is extended beyond that of the drop out itself by use of a short monostable.

In Figure 8.18 the FM carrier signal is demodulated then passed through a 100 kHz low pass filter (to remove residual FM carrier) before being applied to the hold circuit. Figure 8.19 shows the basic circuit and waveforms involved. At the same time the FM carrier is monitored by a drop out detector to generate a drop out pulse which via an inverter provides one input to the OR2 gate. A trigger pulse (generated from the trailing edge of the DOC pulse) sets a short-duration monostable via the OR2 gate. The

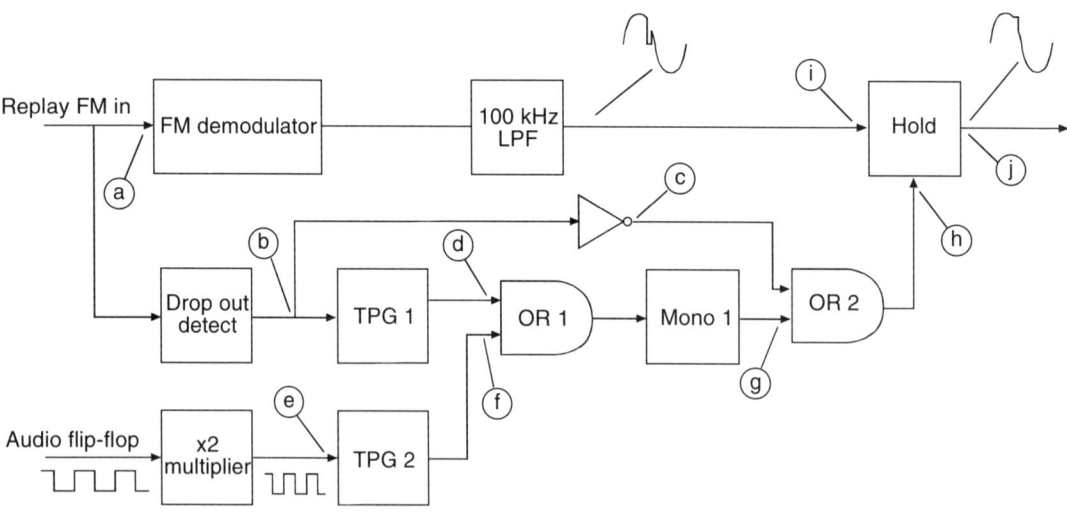

Figure 8.18 *Drop out and head switching compensation*

Hi-fi audio

inverted drop out pulse is applied to the second gate of OR2 so the output pulse is thus prolonged by the effective addition of drop out and monostable as shown Figure 8.19.

Audio head switching pulses into the IC on pin 11 are frequency-doubled to facilitate the generation of trigger pulses in time-coincidence with both positive and negative edges of the head-switch signal. These pulses (f) trigger the monostable via the OR1 gate so that a 'hold' pulse is generated at the instant of each audio head switchover.

As waveforms (j) in both diagrams show, the hold circuit puts a 'lump' on the audio signal, which can be ironed out by a following de-emphasis filter. Another method is to differentiate the audio signal in a separate process, and add the resulting spikes back to the hold section in inverted form. This has the effect of softening the edges of the hold pulse and rendering a smooth and harmonic-free output from a following 20-kHz de-emphasis filter. An idea of this process is given on p.49 in the FM playback description.

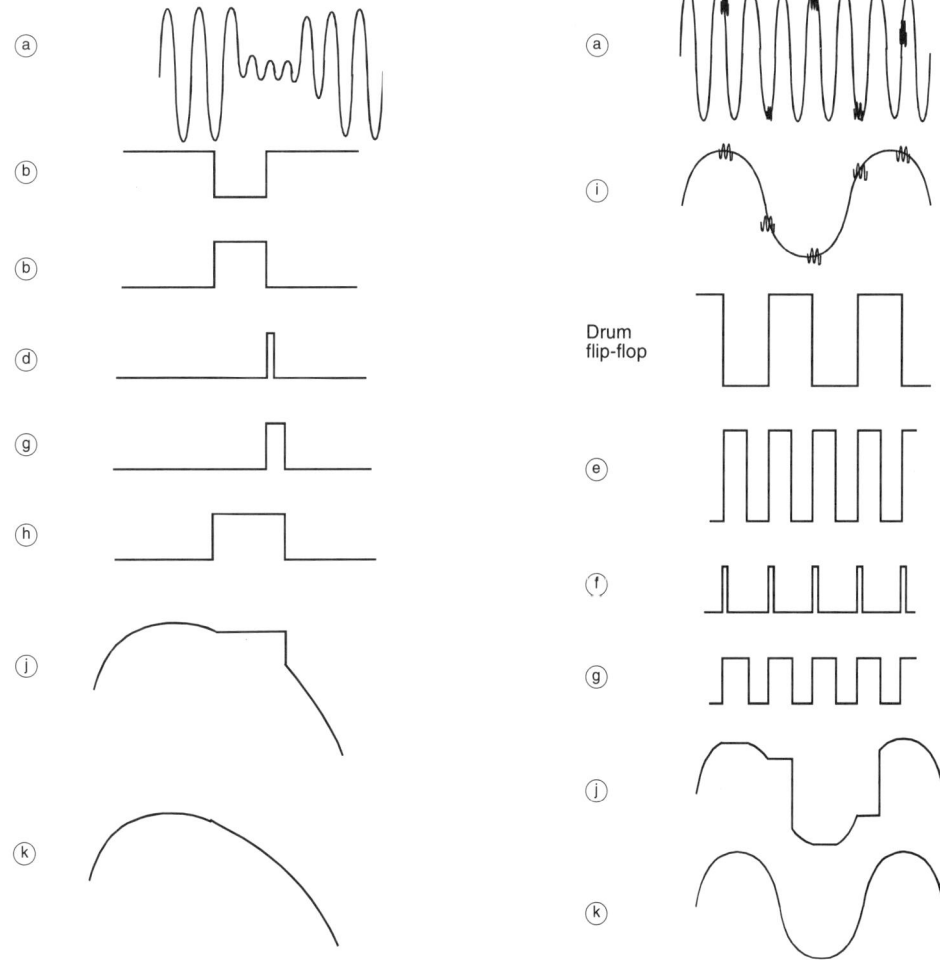

Figure 8.19 *Drop out compensation waveforms*

9
Still pictures and slow motion

The main problem encountered with video recorders, when a still frame is to be displayed, is that the video heads do not follow the same path across the tape when it is stationary as was recorded by them when it was moving. The magnetic tracks are recorded on the moving tape, and providing the tape was moving during replay the video heads will read a path parallel to the one recorded. If the tape is then held stationary the replay path is not parallel to the recorded path and the video head will wander off track somewhere in the replay field. Add to this the fact that each video head will only reproduce its own recorded track, due to the azimuth slant between the gaps, and a problem occurs. When the tape is stationary both video heads follow the same path, so if the path is over Ch.1 track then Ch.2 head will only replay noise. The change in video head paths between stationary and moving tape is shown in Figure 9.1.

Consider that the tape is stationary and that both video heads will follow the path AC. The angle of this track is the same as the angle of tilt of the ruler edge around the drum assembly upon which the tape travels. The distance BC is the amount that the tape travels, when moving, during the time it takes a video head to travel across the tape, following the path AC, during one field scan. When the tape is stationary the video heads contact the tape at point A and travel up, across the tape to point C. If the tape is moving, the video head still contacts the tape at point A and travels towards C. However, during this time, the tape moves the distance BC. When the video head has completed the scan, C is no longer at the end of the track, B has moved into its place. The path of the video head when the tape is moving is AB.

We therefore have two possible video head paths across the tape. First, AB when the tape is moving, which is the normal record/replay path. AB is the standard track angle and length for each TV field during normal operation. Second, AC is the path that both video heads will follow across the tape when it is stationary. When a standard recorded tape is halted during replay then a path of AC is followed and it will cross over at least one and, at worst, possibly two of the recorded tracks AB. It is not possible for a video head to replay its track without crossing over to an adjacent track and producing a band of noise. The noise is referred to as tracking error, or more correctly crossover noise.

Crossover noise is not the only distortion that occurs in the replay of a still picture. It can be seen from Figure 9.1 that the track AC is longer than AB (the drawing is an exaggeration for clarity, and in fact the increase in track length is 0.48%, or about three TV lines). If you think about this point carefully you can rationalise that the video head will travel this slightly extended track in the same time period. In other words AC is scanned in the same time period as AB. So 315½ lines are reproduced in 20 ms, indicating a small

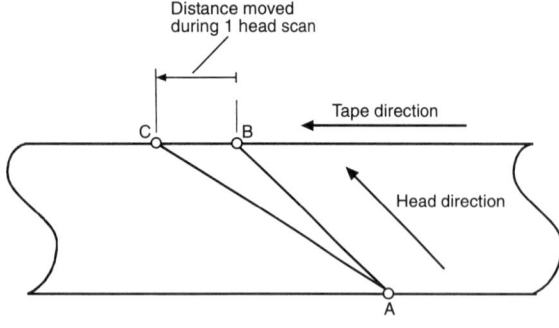

Figure 9.1 *Difference in ape path length and angle between tape stopped and moving*

increase in the replayed line frequency rate.

For a video head to replay a longer track within the same time period means that the head to tape speed has increased by 0.48%. All replayed frequencies off the tape are therefore subject to a 0.48% increase. Such a small increase in line sync frequency and replayed colour under carrier can be accommodated by the AFC loop within the colour circuit, but the most common method is to increase the drum motor speed.

Why standard heads cannot reproduce a still picture

In Figure 9.2 the shaded tracks are Ch.1 tracks and the light tracks are Ch.2 tracks. In (1) the tape has stopped in a position where two video heads cover a Ch.1 track. This allows Ch.1 video head to reproduce its track to almost full level, as the FM output shows. Ch.2 video head, following the same path, reproduces only a small amount of FM at the beginning of the scan and at the end with nothing in between, except noise.

In (2), the replay path covers both Ch.1 and Ch.2 recorded tracks; this is still insufficient to enable a noise free replay. Ch.2 video head picks up no FM at the start of scan and almost full level at the end whereas Ch.1 replays full level at the beginning of scan and little or no FM at the end. Ch.2 reproduces noise at the top of the screen and Ch.1 reproduces noise at the bottom of the screen.

In (3), again the replay path covers both Ch.1 and Ch.2 recorded tracks. Ch.2 provides an output at the start of scan and nothing at the end. Ch.1 has no output at the start of scan and full output at the end; a situation that will reproduce noise bars at the top and bottom of the picture as in (2).

It proves therefore that stopping the tape during a replay cannot provide for a noise-free still field or frame, not only due to the fact that the replay video head path will cover any recorded tracks at random, but also that there is no position that can give a noise-free replay.

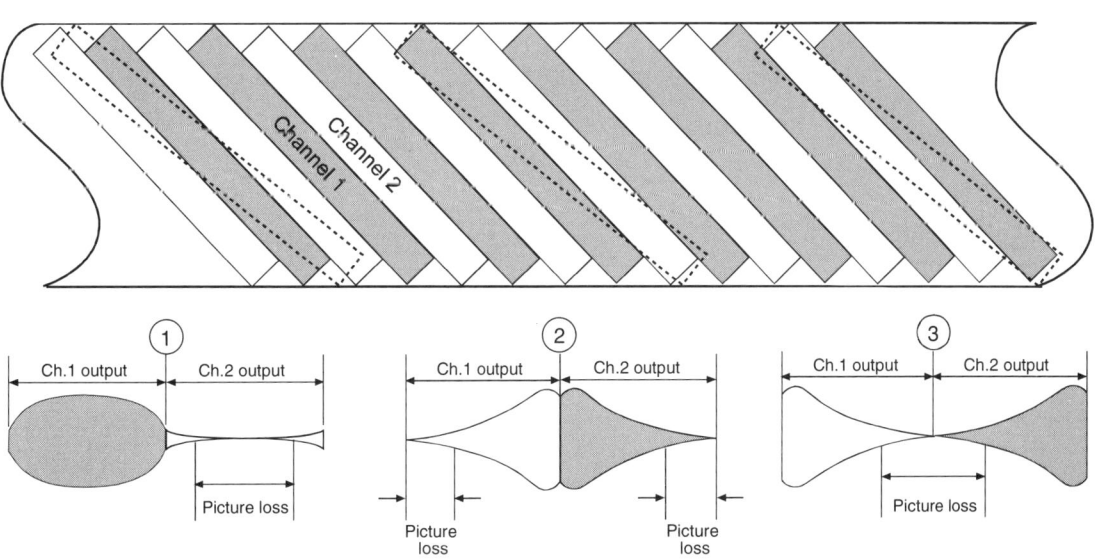

Figure 9.2 *FM crossovers due to head tracking problems in still picture mode*

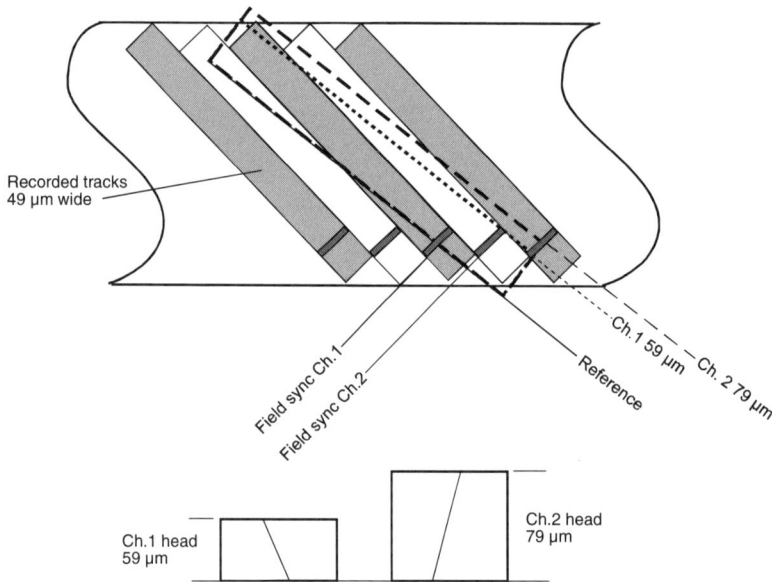

Figure 9.3 *Track scanning for still picture*

The still picture solution

A solution to the problem is to modify the video heads. This can be done in three ways. (1) Increase the thickness of the video head chips so that they can cover a wider area of tape track in still picture replay. (2) Modify the action of static video head chips so that they can move in the vertical plane and follow the original recorded tracks. (3) Add extra heads.

The first option is the one adopted by most manufacturers. The width or thickness of Ch.1 head is increased from 49 μm to approximately 59 μm and Ch.2 head is increased to approximately 79 μm, as can be seen in Figure 9.3.

An important point to note is that the lower edges of the video heads are on a virtual reference plane. This reference is the same height above the main tape deck plane as the 49-μm video heads in a standard recorder. In a still frame servo tracking control system the tape is halted under strict control such that the replay track path sits over Ch.1 recorded track. Alignment of the replay still frame track path is adjusted by a slow motion tracking control. The user adjusts the slow motion tracking control for a noise-free picture. In doing so the replay track path is set upon Ch.1 track.

The reason for this alignment is illustrated in Figure 9.3. The replay path is shown in dark outline upon the Ch.1 (shaded) and Ch.2 recorded tracks shown. Each video head follows the same path, aligned on the reference line; Ch.1 head has a replay track 59 μm wide and Ch.2 head has a replay track 79 μm wide. The result is that Ch.1 head replays the shaded recorded track. It is well covered and sufficient FM is picked up to satisfy the minimum input requirements of the replay limiter amplifiers.

Ch.2 video head follows along the same reference line; its extra thickness provides for a very wide replay track, the whole of the area within the dark border. It allows Ch.2 video head to replay the Ch.2 recording and again sufficient FM is obtained to allow a noise-free replay. However, the limits of the system design are reached at the end of the scan. Generally this will not necessarily allow still frame replay of a tape made on another recorder.

Techniques using extra thick video heads are used on most makes of VHS recorders that have still frame or slow motion facilities; some early exceptions had three video heads. An extra Ch.1 head produced a still field playback.

Still pictures and slow motion

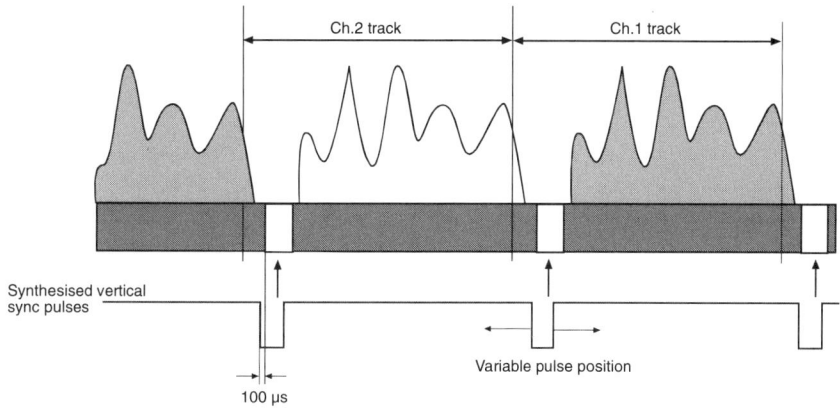

Figure 9.4 *Synthesised vertical sync pulse insertion*

Synthesised vertical sync pulse

When the video recorder is in normal play each video head, in turn, replays the FM signal that contains the field sync pulse. The drum servo, in record, has ensured that each field pulse has been consistently recorded at the same point from the start of the track, as shown in Figure 9.3.

In replay, the field sync pulse is replayed with consistent timing after the start of each track. However, in still frame operation both video heads follow the same path and as such they commence the start of scan at the same point, the start of the track in Figure 9.3.

Ch.1 replays its channel 1 field sync and Ch.2 replays its channel 2 field sync. It can easily be seen that the sync pulse replayed by Ch.1 is later than that of Ch.2. The difference in timing of the replayed field sync pulses, if allowed, will cause the replayed picture to jitter in the vertical plane, i.e. frame bounce. In order to compensate for this synthesised field sync pulses are injected into the video signal during all trick replays — still frame, slow motion, X2 and cue and review.

The timing of the injected field sync pulse is 100 ms before the video sync pulse in Ch.2 replay.

Figure 9.4 illustrates the timing of the replayed pulses and synthesised vertical sync pulses. The timing of the Ch.2 synthesised sync pulses is 100 ms before the replay field syncs. TV sets will lock to the first synthesised pulse and ignore the secondary replayed one.

Synthesised vertical sync pulses are derived from the flip-flop signal that is in turn derived from the rotating video head drum. The synthesised sync pulses will therefore provide for a stable replay in still frame operation. To ensure a bounce-free picture the vertical sync pulse for Ch.1 replay is variable in timing, from an external pre-set control.

Recording with different width video heads

Once it has been accepted that the basis of video recorder still-frame techniques is increased width of the video heads, the question that will be asked is 'how can these video heads record normal 49-μm wide magnetic tracks?' The secret is in the fact that the bottom edge of each of the video

Video and Camcorder Servicing and Technology

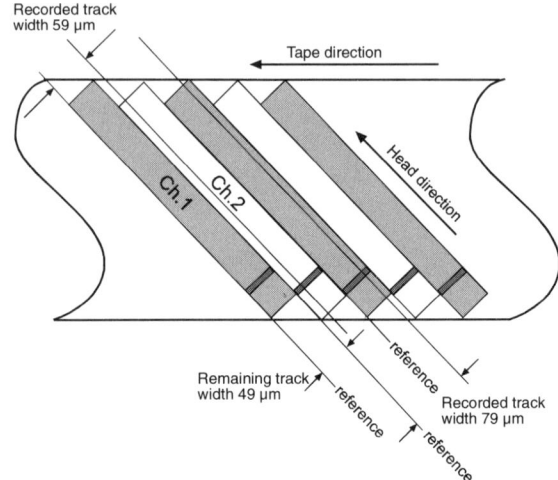

Figure 9.5 *Recording 49 μm tracks with larger width video heads*

heads is on the same reference plane.

In Figure 9.5 the first magnetic track (shaded) is recorded by Ch.1 video head; this track is initially 59 μm wide. Ch.2 video head subsequently scans the tape to record Ch.2 track, which is 79 μm wide. However, when Ch.2 video head lays this track its bottom edge is on the reference line that is 49 μm from Ch.1 reference due to the speed of the tape. The extra 10 μm of Ch.1 track is over-recorded by the subsequent Ch.2 video head. Next, Ch.1 video head scans for the second time laying another Ch.1 track. The bottom edge of Ch.1 video head is still on the reference plane and the reference line is 49 μm from Ch.2 reference line. It records a 59-μm wide track, and in doing so, over-records the excess 30 μm of Ch.2 track.

The sequence then repeats throughout the recording, each video head sits on the reference plane and on each subsequent scan the reference edges are 49 μm apart. Each video head thus over-records the excess recorded by the previous one. The magnetic tracks are recorded 49 μm wide leaving the same recording pattern as a normal head assembly with 49-μm wide head chips.

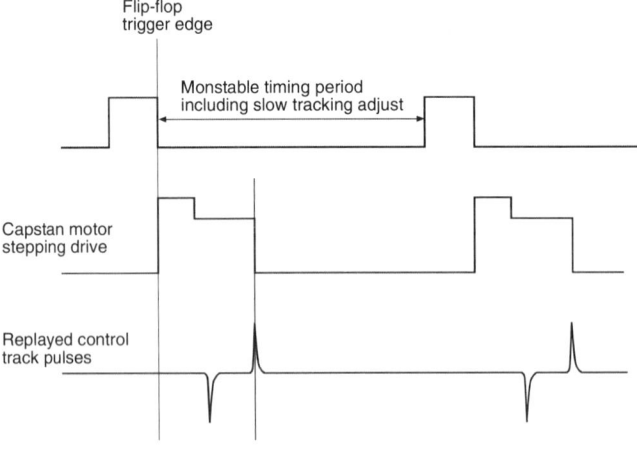

Figure 9.6 *Derivation of the capstan motor stepping drive pulse*

Slow motion

Slow motion is a series of still frames in sequence under the control of the servo and systems control. The rate at which still frames are sequenced is controlled by a slow motion speed control, in early video recorders, which is part of a monostable time constant. At the end of a monostable timer period, when the monostable is reset, the drum flip-flop is used to construct a stepping pulse for the capstan motor.

Figure 9.6 shows the basic pulse timing although in practice a few more monostable delays are used, including slow motion tracking. The small pulse is the short monostable reset period, short by comparison because once the monostable is reset, the next available flip-flop pulse will set it again. In doing so, the flip-flop is used to develop a castellated stepping pulse.

A motor requires a higher starting torque than it does a maintaining voltage, hence the castellated shape of the drive voltage to the motor. Once the drive pulse is initiated it is maintained until a control track pulse causes it to reset. The tape then rests for the remaining monostable timing period until the sequence is repeated and the tape is stepped again, to the next control track pulse.

Slow motion is therefore a series of still frames, and each still frame is displayed for the monostable timer period. At the end of the timer period the capstan motor steps to the next control track pulse. A sequence is shown in Figure 9.7 where the first frame consists of fields A and B, then A1/B1, A2/B2 etc.

After stepping the video heads will read field A1 and field B1. The sequence will be repeated and the capstan motor will step the tape from control track pulse to control track pulse. The slow motion replay will be formed frame by frame and will follow the field sequence AB, A1B1, A2B2, A3B3, A4B4, A5B5 etc. These are true still frames consisting of an odd and an even field and maintaining an interlaced still picture.

The slow motion tracking control affects the start and stop edges of the stepping pulse and in consequence ensures correct 'framing' of the replay video head tracks upon the recorded video tracks; reasonable noise-free slow motion is therefore obtained.

The slow motion tracking control also ensures a noise-free still frame. When still frame or play pause is selected during play, the tape does not stop immediately. The system stops the tape transport in four little hops, these hops helping the slow motion tracking control to place the replay tracks over the recorded tracks correctly so that the still frame picture is noise-free.

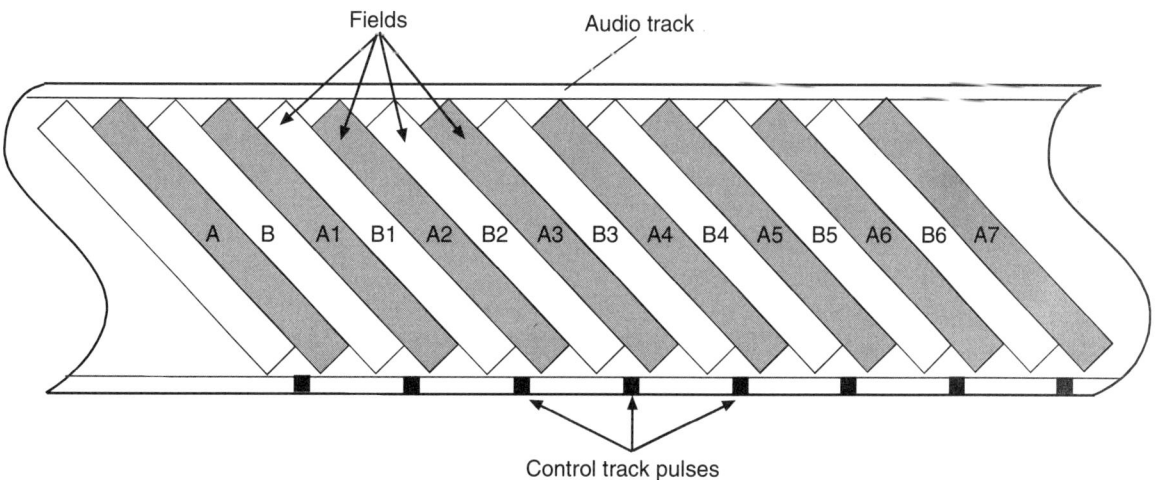

Figure 9.7 *Slow motion track replay stepping between control track pulses*

Video and Camcorder Servicing and Technology

Microprocessor controlled slow motion and still frame automatic tracking

As previously shown in Figure 9.6 a capstan stepping pulse is used to ensure a noise-free still frame picture. Adjustment for such a noise-free picture is achieved by microcomputer controlled variable slow-motion tracking control.

In Figure 9.8 the operation of the system is shown by the timing pulses. In (1) after the pause (still frame) is selected the microprocessor counting delay period (a) is initiated by the first positive edge of the flip-flop waveform. This corresponds to Ch.1 video head commencing replay of its track, after the video head crossover point. After the delay count period (a), a motor reverse pulse (b) is used to stop the capstan motor dead in its tracks by applying a reverse pulse voltage. There then follows a further flip-flop counting period whilst the microprocessor checks the FM signal. If all is well a standard motor drive castellation pulse is used to step the capstan motor.

A 'standard' drive pulse has a start edge, period (c), after a Ch.1 positive flip-flop transient. This is followed by a standard run period (d), which is terminated by a braking pulse (b). However, if during a subsequent checking of the FM signal, the crossover point is prior to the Ch.1 positive transient, then the tape is judged not to have travelled far enough. The crossover point is converted to a pulse from the FM AGC circuit, and a measurement (f) is taken between the crossover point and Ch.1 video head start of scan. The period (f) is then subtracted from the (c) period and the start of the motor drive pulse is advanced. The run period (d) is therefore longer and tape travel is increased to make up for the shortfall. See Figure 7.9 section (2).

In section (3) of Figure 9.8 the crossover point is found to be after the Ch.1 reference edge by

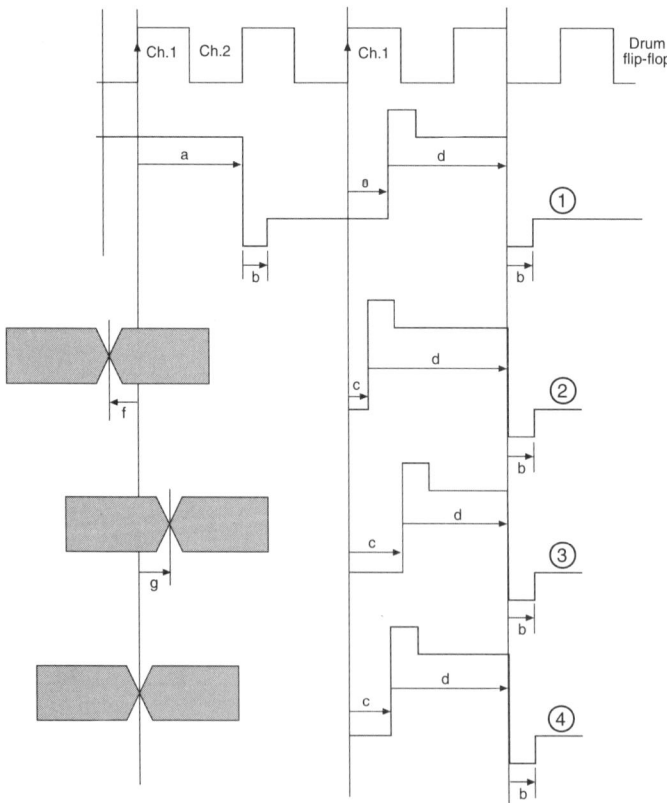

Figure 9.8 *Timing diagram for microcomputer controlled slow motion tracking*

Still pictures and slow motion

period (g). In this instance the tape has travelled too far and a smaller motor drive step is required to correct for it. The microcomputer therefore adds the period (g) to period (c) in order to lengthen it. Run period (d) is therefore reduced and the motor drive pulse is shortened.

In section (4) the crossover point coincides with the Ch.1 video head transient which means that the tracking noise is off the screen at the bottom of the picture. A noise-free picture is therefore obtained and the stepping pulse is then re-adjusted to standard.

When a still frame is selected, the microcomputer typically has a maximum of eight stepping pulses to set up a noise-free picture. If it fails to do so it will stop on the eighth anyway.

In slow motion mode, continuous checking and correcting is carried out to maintain a noise-free picture.

The system is automatic for still frame and slow motion tracking and it will operate for interchanged tapes as well as the replay of the video recorder's own tapes.

In the digital domain all timer periods are counters and not monstables, computation of errors due to on-screen crossover noise are easily calculated resulting in precision still picture performance and slow motion.

Where a digital field or frame store memory is incorporated - the fields are read off tape while it is pulsed at play speed to maintain accurate scanning of the video tracks. Viewed pictures are read out of the memory at the same rate that the tape is pulsed.

Combined four head drum

In high speed picture search a video head will cross over 9 or 10 tracks during a single field scan in picture search when the tape speed is 9–10 times normal play. High performance video recorders improve the still picture and slow motion performance and the noise bands are reduced to just thin lines that do not greatly detract from the picture information. This is done by the use of four video heads, two for normal record and playback, referred to as the SP (standard play) and two for LP (long play). As the LP heads are used to 'support' the two main video heads as the

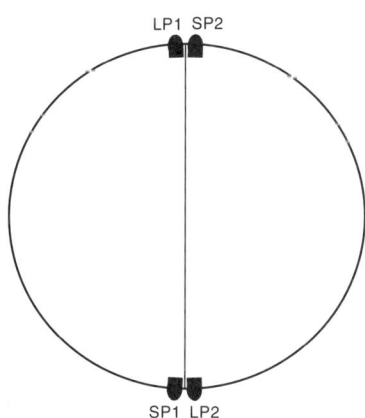
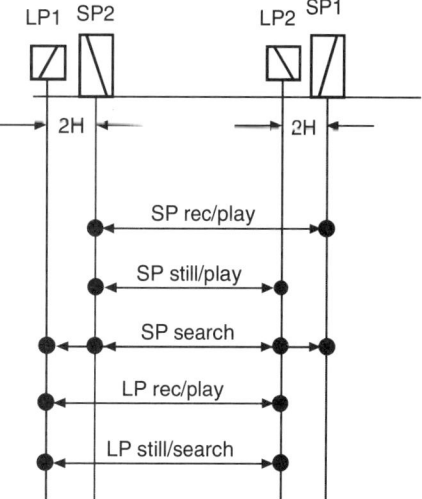

Figure 9.9 *Four head drum for noise-free picture search*

Video and Camcorder Servicing and Technology

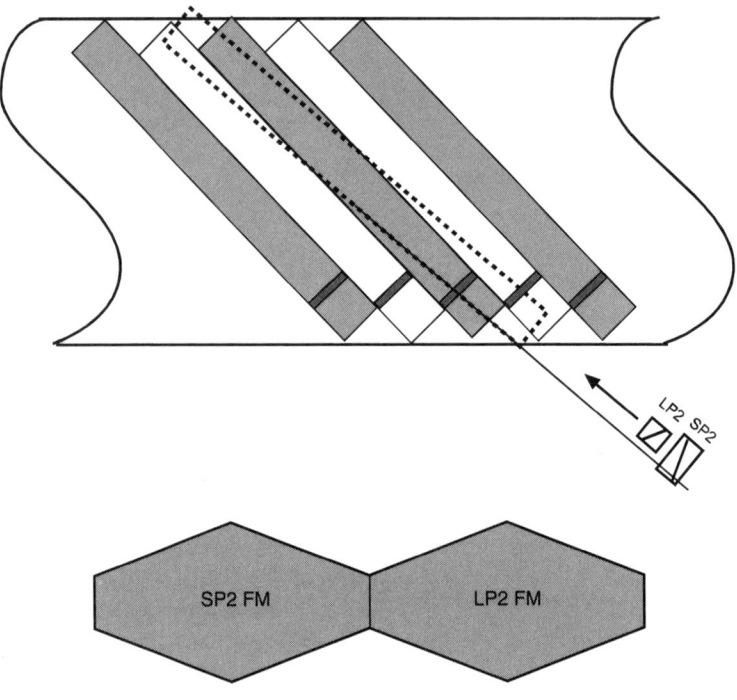

Figure 9.10 *SP/LP heads reproducing a good replay FM carrier symmetry*

noise-free picture search applies only to standard speed.

Figure 9.9 illustrates the head mounting configuration. SP1 and SP2 heads are the standard record and replay heads and are spaced 180° apart. SP1 head has a track width of 59 µm and SP2 is 79 µm wide. The LP1 and LP2 heads are also 180° apart but displaced from the main heads by a distance corresponding to 2 TV lines, 2H. SP1 and SP2 are used for standard speed normal recording and playback; in picture search or still picture mode all four heads are active. It is possible to do this as the azimuths of the LP heads are opposite to the SP head, with which they are co-mounted so that SP1 and LP2 are together, and SP2 and LP1 together. This puts SP2 and LP2 180° apart except for the 2H displacement. The reproduced picture is a still field (312.5 lines) from SP2 and LP2 heads reading the same video track as shown in Figure 9.10. Picture vertical jitter created by the 2H displacement is compensated for by a 2H shift in the vertical synthesised sync pulse timing periods between the drum flip-flop switching signal and the synthesised vertical sync pulse.

A more interesting feature of the use of four video heads is in the SP picture search mode of operation when all four of the video heads are brought into use. By switching between the four video heads selectively it is possible to replay sufficient FM carrier on a continuous basis to eliminate noise bands, as shown in Figure 9.11. The Ch. 1 video head signal starts off low and increases, whereas the corresponding Ch. 2 starts off high in level and decreases. By selecting a level of about half maximum and then switching from Ch. 2 to Ch. 1 sufficient output signal level can be maintained. In this example Ch. 1 and Ch.

Still pictures and slow motion

2 signals can be from either the SP or the LP heads.

Figure 9.12 shows the block schematic of a microcomputer controlled switching system for noise-free picture search.

SW1 switches the outputs of the SP video heads and SW2 switches the LP heads. When the drum flip-flop output is low in level, SP1 and LP2 are 'active'. When the flip–flop output is high then SP2 and LP1 are selected. The selected outputs from SW1 and SW2 are passed on to SW3. Switching between the SP and LP heads by SW3 is affected by a 'head select' signal that is high for LP and low for SP and so either output of SP heads, or LP heads, is output as playback FM carrier.

With a combination of head select and flip-flop switching pulses any one of the four video head outputs can be routed through the playback system.

The head select switching signal is synchronised to line within the microcomputer as switching the video heads at the beginning of a TV line minimises picture disturbance. This is of little use in normal playback but is significant in the noise-free picture search performance. Also part of the microcomputer is a colour rotate switching signal, derived from the head switching signals, to ensure the correct colour phase. Once again it is more important that it is synchronised to the head switching points and line syncs during picture search.

Flip-flop switching signal is generated from the drum PG pick-up head as discussed in Chapter 4, and it is inverted between SP and LP. The inverted signal in LP mode is also delayed by a 2-line period to compensate for the physical displacement between the two sets of heads upon the mounting assemblies.

When picture search is selected the FM envelope

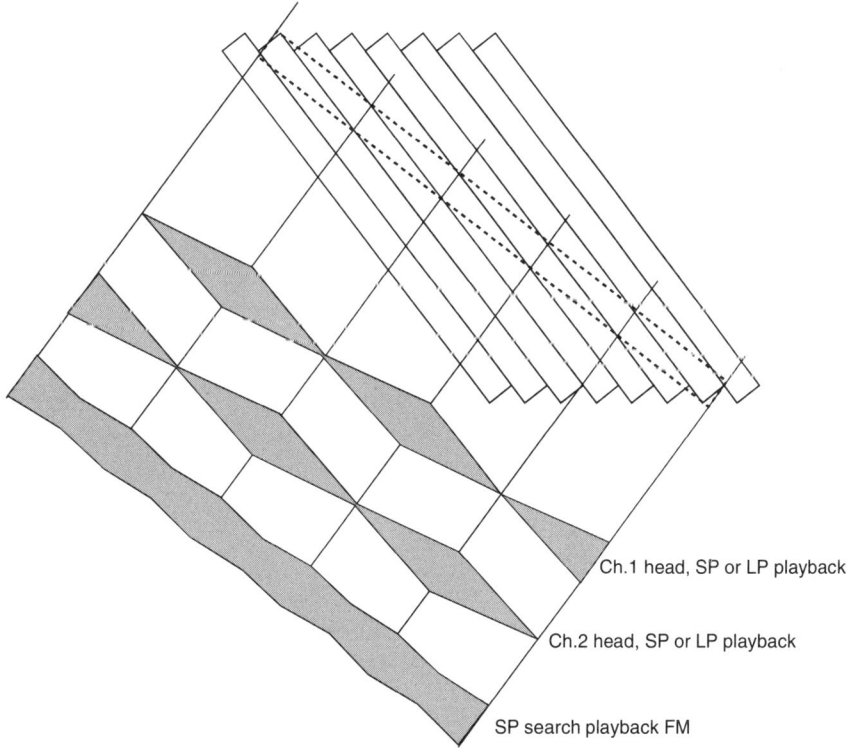

Figure 9.11 *FM replay by multiple heads and combined output FM carrier*

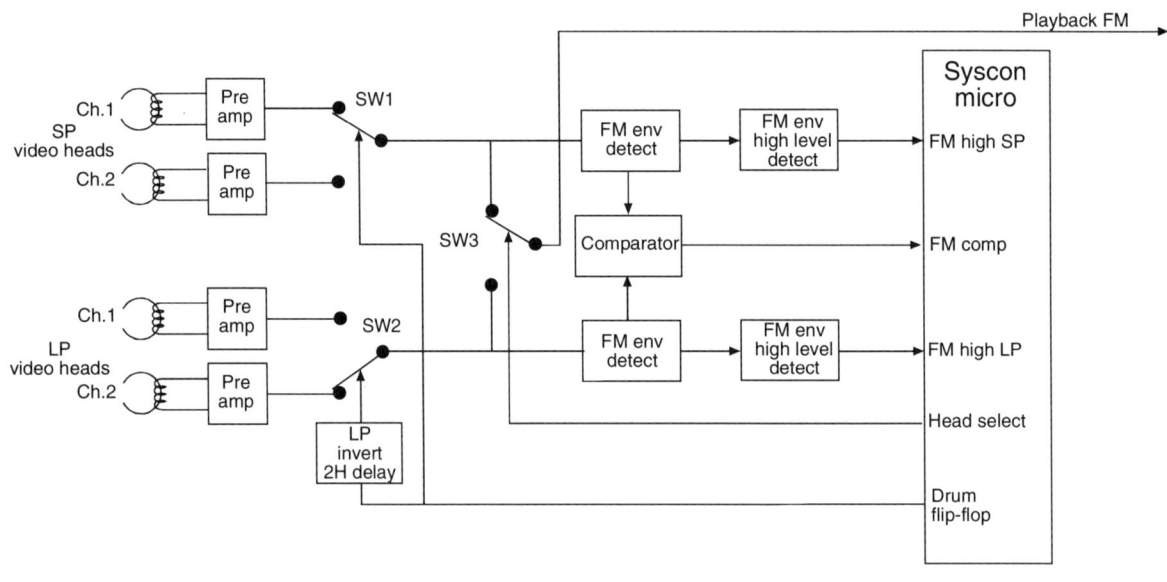

Figure 9.12 *Microcomputer controlled multiple head switching system*

levels of the SP and LP head outputs are compared and fed to the micro as logic levels. The output of the comparator is low if the SP FM output is higher than that of the LP heads, and high if the FM level is lower than the LP heads.

System control microprocessor controls the switching functions and is programmed for the noise-free picture search. In slow motion or still picture modes it controls the stepping of the capstan motor by determining the noise bar position within picture area. This is done by FM measurement from the comparator, and by comparing the event time within the confines of the drum flip-flop period before stepping the noise bar to the vicinity of a flip-flop transition, i.e. out of picture area.

In picture search mode the output levels of all four heads are effectively checked and the highest output is switched through with switching synchronised to line syncs and colour phase rotated through 180° to provide for the minimum of picture disturbance at the switching point.

Dynamic drum system

Previous methods of improving picture performance have reduced noise bars to smaller proportions of the total picture area but not succeeded in eliminating them. The principle of the dynamic drum is to alter the head scanning

Still pictures and slow motion

angle and path in still and search modes so that it follows that recorded onto the tape. In this way the replay FM carrier will not have any crossovers and the picture is clear of noise bars.

In search modes the entire drum assembly is tilted to match the head scanning path with the faster moving tape tracks. In fast forward the tape is moving in the same direction as the head, and the replay track lengthens. Refer back to the explanation of Figure 9.1, which shows the difference in track angle between still and forward motion. The distance BC lengthens as the tape speed increases making the track AC longer with a more acute angle. In reverse search the opposite occurs and the effective track is shortened producing a more obtuse angle.

The dynamic drum system incorporates a precision video head drum mounted upon eight pivots and tilted by an arrangement of motors, helical gears and a positional pulse counter.

Around the drum is the ruler edge or lead that the tape runs upon that determines the compatibility. It is not fixed to the dynamic drum or part of the aluminium lower drum casting. By not being fixed it allows the upper drum assembly to tilt without affecting the tape transport path so that the path along which the heads scan can be tilted with respect to the tape path. The tilt is very small, measured in microns, so pulse counting techniques are employed as tilt measurement.

Figure 9.13 shows the drum assembly for normal record and playback. Figures 9.14 and 9.15 show the positions in forward search and reverse search.

Where play or record are selected the drum sits in the normal position held firmly upon the pivots by strong springs. For other modes, i.e. search, fast play forward, fast play reverse, slow motion and still, the drum is first set to a nominal position for that mode. The microcomputer then detects the replay FM envelope shape and adjusts the tilt of the drum assembly to minimise crossovers until the envelope has a rectangular shape, i.e. the heads follow the track path and there are no visible noise bars.

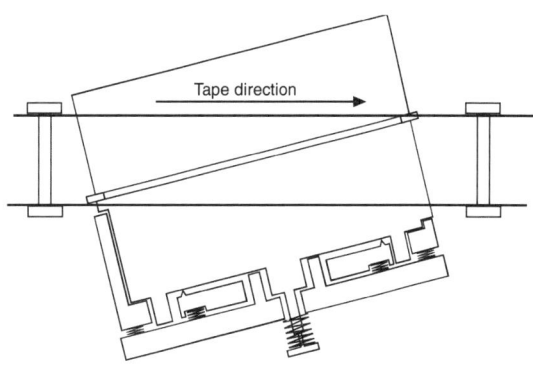

Figure 9.13 *Normal record/playback*

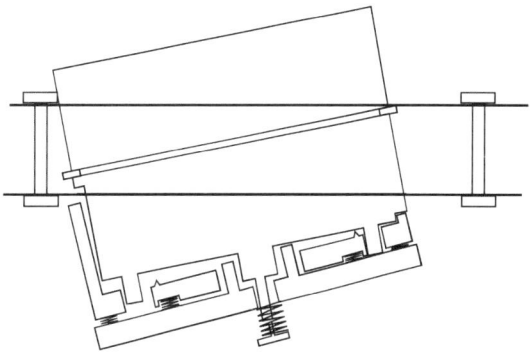

Figure 9.14 *Reverse search, tilt down, longer path*

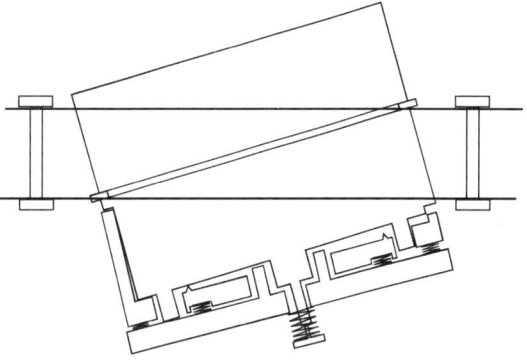

Figure 9.15 *Forward search, tilt up, shorter path*

10
Camcorders

The VHS-C format is a small version of the standard VHS format. While cassette size is only about one third of that of a standard cassette the recordings are VHS compatible. It first appeared as a small compact portable video recorder, part of a camera/video recorder package to meet the demand for a lightweight VCR that could be carried on a shoulder strap comfortably. The small VCR was not as popular as the standard size, it seemed that the requirement for up to 3 hours recording time outweighed the lighter VHS-C recorder with only 30 minutes capability.

The VHS-C format system was adopted by JVC for its range of camera/recorder combinations that became known as the camcorder or video movie, and provided for up to 30 minutes of recording time. Other manufacturers also produced full size VHS camcorders but the cassette housing determined the size of the unit, which was not heavy, but nevertheless it was bulky and difficult to balance for long periods.

To overcome the criticisms of its camcorders with regard to recording time JVC designed a VHS-C format camcorder, model GR-C7, with a CCD pickup tube and half-speed long play facility. The ability to record for up to an hour with a very lightweight camera/recorder was popular with other Japanese manufacturers and some European manufacturers also. JVC produced clones for many other brand names. Later in 1986 a record-only model was announced which, with cassette and battery, weighed less than 1 kg. At that period of development the 45-minute cassette was not available, even so it is unlikely that the mechanisms of that period could reliably handle the thinner tape.

VHS-C system cassette

The supply spool is a small version (Figure 10.1) of the standard VHS cassette spool, and can sit upon an open turntable within the video recorder. The spool socket is the same as a standard cassette despite the small diameter of the spool. The take-up spool is mounted internally on a small spindle and it is not driven by its axial but by a toothed gear on its circumference, as shown in Figure 10.1.

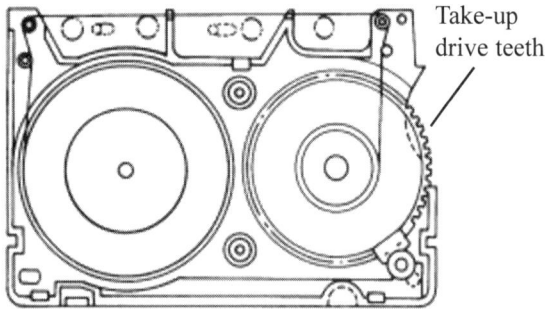

Figure 10.1 *VHS-C cassette construction*

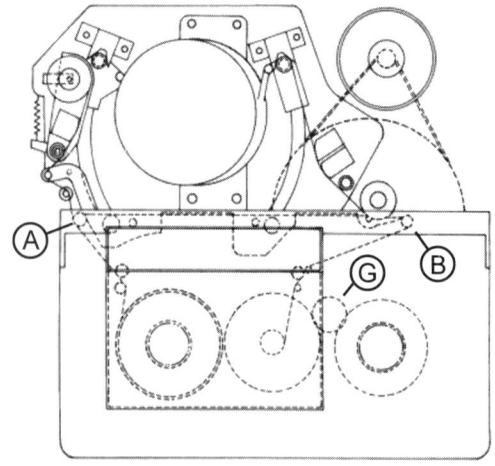

Figure 10.2 *VHS-C cassette in an adapter on a standard VHS deck*

Cassette adapter

When the VHS-C system cassette is mounted within a cassette adapter then the recording can be played on a full size VHS video recorder. The supply spool sits upon the VCR turntable and the take-up spool is driven from the VCR take-up turntable via a small intermediate gear (G) arrangement within the adapter (Figure 10.2). Earlier versions of the cassette adapter were 'manual' in that the loading arms A and B had to be moved outwards and forwards to match the tape path, manually, by a side mounted wheel and internal gears. Later versions use a battery driven motor actuated by the VHS-C cassette being inserted into the adapter. In either type of adapter the tape is pulled out of the VHS-C format cassette to the front of the adapter thus matching it to a standard VHS cassette. A small point worth noting when putting a VHS-C format cassette into an adapter is that a few seconds of the beginning of the VHS-C cassette recording are 'lost' within the adapter due to the extended tape path of the tape looped inside.

VHS-C format video head drum

The VHS-C camcorder utilises a very small, lightweight, video head drum assembly and various arrangements have been made to maintain VHS compatibility. It is only two-thirds the size of a standard VHS video head drum and has four video heads mounted upon it, these are designated A, A', B and B' and they are used in turn to play or record the video tracks upon the tape. In order to follow the standard VHS signal tracks, track Ch.1 and track Ch.2 (see Figure 4.12 p.70), use is made of the four video heads and their switching sequence, determined by the extra amount of tape wrapped around the drum.

A comparison of the two video head drums looking from the non-recording side, is shown in Figure 10.3. Figure 11.1 compares specifications. The slanted track is laid down in the VHS format with an azimuth tilt of $\pm 6°$ between heads. Head A (Ch.1) has an azimuth of $+6°$ and head B (Ch.2) has an azimuth of $-6°$ and it follows that the tracks can be labelled A and B. From Figure 10.3(a) in the standard two-headed VHS system it can be seen that head A contacts the tape upon its lower edge just as head B is leaving at the upper edge, hence head A will follow an A track, and head B a B track. As the video head travels around the drum the tape moves very slowly by comparison in the same direction. The head travels across the tape as a slant track and off the top edge having gone around by 180° plus the overlap margin of about 3° at the start and 3° at the end making a wrap around of 186° in all.

Let us assume that the part of the circumference of the 62-mm diameter drum where the head is in contact with the tape is L, and that it is equivalent to the length of the track for the purpose of this comparison. Ignoring the extra wrap around we can say that the length of the path L is:

$$L = \pi \ (3.142) \times D/2$$

Where D=62 mm therefore L is 97.4 mm.
In the case of the VHS-C format shown in Figure 10.3(b), the video head drum path L has to be the same 97.4 mm for compatibility. As the diameter of the VHS-C head drum is 41 mm then the total circumference is pi (3.142) x D, which is 128.8 mm. Path L is 97.4 mm, which represents 97.4/128.8 or three-quarters of the total circumference. So in order to be compatible the tape has to be wrapped more than 180° around the VHS-C head drum. A wrap around of 270° is used as the result of ($\frac{3}{4} \times 360°$) = 270°, plus an overlap margin of 7°, totalling 277°.

We have now established that a smaller VHS-C video head drum requires a greater degree of wrap around in order to keep the same length of tape in contact with the drum to record or replay the standard video track length.

Ingenious use is made of the four video heads on the drum. They are mounted in the order A, B, A', B' and record or playback in that order. An easy way to work out the sequence is to remember that whichever head is contacting the tape then the next track is recorded or played back is by the head that precedes the current one by 90°. This

Video and Camcorder Servicing and Technology

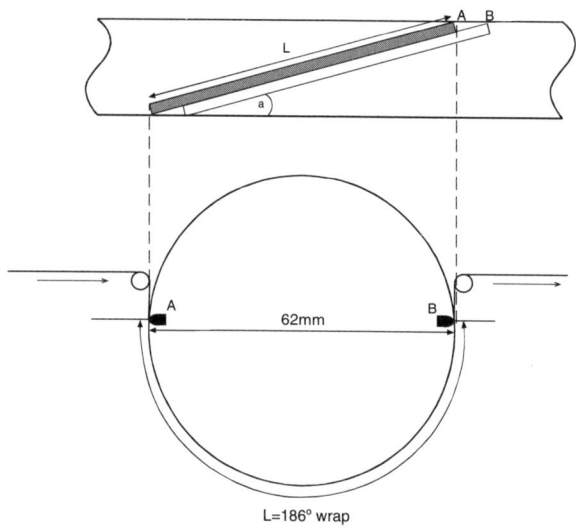

Figure 10.3(a) *Standard VHS tape path length*

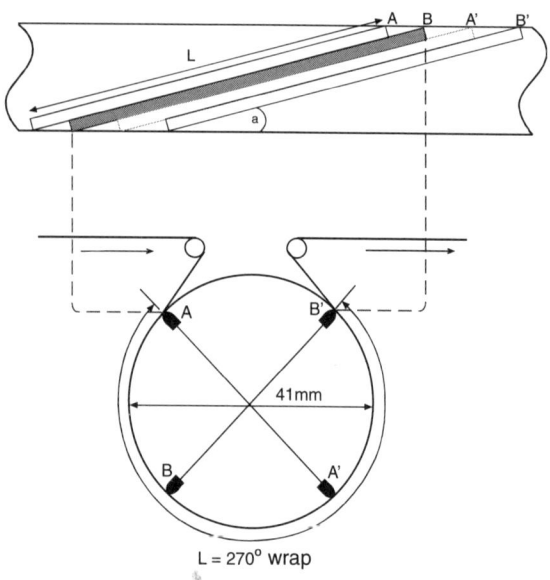

(b) *VHS-C tape path length*

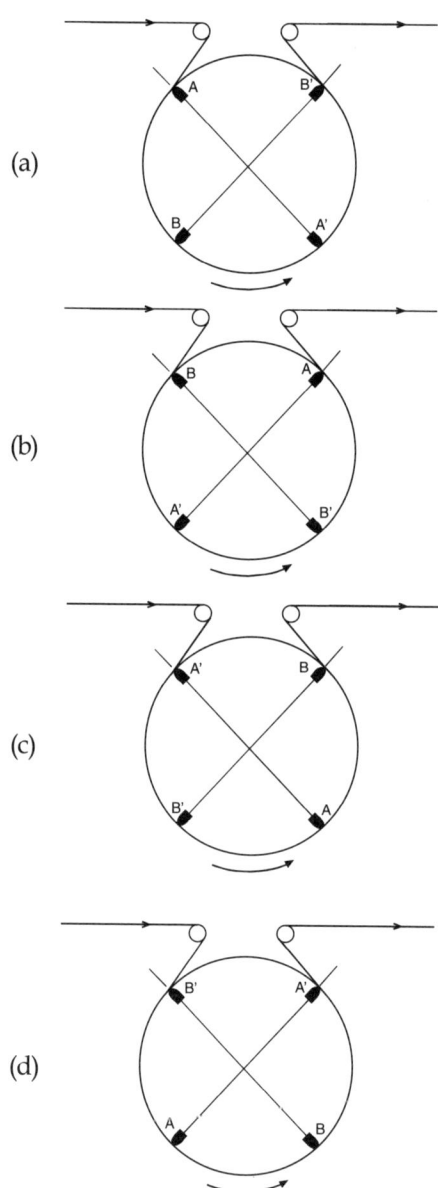

Figure 10.4 *VHS-C head scanning sequence*

is shown in Figure 10.4 . Initially head A contacts the tape (a) and records a track As it leaves, head B contacts the tape (b), and as head B leaves the tape then head A' takes over to record (c). As A' leaves the tape then head B' records (d) and as head B' leaves the tape head A takes over once more (a) to repeat the sequence. The tracks are recorded in the order A, B, A', B'. Heads A and A' have an azimuth of +6° whilst heads B and B' have an azimuth of -6°. Therefore the VHS tracks are recorded and replayed in the standard A, B sequence.

Drum speed

When servicing a VHS-C format camcorder there is a trap that even experienced video engineers can easily fall into. The standard VHS recorders have servo timings which are equivalent to field rate, that is 40-ms periods. However, careful consideration of the VHS-C format head drum will show that its rotational speed is different, resulting in different servo waveforms.

The standard VHS video head drum rotates at a speed of 1500 rpm to meet the requirement that a TV field is recorded onto a single track in half of a revolution. A TV field has a period of 20 ms, it follows then that the path L is traversed in 20 ms, so a full revolution takes 40 ms, equivalent to 25 rev/s or, 1500 rpm.

In the case of the VHS-C format head drum the path L, now 270°, is still traversed in 20 ms to maintain compatibility. If 270° is equal to 20 ms then 360° is 26.6 ms, which is equivalent to a rotational speed of 37.5 rev/sec or, 2250 r.p.m.

To summarise, the VHS-C format video head drum is VHS compatible, it is two-thirds the size of a standard VHS video head drum, it has four video heads and rotates at a speed which is 50% faster.

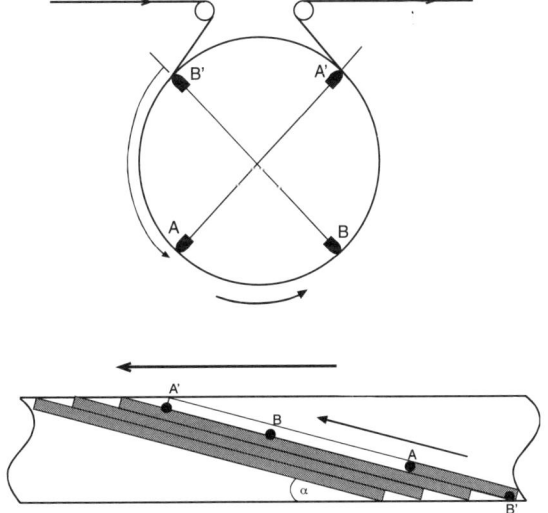

Figure 10.5 *Four heads in contact with the tape only one is active.*

Video head switching

In a standard VHS video recorder the video heads are switched in playback in order to prevent noise pick-up from the video head that is not in contact with the tape, so it is switched off. In record mode switching is not necessary. In the VHS-C format there are usually three video heads in contact with the tape at any given time, but as only one is active, it is essential to switch off the others to prevent any sort of cross-talk between heads.

This is illustrated in Figure 10.5 and shows the position of head A around the circumference of the head drum after it has contacted the tape and travelled along 30% of its track. The shaded portion represents recorded tape and head A has started at the lower edge of the tape and is travelling up and along the centre of its track. The position of the other heads at this point in time is determined by the fixed tip position in relation to both angular position around the drum and relative tip height. As you can see from the illustration at this point, all four heads are in contact with the tape. As this is a recording situation the need to switch A', B and B' fully off is obvious to prevent A' and B from damaging a previous track recording and B' from altering the A track currently being recorded. On the other hand, in playback mode severe cross-talk would result from heads in contact with the tape and replaying residual FM, particularly the head that has the same azimuth as the current replay head. The video head to record after A will be B and by the time it has travelled from its current position to the start of its track then the tape will have moved along correspondingly.

Overlap

In the case of a standard VHS two-head drum, both heads are simultaneously driven by the FM recording signal as they are 180° apart, and as the wrap around is 186° then there is duplicated information at the start and end of each track, within which replay head switching occurs to ensure signal continuity and interference-free

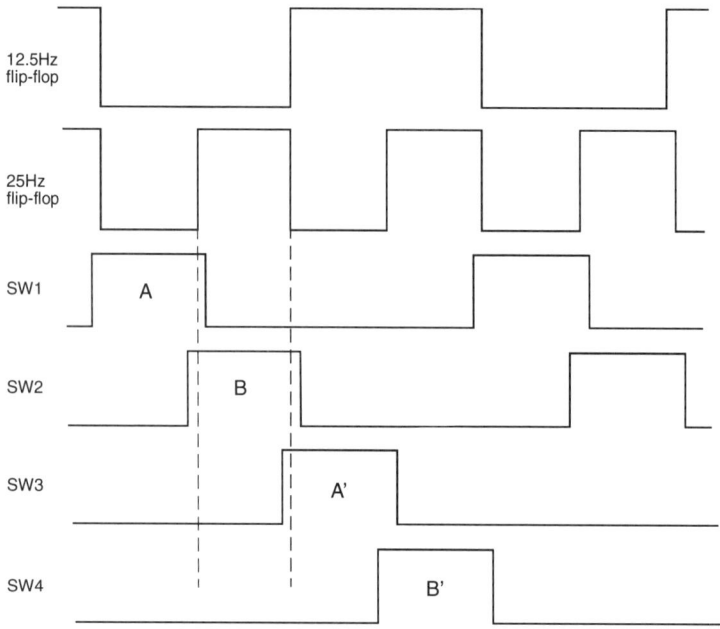

Figure 10.6 *Timing waveform showing SW1–SW4 signal enable pulses*

switch-over points (refer to Figure 4.2). This duplication period is the overlap region and it is not so easy to obtain it by the same method in the VHS-C format because the video heads are switched in record as well as playback. The overlap period must be constructed in order to provide for interference-free head switching in playback and it is achieved by the relationship of the head switching signals, by switching one head on slightly before its predecessor is switched off.

In the example in Figure 10.5 when head A nears the end of its track and head B has contacted the tape, a co-existence occurs as the heads are 270° apart. There is a 277° wrap around and so for 3° both heads are contacting the tape. Only the head switching is required to be modified so that head B is switched on for about 1 m before head A is switched off. This corresponds to an overlap period of about 16 lines and as the replay RF switching signal is centred within the overlap period it gives a margin of ± 8 lines. Figure 10.6 shows the main switching waveforms of 12.5 Hz and 25 Hz. As in the standard VHS video recorders the 25 Hz signal is the RF switching signal or flip-flop and the 12.5 Hz signal is obtained, by division, from it. The four head switching signals SW1–SW4 are synchronised to the RF switching 25 Hz but the timing is modified, so that each one switches 'on' (high level) 8 lines (518 μs) before the RF transition and 'off' 8 lines after the RF signal transition. The recorded overlap period is therefore centred within a 8-line overlap period leaving sufficient margin of error for interchange tolerances in replay. In practice the timing signals are accurately developed digitally within a single IC and are not dependent upon capacitors or susceptible to drift.

Timing

Figure 10.7 shows a block diagram of the video head record and replay circuits. The recording FM carrier is applied to the video heads via the head switching signals. This ensures that excess signal for the overlap period is recorded. In replay the head switching signals are still active to prevent cross-talk by switching unused heads off. Note that the replay signal path also embodies the 25 Hz and 12.5 Hz switching signals. Each video head is enabled in turn by the individual head switching signals. The precise switching point and video head is selected by the combined logic of the 12.5 Hz and 25 Hz.

The head switching waveforms are derived within the main servo IC and all switching and servo timings are determined by a 32.7 kHz crystal clock. Clock pulses are counted down and all pulse timing determined in a manner similar to the examples given in the section on digital servos in Chapter 4.

The timing diagram is shown in Figure 10.8. Two inputs are used: a 32.768-kHz clock with a pulse period of 30.53 µs, and the drum pick-up PG pulses generated once each revolution of the drum cylinder. PG pulses are at a frequency of 37.5 Hz so they are divided by three to 12.5 Hz (or 80 ms). The pulse developed from the drum PG monostable is about 1.9 ms and is pre-set. It provides a delay between the drum PG and the start of the 25 Hz flip-flop, and it can be used to set the replay switch phase adjustment. To set the RF switching point to 6.5 lines prior to field sync, this adjustment is now restricted to a single adjustment (PG1) and not two (PG1 and PG2) as in standard machines as the rest of the timing is digital. Note that it takes three full revolutions of the drum cylinder to play or record with all four video heads.

Upon the falling edge of the PG pulse the 32 kHz oscillator is reset and a single clock pulse is generated; this is the preset pulse. The reset pulse resets all counters and initiates the start of the 80-ms timing period. Two counter networks are responsible for the overlap pulse delay and the 25-Hz delay pulse trains. The overlap pulse is a count of 33 clock pulses high and 622 clock

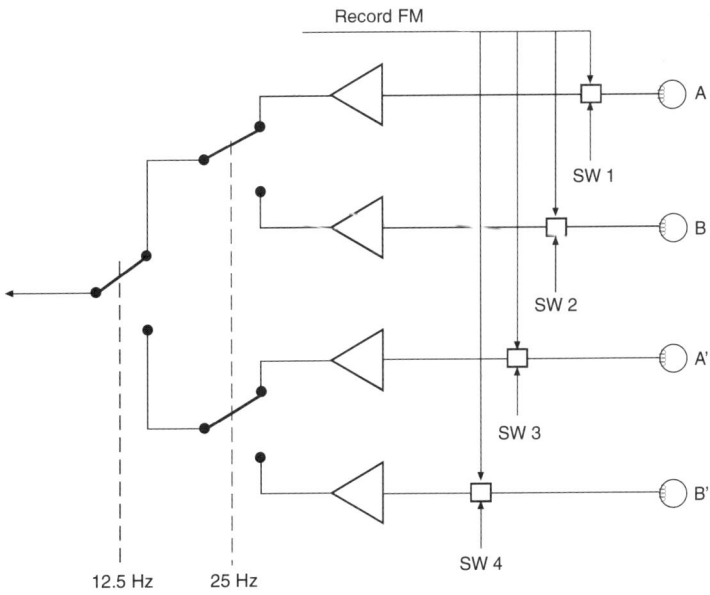

Figure 10.7 *Replay signal path switching and record SW1-SW4 enable paths*

Video and Camcorder Servicing and Technology

pulses low and four of these occur between each reset pulse. It can be seen from the diagram that the 33 count is in fact the overlap period. The 25 Hz flip-flop delay signal is high for a count of 17 clock pulses and low for 638 clock pulses and there are four of these between each reset pulse. In both cases the total count is 655 where 655 × 30.53 μs = 20 ms. In order to time the overlap period each video head switching pulse (SW1–SW4) is set by the rising edge of the overlap pulse and reset by the falling edge, this extends each head switching pulse to 688 counts or 21 ms. The 25-Hz delay period is 17 counts and the 25-Hz RF switching signal corresponds to the falling edge of this pulse. The effect is to locate the 25-Hz RF switching pulse transitions precisely central inside the SW1–SW4 head enable periods. In record the FM signal is enabled for the full SW period, recording the extra FM overlap. In replay the heads are each enabled by the SW period and within this period the 25-Hz head switching gates the replay signal, switching in the overlap sections of the tape.

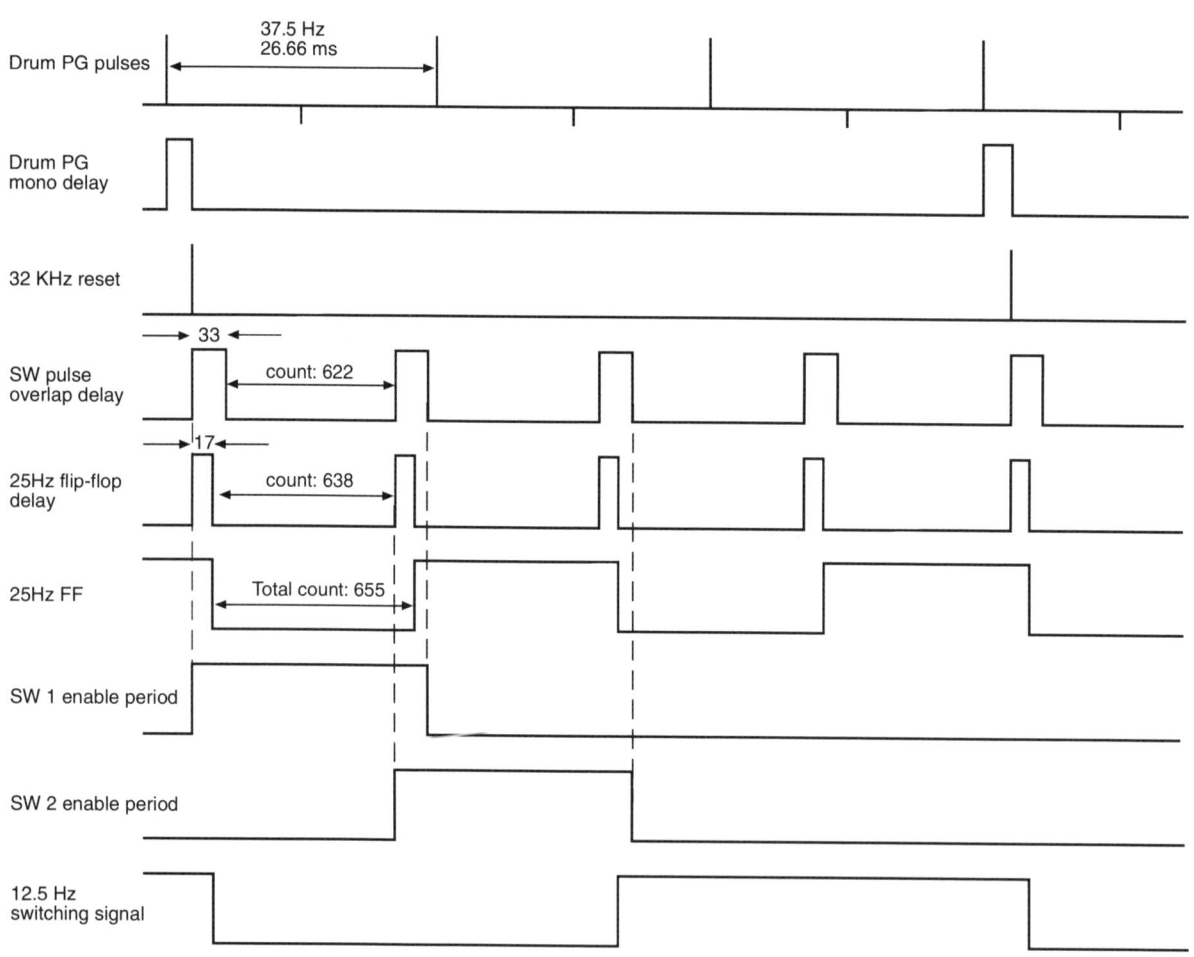

Figure 10.8 *Timing diagram for derivation of SW enable pulses for overlap area*

8mm format

8mm video movie is a nomenclature for a video format which is derived from the tape width and is marketed towards replacing 8 mm cine film. It was heralded as a new technology format to replace all existing video formats, including VHS. However, this did not prove to be so.

Nevertheless, there were advantages in specifying a new video format as the latest technologies could be incorporated. For instance, the 8mm video system accommodates linear, FM and pulse code modulated (PCM) audio, as well as automatic tracking.

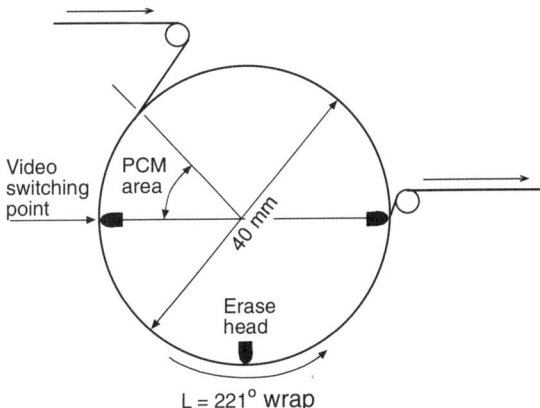

Figure 10.10 *8mm tape wrap around the drum*

The basic mechanical data is as follows.

Drum diameter	40 mm
Speed, head disc	1500 rev/min
Scanning speed (relative speed, video head/tape)	3.12 m/s
Tape speed	2.0051 cm/s (1.0058 cm/s LP)
Video head gap length	0.2 mm
Video track width	34.4 mm (17.2 mm LP)
Distance between two video tracks	0
Audio track width (auxiliary)	0.5 mm
Sync track width (cue track)	0.5 mm

The magnetic tape format is shown in Figure 10.9, with the cue track on the top edge of the tape, the auxiliary audio track on the bottom edge, and with the video and PCM tracks in between.

The total tape wrap around (Figure 10.10) is 221°. The video tracks are recorded in the 180° sector and follow normal video technologies, with the video heads taking 20 ms to record/replay because the drum is rotating at 1500 rpm, which is equivalent to 40 ms per revolution – as in VHS.

In the 8mm format, however, the band of tape used for video tracks is only 5.351 mm wide so the area allocated to analogue recording is by comparison quite small. Higher coercivity tape is recommended to maintain good signal-to-noise ratio for the replayed FM carrier.

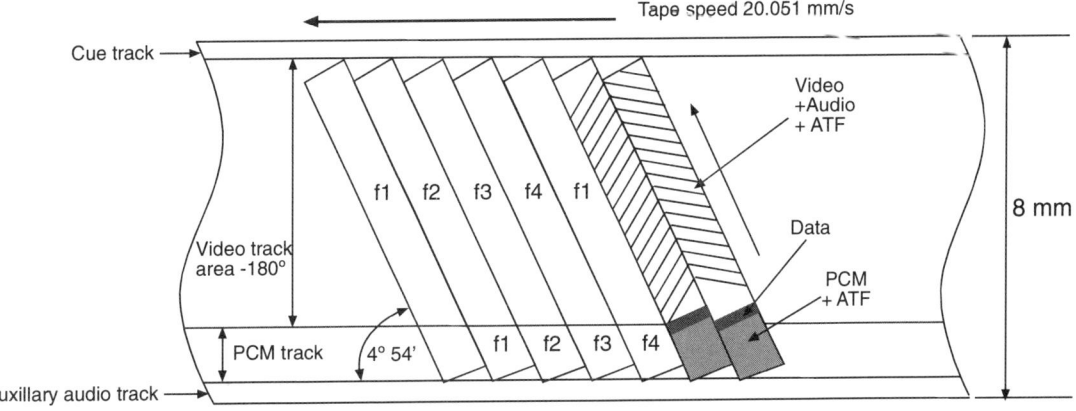

Figure 10.9 *8mm tape format*

Video and Camcorder Servicing and Technology

An extension of the video track is used for PCM audio which is recorded and replayed by the video heads in an extra 31° of wrap around before the video tracks. The overlap margins 5° at the start and finish of the head scan of each take up the remaining 10° of the total wrap around of 221°. Automatic track following (ATF) tones are also recorded on the video tracks, as described later.

The two video heads have a thickness of 27 mm with an azimuth tilt of ± 10°. Standard play video tracks are 34.4 mm wide but with video heads of only 27 mm there is an effective 7.4-mm wide guard band gap between each video track. In long play mode the video tracks are only 17.2 mm and this is obtained by successively trimming alternate 27-mm wide tracks as discussed in Chapter 9, Figure 9.5. A flying erase head is mounted at 90° to the video heads and is 68 mm thick. This thickness allows it to erase two tracks simultaneously and provides the 8mm format with a high grade of editing, without the colour flutter on assemble edits, found in formats and models that edit on a single field basis.

Recording frequency spectrum

For later generations of camcorders the 8mm frequency spectrum has been altered to accomodate stereo FM as PCM became too costly in a competitive market (see Figure 10.11). The luminance sync tip frequency is 4.2 MHz and peak white is 5.4 MHz, with the upper side band extending well over 6 MHz giving improved luminance frequency response. The lower luminance sideband is filtered to just less than 2 MHz to avoid beats with the 1.5 MHz and 1.7 MHz audio FM carriers. Initially there was only a mono audio FM track; stereo is accommodated by the PCM audio system and is not used in the standard video movie camera. Later an additional FM audio carrier was added at 1.7 MHz and was designated a difference carrier. The centre 1.5 MHz has a maximum deviation of 100 kHz but in practice it is restricted to 60 kHz. This is now designated left (L), with the additional difference carrier at 1.7 MHz designated (L-R).

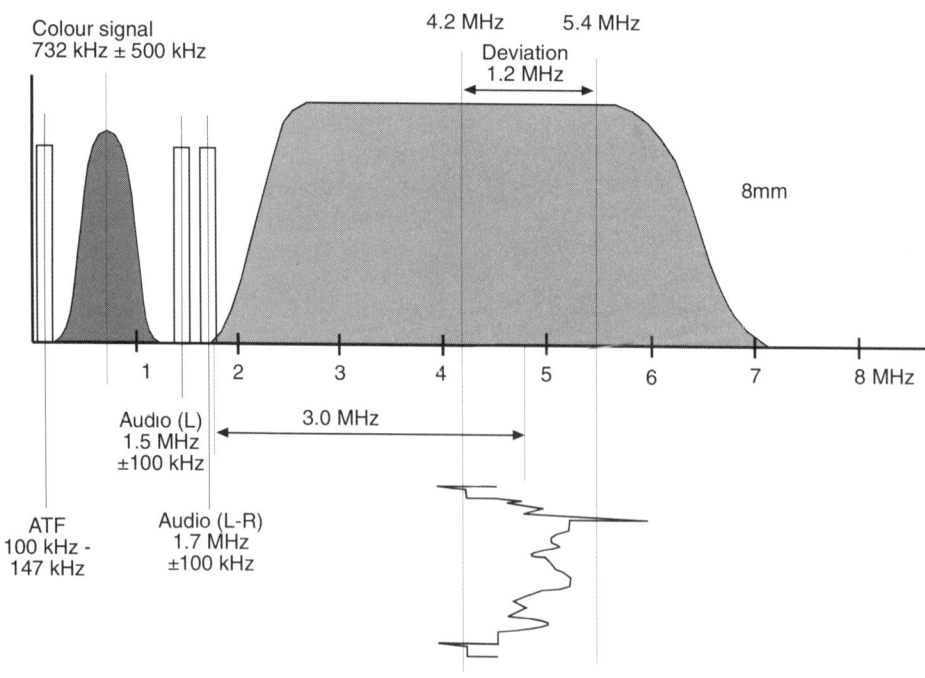

Figure 10.11 *Upgraded 8mm recording spectrum to accomodate a second stereo audio channel*

Automatic tracking

Automatic tracking for the 8mm format has a system that uses four automatic tracking frequencies (ATF frequencies), as shown below.

	Precise frequencies (Khz)	Rounded off (kHz)
f1	101.024	101
f2	117.199	117
f3	162.760	163
f4	146.484	146

The different frequencies used for ATF are 16 and 45 kHz and these are obtained in playback by the subtraction of pick-up from adjacent tracks. During record the ATF reference frequencies are recorded in the sequence f1, f2, f3, and f4. During replay, the reference sequence is changed to f4, f3, f2, f1, in order to determine the correction range of video tracking by use of the capstan servo. When the recorded ATF frequencies for ATF are replayed, they come off the tape in the original sequence f1, f2, f3 and f4. Both replayed and reference frequencies are mixed, as shown in Figure 10.12(a). Reference frequencies are generated by the control system, and their frequencies and sequence are programmed for different modes of operation such as cue, review and reverse play. The mixed result is fed to 45 kHz and 16 kHz filters, the outputs of which are detected and provide inputs to balance an amplifier.

When the video head is 'on track' the operational amplifier is balanced as both inputs are at equal level. If the video head is advanced so that the 16-kHz signal is at a higher level than the 45-kHz signal then the op-amp output falls. The reverse happens if the video heads is lagging, i.e. it rises. A sample and hold circuit reads the error signal and holds it for the capstan servo. A number of sample pulses are used at different times to detect phase lock to ensure the correct sequence of reference pilot frequencies as well as the phase error.

The matching sequence that determines correct tracking operation is shown in Figure 10.12(b). Note that f4 is referenced to f2 resulting in a centre frequency of 29 kHz which is ignored as it does not match any filter, only 45 kHz and 16 kHz difference beats are utilised.

An advantage of this type of ATF servo is very accurate editing, as any edits must maintain the correct replay sequence to avoid a disturbance at the edit point. This is achieved by using f1 as the edit switching reference signal. There is also the future possibility of editing over a PAL four field sequence, which is normally only found on broadcast standard video recorders.

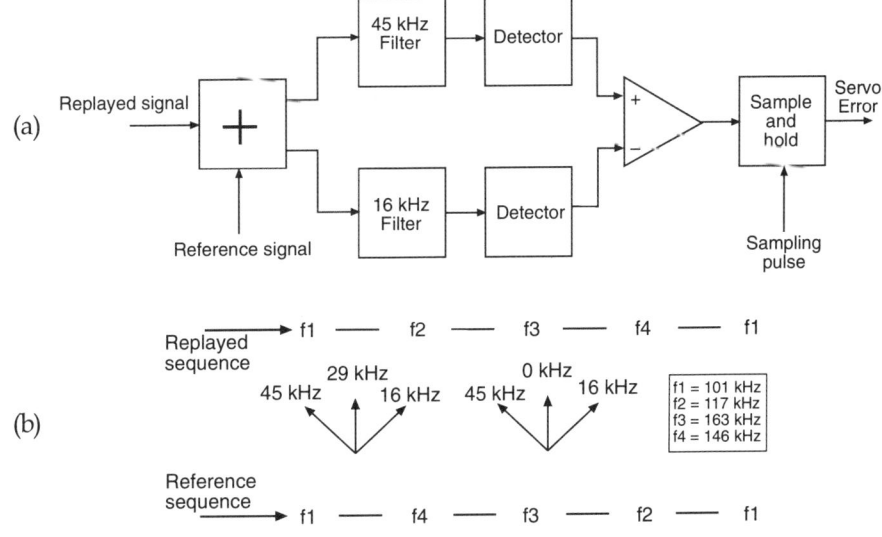

Figure 10.12 *ATF replay frequencies and filter system*

Video and Camcorder Servicing and Technology

Videomovie

The videomovie consists of two main units, the camera and the video recorder. As the two units are contained within the same housing then some simplification and shared facilities allow a reduced component count as well as lower power consumption, compared to separately housed equipment. For example, supplies are shared and luminance and chrominance signals are maintained separately prior to recording. Figure 10.13 gives an overall block diagram of the camera and recorder sections. Signals from the ½-inch high band saticon pick-up tube are decoded and processed to PAL colour and luminance, and passed to the recording circuits.

All timing signals are derived from the sync signal generator (SSG) and these drive the tube deflection circuits and provide the very high voltages required for correct tube bias. Syncs, clamping and colour carrier signals are fed to the video process circuits to decode the tube signals and produce separate luminance and chrominance signals. Luminance is passed through a low pass filter and then a dynamic aperture corrector (DYAC) before being frequency modulated for recording. Chrominance is passed through a band pass filter (BPF), not so much for filtering as it is already at 4.43 MHz, but for timing purposes as a small delay to match the luminance signal timing, slowed down by its LPF, during record. Down-conversion and colour level control are as for standard VHS recording. Luminance FM recording carrier and low frequency colour signals are mixed for recording and then switched to the video heads by the matrix switching circuits.

Luminance and chrominance signals are simultaneously mixed to a composite colour signal and fed to the output for E/E monitoring. A monochrome viewfinder monitor allows monitoring via a low pass filter (to prevent colour dot patterning).

In replay mode the signals from the video heads are switched again on playback from each video head, the luminance FM signal is limited and demodulated while the colour signal is up-converted and phase corrected, and the two are mixed to a composite signal for the output.

Audio recording and playback are fairly standard and recording is only available from the inbuilt microphone and can be monitored from the main output socket or by earphone.

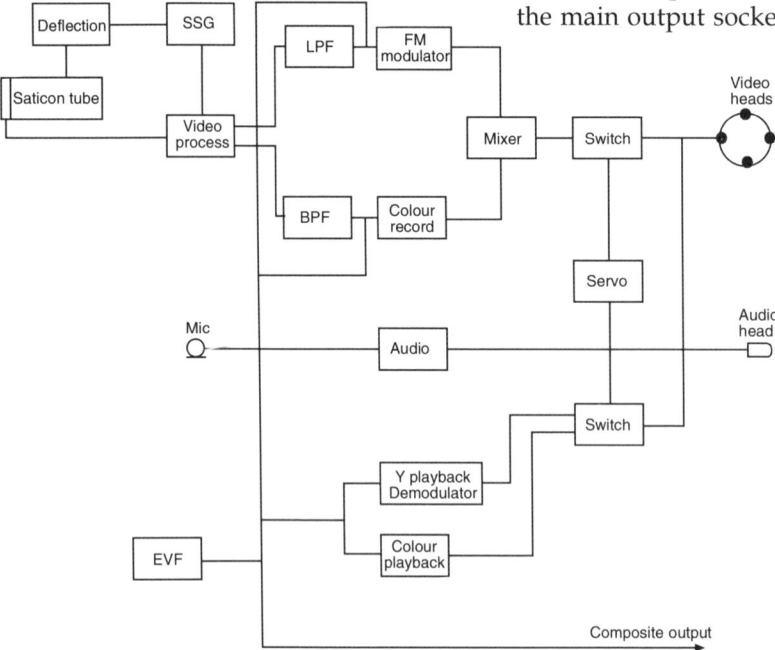

Figure 10.13 *Basic camcorder block diagram*

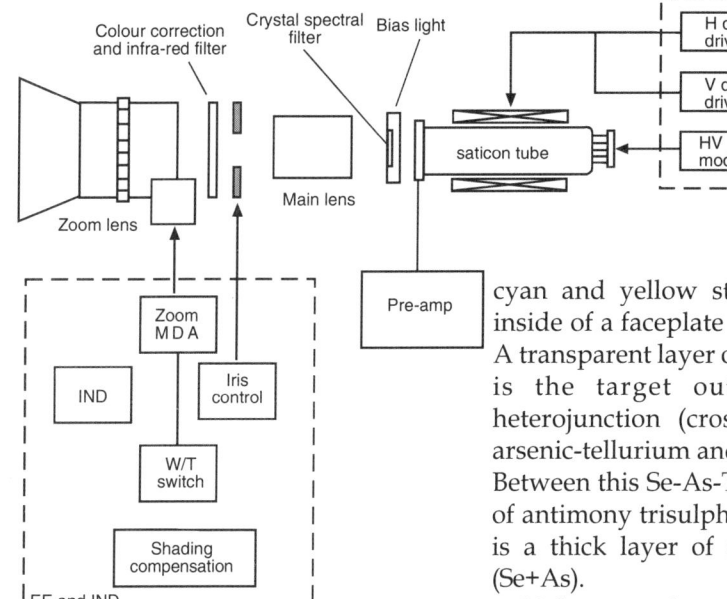

Figure 10.14 *Tube camera head and lens components*

Camera tube

Light passes through the camera zoom lens, colour temperature filter and a crystal filter focusing the image onto the saticon tube faceplate and photosensitive target area. Within the tube an electromagnetically deflected electron beam scans the inside of the target area and conforms to the specific TV standard of 625 lines and 50 Hz field scan. Variations in light intensity upon the target cause variations in the electron beam's current as it scans the target. Without any light at all some beam current still flows, although it is very small (in the range of pico amps) and is referred to as the dark current.

The tube's target is a high impedance source and the small signal current is very susceptible to external interference. The pre-amp is therefore mounted upon the tube within a screened container embodying the target connection. The target is a complex sandwich layer of selenium-arsenic-tellurium (saticon) film, bonded to other semiconductor layers in a sandwich. A structural diagram is shown in Figure 10.14.

Tube construction is quite complex. Diagonal cyan and yellow stripes are deposited on the inside of a faceplate and over layered with glass. A transparent layer of stannic oxide (SnO_2), which is the target output electrode, forms a heterojunction (cross-coupled) with selenium-arsenic-tellurium and chalcognide glass material. Between this Se-As-Te (or SAT) layer and a layer of antimony trisulphide (Sb_2S_3) on the beam side is a thick layer of a selenium arsenic mixture (Se+As).

Light passes through the faceplate, stannic oxide and SAT layers to the selenium arsenic layer where electron and hole charge carriers are produced. Electrons flow towards the faceplate and into the stannic oxide via the SAT layer. Holes, on the other hand, cannot penetrate into the stannic oxide layer due to the selenium film deposited upon it by the SAT layer. The holes travel the opposite way and couple with the scanning beam electrons through the antimony trisulphide. Electrons from the scanning beam cannot pass through the antimony trisulphide due to a film of selenium on the opposite side. The scanning beam current is therefore dependent on the quantity of holes available for coupling.

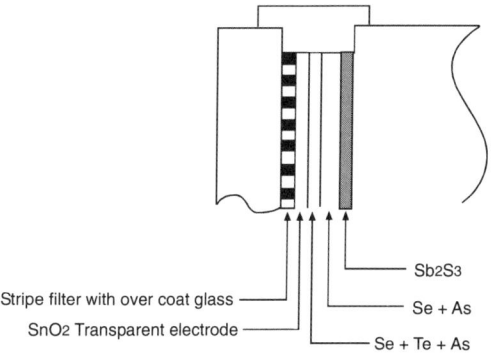

Figure 10.15 *Target layer construction*

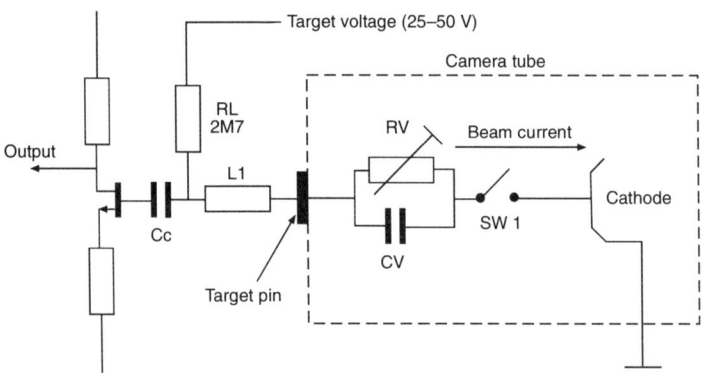

Figure 10.16 *Target electrical circuit*

So much for theory, but a much more practical approach can also be realised. An equivalent circuit representing a single pixel element of the photosensitive target layer is shown in Figure 10.16. The saticon target layer effectively forms a parallel circuit RV and CV. RV is varied by the intensity of the image light falling upon it. With no light, RV has a finite value and is designed to be as high as possible to minimise the standing dark current. As the target is scanned by the electron beam each pixel is hit by the beam once every frame. Contact of the electron beam on the target pixel is represented by the switch SW1. When the electron beam hits the pixel, SW1 closes and CV is charged up between the target and the cathode as the electron beam effectively connects one end of CV to the cathode. After the electron beam has passed, CV begins to discharge via RV whose value is light dependent. If there is no light, RV is a high resistance and the amount of discharge from CV is very small. If the light level is high then RV is small in value and the discharge of CV is high. The next time the electron beam passes over the pixel the amount of beam current that flows through the target load resistor RL will be that amount required to recharge CV. Hence if the light level is high then the charge current through RL is high and the signal voltage across RL increases.

The target output signal is therefore dependent upon the refresh charging current of each pixel capacitor CV and the light variable resistance of each pixel resistance RV. As the saticon electron beam scans a TV line, many hundreds of small capacitive elements charge up. The variation of charging currents through the target load resistor, RL, develop the signal voltage which is connected to the camera pre-amp via Cc.

The signal voltage across RL is very small but it has a very good signal-to-noise ratio, and this must be maintained. A very low noise junction field effect transistor (JFET) is used for the first stage and is selected for its low noise characteristics. L1 is called a Percival coil. It is used to resonate with the capacitive output of the target and lift the output at high frequencies.

Tube construction

At the faceplate end of the tube, shown in Figure 10.17, is the external connection of the target, the target flange or rim. A small leaf spring contacts the target rim and extends into the pre-amp to connect by soldering to the Percival coil.

One danger when servicing cameras is the presence of very high voltages on innocent looking printed circuit boards. While a shock may not be lethal it can cause the demise of the camera if it is dropped. Some reasonably high voltages are found around the camera tube.

The highest voltage is the mesh voltage of G6 and G3, which is around 1400–1600 volts. Grid G5 is a collimator lens to narrow the beam as it approaches the target and is at 700 V, functioning in conjunction with G4, which is variable between 200, and 300 V, as the electrostatic focus voltage.

Camcorders

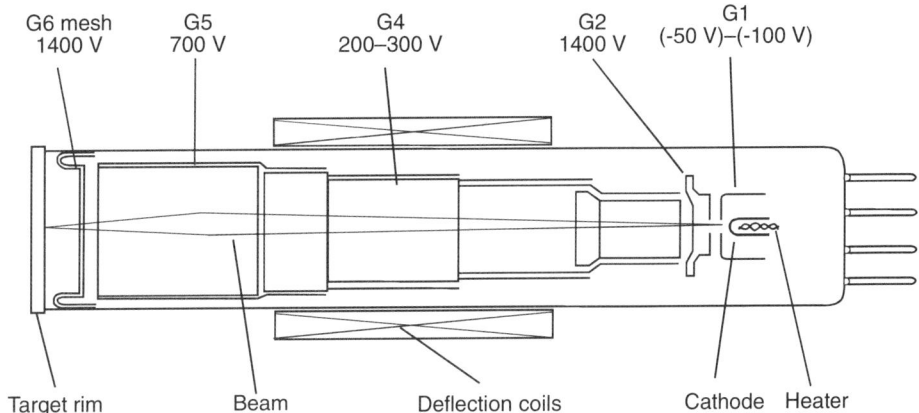

Figure 10.17 *Saticon tube construction*

If the beam is not focused the tube resolution drops and the R/B colour carrier reduces to zero leaving a green picture.

G2 is the first high voltage electrode to accelerate the electrons leaving the cathode. G1 is a reverse-biased grid and is variable so that the beam current can bs set to a value sufficient to charge CV at full white, without saturating the target. The visual effect of reducing the beam current below the nominal value is to clip the peak whites, as CV cannot fully charge back up. Hence the beam current is set to a level just above peak white maximum voltage at the signal output. Too much beam current will adversely affect focusing as the 'spot' size of the beam widens out, effectively shorting adjacent pixels.

Cathode DC standing voltage is about 6 V above chassis, but the tube needs to be turned off during field and line blanking, so blanking pulses are applied to the cathode at about 64 V p/p. An oscilloscope on the beam current grid, G1, will show negative blanking pulses down to -70 V with a black level of about -50 V upon which the video signal can be seen. This measurement indicates a functioning tube even if the output pre-amp has failed, and no apparent output signal is present from it.

While focusing is electrostatic, deflection is electromagnetic and is achieved by a deflection yoke around the tube. Shorted turns within the yoke can stop deflection resulting in no output from the target and no grid signal.

Bias light

To reduce lag or after-image effects caused by a too small dark current, a bias light illuminates the target. If, say, there is a rapid change from bright light to dark it is possible for CV elements to be left with residual charges which could not be discharged through RV between electron beam scans, as RV is too high in value. These charges show up as an after-image. The bias light is used to maintain lower values of RV and hence increase the discharge of the CV elements.

Crystal filter

The crystal filter is used to reduce cross-colour effects or moiré patterning from scenes with small chequered patterns or fine stripes. If, within a scene, luminance signals of fine patterns around the 4.433-MHz chrominance subcarrier frequency occur, then random blue/purple moiré or herringbone patterns are seen. A similar effect occurs in a camera when scenes which create luminance frequencies around the R/B carrier frequency of 3.9 MHz are viewed. The crystal filter is an optical spatial filter with optical gratings effective at the 3.9-MHz frequency, which effectively block fine patterns at the filter.

191

Colour signal

A colour television tube has coloured phosphors of red, blue and green to enable it to display colours. A colour camera tube has coloured stripes upon its faceplate in order for it to distinguish colours.

Although the target area covers most of the tube faceplate the main area of use is a rectangle of 5.6 by 7.2 mm wherein the purity of material and uniformity of sensitivity are kept within fine limits (Figure 10.18). The image from the lens is focused onto this area and in order to suppress signals outside this area the rest of the tube is masked off by a black mask. A precise fine stripe filter covers the image area to provide for colour recognition. The main tube signal is green, and stripes on the target face are yellow and cyan, so within the decoding system red, blue and white can be obtained. As the light transmission of red and blue filters is low the complementary colours of yellow and cyan are used, cyan being -R and yellow being -B. The stripes cover the target faceplate in diagonals at an angle of 25.40° with a width of 13.95 μm and a pitch of 30.89 μm, as shown in Figure 10.19. The stripes act not only as colour filters but the physical dimensions and spacing are used to generate specific signal parameters. As the electron scanning beam passes over the stripes it generates one output per stripe and as the number of stripes along the horizontal is fixed then a carrier signal is generated. An analogy is that of a child running along some railings with a stick to generate a tone. In the videomovie the number of stripes in a line period generates a carrier of 3.9 MHz. It can be noted then that the horizontal width is critical. The green signal is present in all three stripe colours, white, cyan and yellow, and so the target video signal tends to be green rather than white. It is complemented, however, by the 3.9-MHz carrier which contains the red and blue colour information as an amplitude modulation with red and blue 180° apart, in phase.

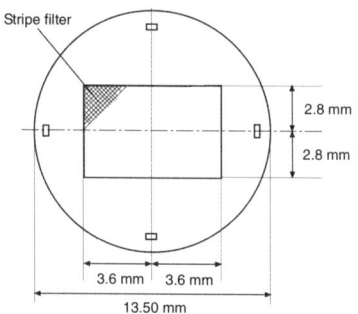

Figure 10.18 *Tube faceplate*

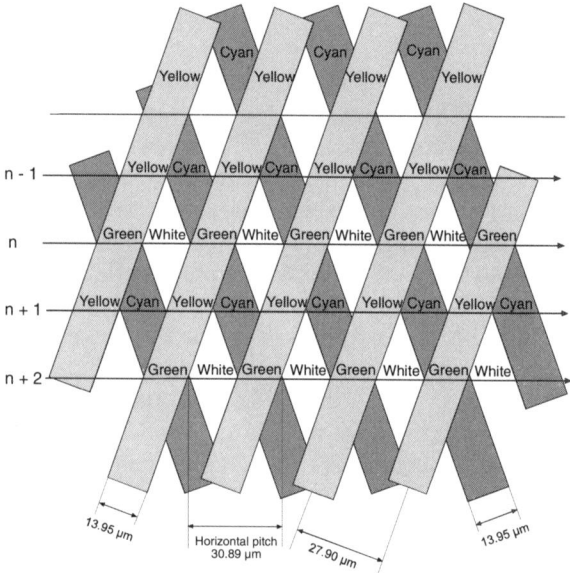

Figure 10.19 *Tube faceplate stripes*

Colour separation

In Figure 10.19 the stripes are shown for line scans (n-1), n, (n+1) and (n+2). The relative tube outputs are illustrated in Figure 10.20. As the TV lines are passed through a 1-line delay line we can relate signals from current line (n) to those of the previous line, (n-1). If we take line n and look at the output as it passes over the stripes we get white/green/white, etc. As white is equivalent to (R+B+G) and green is just G we get:

(R+B+G)/G/(R+B+G).

Camcorders

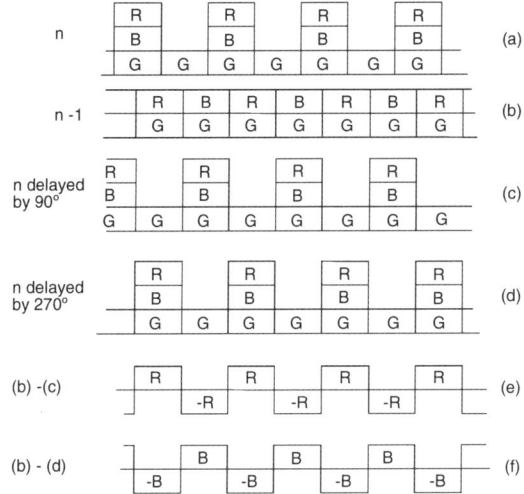

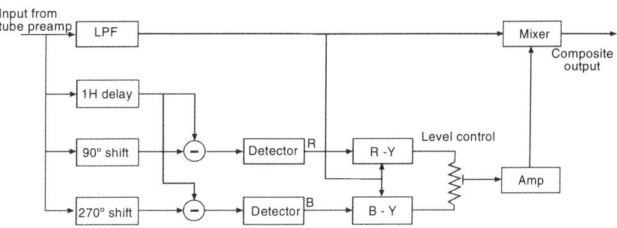

Figure 10.21 *R/B colour separation circuit*

Figure 10.20 *R/B separation waveform*

The electrical signal is as shown in Figure 10.20(a).

The previous line, (n-1), crosses yellow/cyan/yellow, which can be considered to be:

(R+G)/(B+G)/(R+G).

This is shown in 10.20(b), because the camera faceplate stripes are as in the colour bars on a TV display where yellow is made up of red and green, and cyan is made up of blue and green.

In order to separate the red and blue components, line (n) is passed through two phase-delay networks, as shown in Figure 10.21, to give a 90° phase delay, see Figure 10.20(c), and a 270° phase delay, see Figure 10.20(d). By using Figure 10.20(b) as the reference or delay line output (n-1), we can subtract from it (n) delayed by 90° (b - c) to give the red output, see Figure 10.20(e), and (n) delayed by 270° (b-d) to give the blue output, see Figure 10.20(f). The green output is taken as Y and a PAL encoder is then used to provide the composite colour signal.

The tube output is passed through a 3.9-MHz trap and the output green is converted to two components; luminance signal YH and YL by passing through a LPF at 0.9 MHz. YL is subtracted from red and blue to produce (R-YL) and (B-YL). Green will not be recovered until the TV colour decoder and is dependent upon the level of Y with respect to (R-Y) and (B-Y). If the light level is too low and R and B signals are lost then the values of (R-Y) and (B-Y) will fall leaving the TV decoder with a majority output on the green channel.

A later development for camcorders simplified the colour separation circuit, as shown in the block diagram of Figure 10.21. First, green is eliminated by passing the R/B signal (a 3.9-MHz carrier) through a band pass filter. The R/B signal (n-1) in Figure 10.20(b), can be split into the two components of red and blue 180° out of phase as (R-B). Signal (n) in Figure 10.20(a), however, comprises red and blue signals in phase (R+B).

This can be confirmed by inspection of the stripe filter and comparing lines (n-1) and (n+1). The yellow stripe has advanced nearer to the start of the scan, whereas the cyan stripe has retarded further away from the start of the scan. The position of the yellow diamond in line (n+3) is the same as (n-1) so that the pattern repeats over four lines, so it is not difficult to work out in terms of phase that the stripes shift 360°/4, or 90°/line, and that the yellow phase advances whilst the cyan retards.

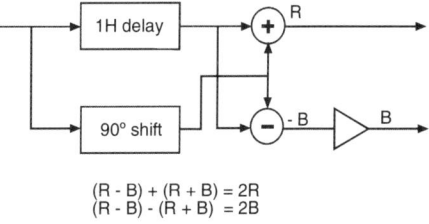

(R - B) + (R + B) = 2R
(R - B) - (R + B) = 2B

Figure 10.22 *Simplified R/B separation circuit*

193

Video and Camcorder Servicing and Technology

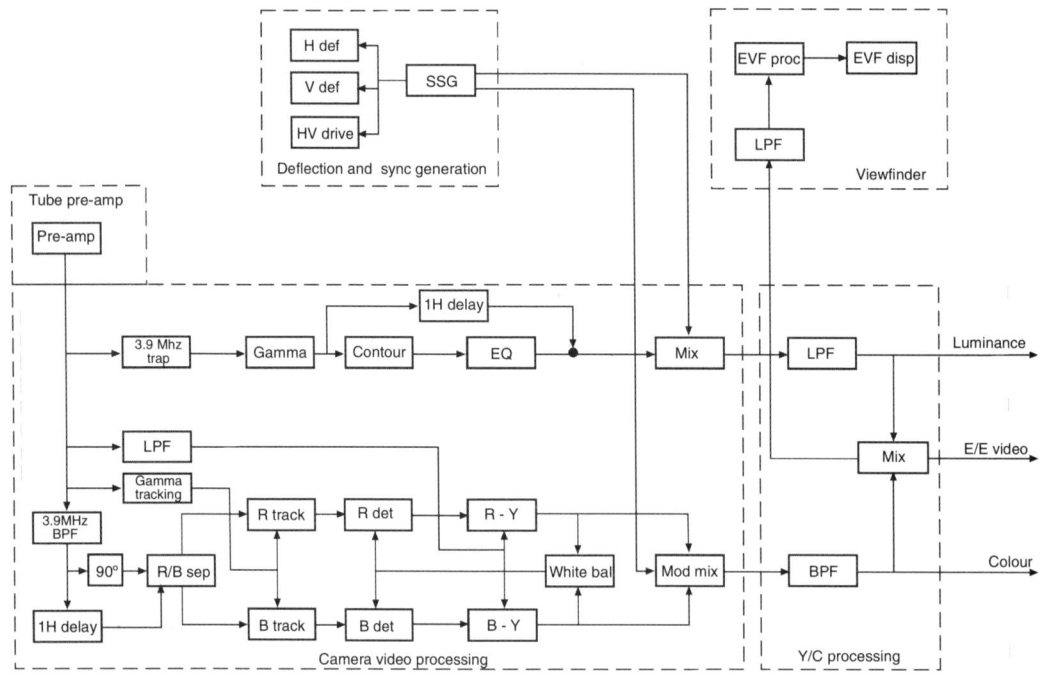

Figure 10.23 *Camera block diagram*

In terms of red and blue then we have a constant 180° phase shift between them which is rotating at 90°/line. It therefore becomes possible by the use of a 90° phase shift and a 1-line delay with a sum and difference network to separate red and blue.

We can now say the signal coming out of the delay line in Figure 10.22 is the difference signal (R-B) while the direct signal, brought into line with a 90° phase shift is the sum signal (R+B). Separation in a sum network is:

(R-B) + (R+B) = 2R

whilst in the difference network we have:

(R-B) - (R+B) = - 2B

where there is a negative sign it indicates a 180° phase shift. This is easily accommodated within the network by a unity gain inverting amplifier to give 2B.

Figure 10.23 illustrates the overall colour video signal handling. In better quality broadcast cameras there is an aperture correction circuit to crispen up the pictures by correcting transition edges in both the horizontal and vertical directions. In the VHS camcorder there is a vertical edge enhancement circuit which by the subtraction of direct and delayed TV lines will enhance the edges, often referred to as contour correction. Horizontal aperture correction is also carried out in the video recorder record circuits and an explanation can be found in the section on dynamic aperture correction in Chapter 3. Also, the colour signals undergo greyscale tracking and white balance correction.

Colour spectrum

From the tube pre-amp comes a signal that has a green signal and an R/B carrier signal, Figure 10.24 illustrates the spectrum of this signal. As green represents the Y signal this is shown as EY

Camcorders

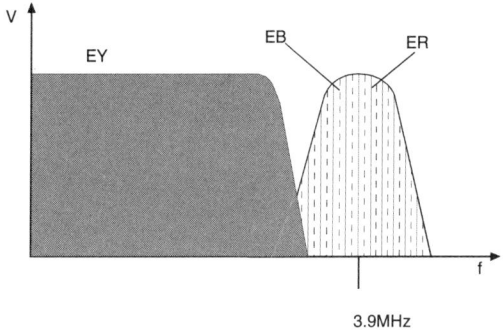

Figure 10.24 *Signal spectrum from tube target*

Gamma correction and tracking

In a single tube camera the levels of the Y signal and the R/B carrier do not remain constant with respect to each other throughout the range of input light. At low light levels the R/B carrier is lower in proportion to the Y video output than at higher levels. This cannot be allowed in high sensitivity colour cameras as there would be a tendency towards dark green pictures as the red and blue outputs fall off if the camera is used in lower light levels. It can be seen in Figure 10.25 that the blue sensitivity is lowest and that red falls between it and the higher Y level. In mid-range light levels the differences remain constant, but at the extremes of low and high light levels the red and blue signal levels fall off first.

It is necessary to ensure that the gains of the red, blue and Y (green) signals remain constant with respect to each other throughout the light range to prevent different tints at different light levels. To this end a triple level tracking correction system is employed as shown in Figure 10.26.

from DC to around 3 MHz. At 3.9 MHz is the R/B carrier signal generated from the coloured stripes, yellow and cyan. In the carrier are the red and blue components 90° phase shifted with a harmonic of fh/2, i.e. half line frequency. As discussed later, in almost all cases the colour signal from the imaging source is line alternate. In this case the line alternate signals are the components of (n) and (n-1).

In the camera colour circuits three components are derived from the colour separation matrix. These are YH, the full bandwidth Y signal, R/B colour component and YL, which is a Y signal but with a much reduced bandwidth to match that of the R/B signals; approximately 1 MHz.

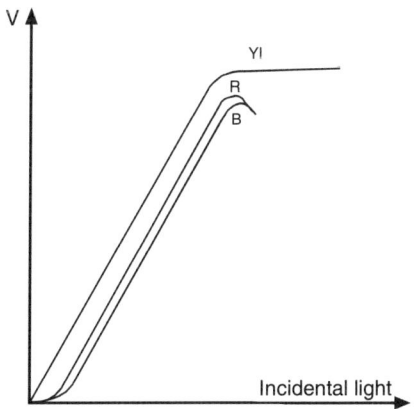

Figure 10.25 *Saticon tube photo/voltage characteristics*

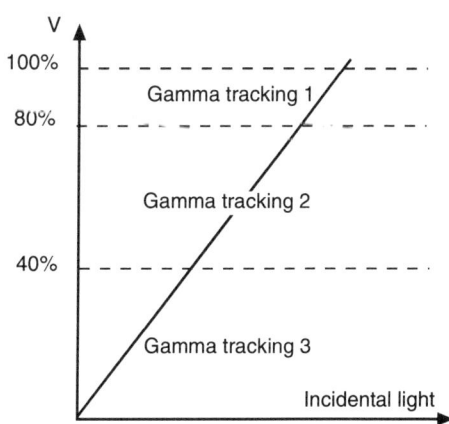

Figure 10.26 *Multi-level gamma tracking correction*

Tracking is corrected by adjusting the gains of the red and blue control amplifiers against the Y reference signal which has already been gamma-corrected against tube gain non-linearity. Tracking (1) works in the range up to 40%, tracking (2) works between 40 and 80%, and tracking (3) operates at high light levels only.

The system used is similar in most camcorders and is incorporated within two ICs, shown in Figure 10.27. IC1 is referred to as the gamma-tracking generator and produces three signals derived from the luminance input. Pin 16 produces an output that is clipped at 40%, hence any luminance signals up to 40% peak white level are sent to the red and blue tracking (1) controls to white-balance the picture up to this level. Pin 17 produces an output of luminance signals only between 40 and 80% to control the white balance between the two levels. Pin 15 outputs peak white signals only to white-balance the peak white levels. It is not easy to adjust these controls without the correct lighting conditions of 3200°K and a grey scale chart. They interact at the crossover levels of 40 and 80% and considerable inter-adjustment is required to get the whole grey scale tracking correctly, and usually some compromise is found. The peak white levels can only be adjusted if control over the automatic iris is possible, or a small white area is illuminated within a black background to fool the iris control and open up the iris to provide the high level of peak white needed for accurate adjustment.

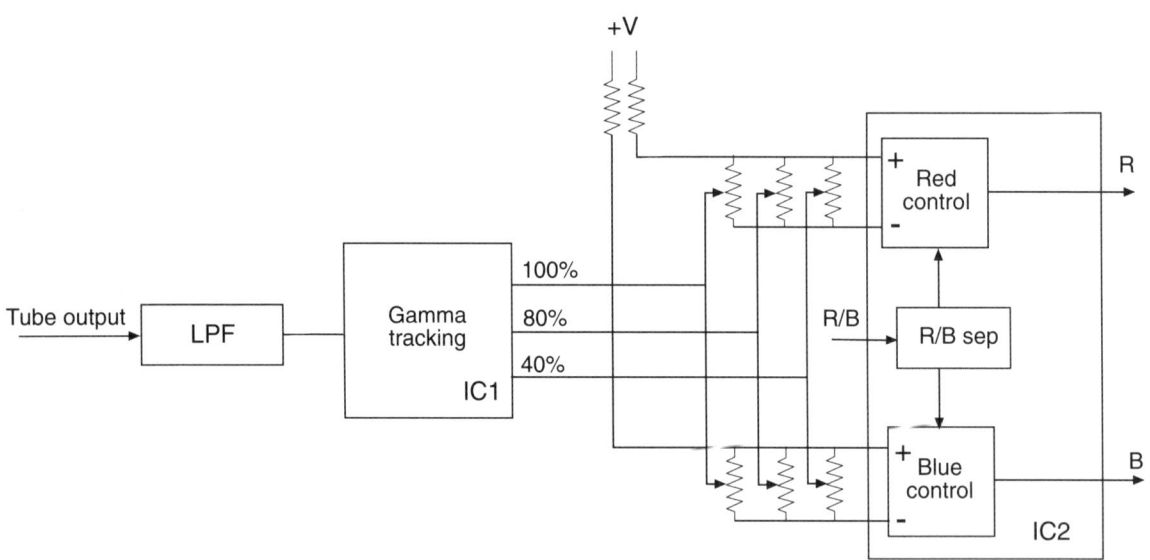

Figure 10.27 *Multi-level gamma tracking correction circuit diagram*

CCD photo-sensitive pick-up device

The solid state photo-sensitive device used in colour camera to replace the saticon tube is commonly know as the CCD imager, or a charge coupled imaging device, a term which refers to the fact that shift registers within the device operate on the charge coupled principle. Advantages are its compactness, low power consumption, long life, and resistance to burning from bright light sources. Unlike a camera tube there is no heater element and as the photo elements are solid state there are no adjustments for installation, temperature variation, or for ageing and usage.

There are two types of CCD imager: frame transfer and line transfer.

In a frame transfer the entire picture is transferred from the sensing area to a storage area from which the picture elements, or pixels, are read out, as shown in Figure 10.28. A single frame transfer imager is more suited to monochrome than colour. In camcorders the line transfer imager is more commonly used and this will be described in more detail.

Line transfer CCD

This is the type of CCD imager used in colour portable cameras and camcorders. A CCD solid-state pick-up unit is a large scale integrated circuit consisting of some 300000 photodiodes for picture elements, or pixels, on a small rectangular substrate. There are in excess of 500 vertical shift registers and one horizontal shift register and accompanying pulse control circuits all on the same chip!

The scene that is recorded by the camcorder is inverted through the lens and when it is displayed on the surface of the imager it is an inverted mirror image, as shown in Figure 10.29. Scanning of the pixels in the imager starts at the lower right-hand corner and finishes at the top left-hand corner. This explains why the horizontal shift register is at the bottom of the stack: it is there so that the first pixel to be read out is the one at the start position in the lower right-hand corner.

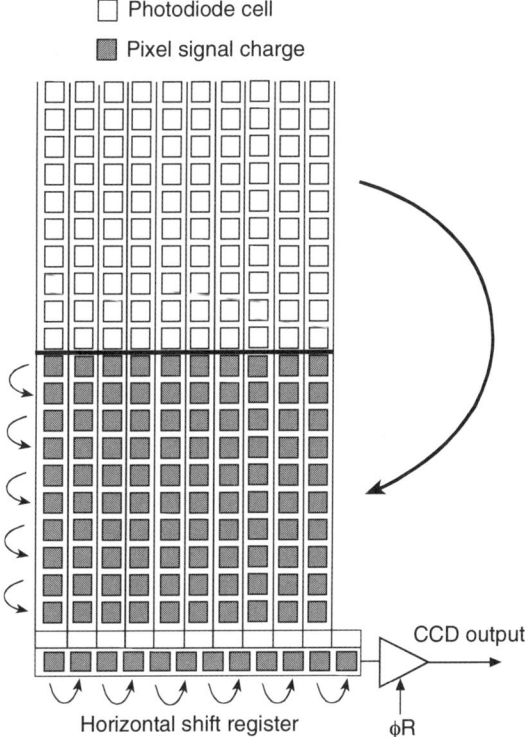

Figure 10.28 *Frame transfer CCD*

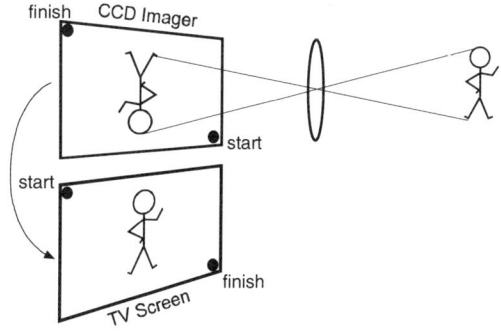

Figure 10.29 *A scene from the lens is an inverted mirror image on the CCD*

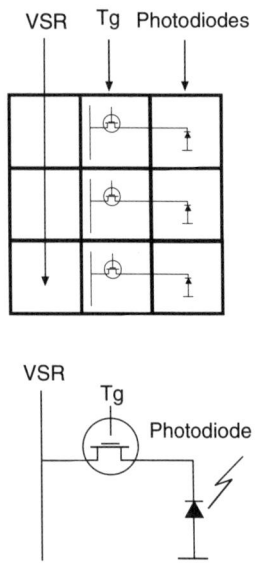

Figure10.30 *Transfer gate*

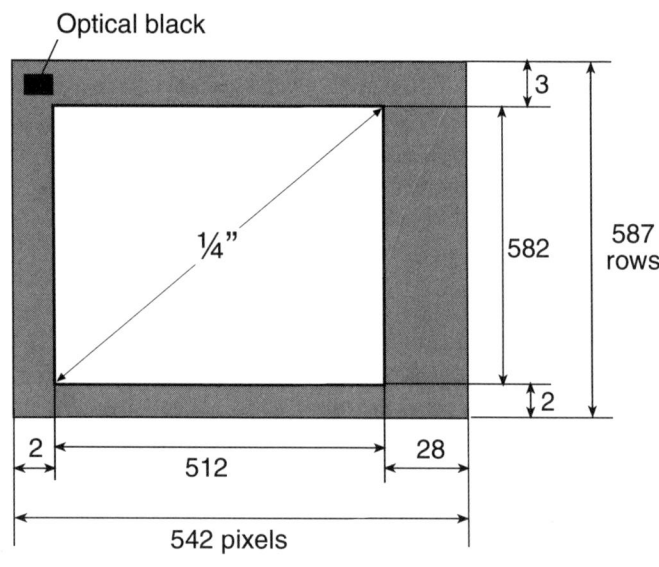

Figure 10.31 *CCD imager construction*

Transfer gate

Charges are developed in the photodiodes according to the amount of light falling upon them during a field period of 20 ms. Each photo diode has a MOS transistor transfer gate attached to it, as shown in the small section of an imager in Figure 10.30. All of the gates in a column are connected together. At the time of a transfer gate pulse the charge on the diodes is transferred over to the vertical shift register and the diodes can then start to acquire a further light dependent charge during the ensuing field period.

CCD imager structure

Not all of the pixels within the imager are active and a small section of pixels are masked for a reference of optical black level. In analogue camcorders the number of pixels that can be manufactured on the CCD substrate and the size of the imager have a bearing upon the picture resolution; the more there are then the better the picture resolution is. For digital camcorders, the number is fixed due to the way in which the data compression is structured to 576 rows of 720 pixels (see the section on digital camcorders). In this example the analogue imager is about 0.25" corner to corner of the active area, see Figure 10.31. There is a total of 587 rows of 542 pixels giving an array of some 318154 pixels, whereas the active area is 582 rows of 512 pixels and the array is 279984 (no doubt that the marketing department will stretch this to 300000!).

Each pixel is a light-sensitive diode that accumulates a small voltage charge, proportional to the amount of light that falls upon it. At a specified time, equivalent to field blanking, the charge is transferred to an adjacent cell forming part of a vertical shift register.

For colour reproduction the light sensitive photodiodes are covered by an array of green, cyan, magenta and yellow filters. Why not used red and blue? Simply that the light transfer characteristics of the complementary colours are better than the primary colours, particularly Blue, and this improves the overall sensitivity of the imager. Complementary colours are yellow filter = R+G, cyan filter = B+G, magenta filter = R+B.

Camcorders

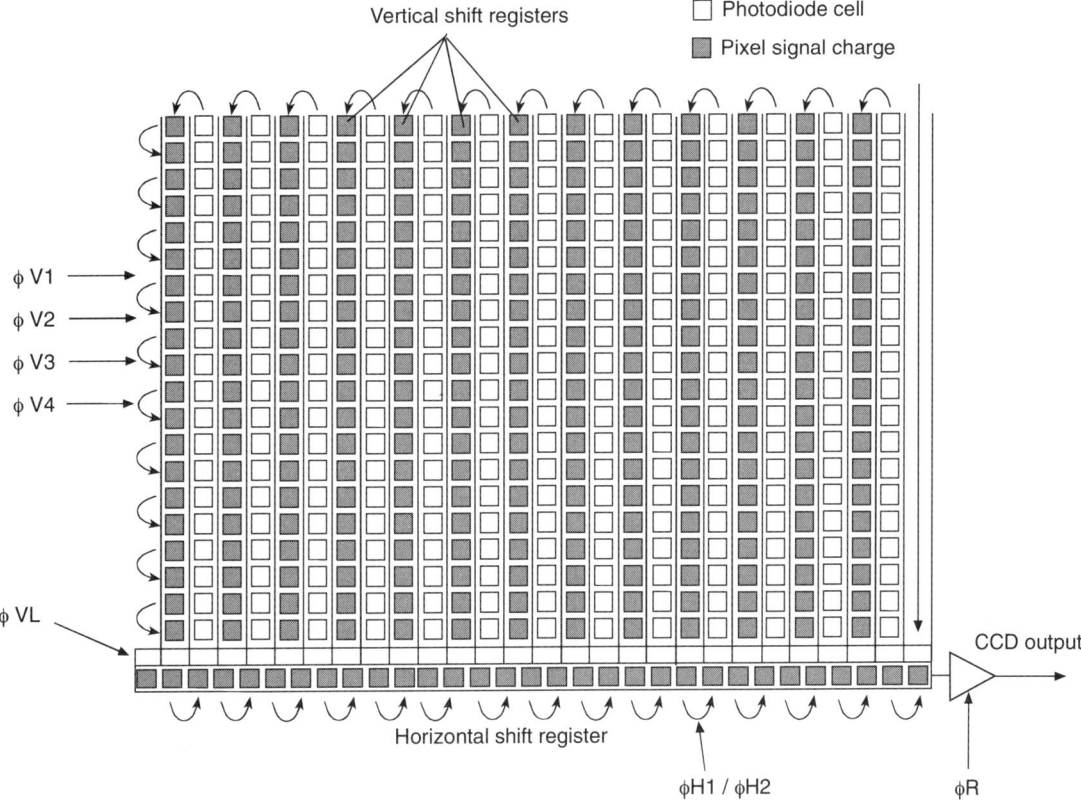

Figure 10.32 *Imager scanning method*

Imager scanning

A CCD image is not scanned in the same way as a TV tube or LCD display as the electrical charges from the light-sensitive diodes are 'marched out' like soldiers leaving a parade ground.

Between each column of pixels (white squares) in Figure 10.32 there are vertical shift registers (dark squares). Each register is to the left of its allocated light-sensitive diode pixel and between each pixel and register cell there is a transfer gate.

During the period equivalent to field blanking, a pulse called the transfer gate (Tg) pulse initiates a total side step of all the light-sensitive diode charges. Each diode's voltage charge is stepped left into a vertical shift register; all 279984 charges are stepped at once. In effect the whole picture shifts from the light-sensitive diode pixels into adjacent shift registers. The diodes may then begin to charge up over the next 20 ms until another Tg pulse is sent from the timing control IC.

Now there are 512 shift registers connected in parallel, each with a column of 582 electrical charges of various levels. Four pulses are now employed to step the charges down the shift registers. The timing is that of a TV line; each charge is stepped down every 64 µs.

At the bottom of the pile the charges are stepped into a horizontal shift register and 512 voltage values are rapidly stepped out during the TV line period of 64 µs by two pulses, φH1 and φH2, through a gate controlled by φR, a high frequency gating pulse with a frequency around 8 MHz.

Video and Camcorder Servicing and Technology

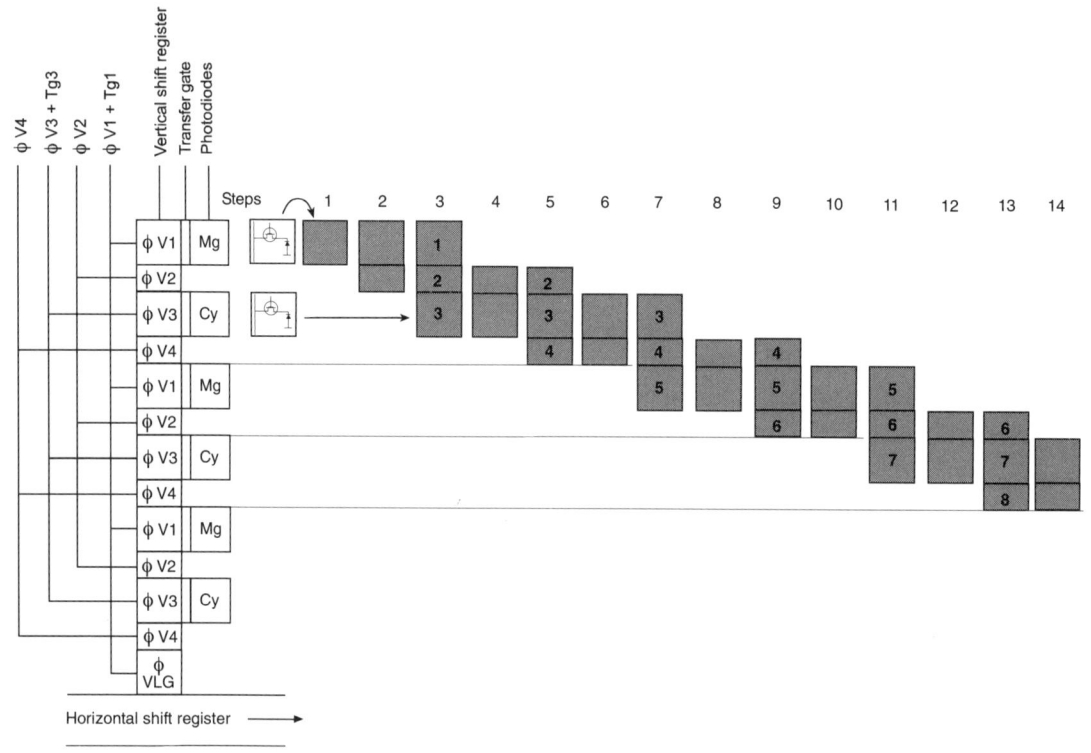

Figure 10.33 Vertical shift register operation

Vertical shift register

The shifting of the pixel electrical charges down the vertical shift register stack is carried out with a four-phase drive, V1, V2, V3 and V4. The transfer of charges from the photo diode pixels into the shift register is initiated by transfer gate (Tg) pulses attached to V1 and V3. Figure 10.33 shows the sequence of charge transfer down the shift register in graphic form.

Step 1. V1 goes high and the attached Tg1 pulse transfers the charge from a magenta pixel into the first shift register cell.

Step 2. V2 goes high and the magenta charge is shared between the first and second shift register cells.

Step 3. V3 goes high and the magenta charged is further shared out into the third shift register cell. Simultaneously, Tg3 transfers the charge from the cyan pixel cell into the shift register. It is added to the charge stored in the other three cells and results in the combined [Mg + Cy] charge.

Step 4. V1 goes low and the total charge is now stored between cells 2 and 3.

Step 5. V4 goes high and the charge spreads over to cell 4, sharing 2, 3 and 4.

Step 6. V2 goes low and the charge is shared between cells 3 and 4.

Step 7. V1 goes high and the charge is further shared out between 3, 4 and 5. There is no Tg pulse now until the next field.

Step 8. V3 goes low and the charge progresses to cells 4 and 5.

This sequence continues with all shift registers working in tandem stepping the charges down to the horizontal shift register and out through a buffer.

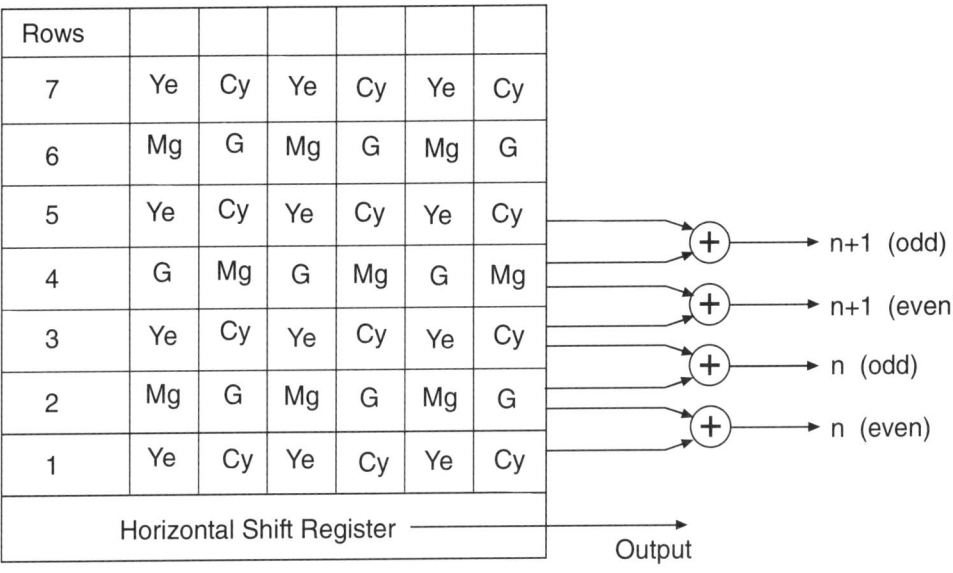

Figure 10.34 *CCD imager colour filter matrix*

CCD imager colour matrixing

In order to obtain the correct colours commensurate with maximum imager sensitivity the light-sensitive diodes are covered by a colour filter array as shown in Figure 10.34. This is in the lower right-hand corner and represents the top left-hand corner of the picture, remembering that the scene upon the imager face is inverted.

The initial stage in this procedure is that the Tg pulse copies all light-sensitive diode charges into the vertical shift registers. There is then one vertical step in two stages, and a second Tg pulse occurs. This second Tg pulse adds the charges of a second TV line to the first line. After that the stepping continues normally until the picture has been stepped out and the next field begins. This additional Tg pulse is important in that it adds the charges from one colour filter to the charges of a different colour.

Referring to Figure 10.34: in the first even field the colour components of rows 1 and 2 are added together, as are rows 3 and 4 in the next line adding [Cy+G] and [Mg+Cy], respectively. For the odd field rows 2 and 3 are added and rows 4 and 5, adding [G+Cy] and [Cy+Mg], respectively. In fact the colour component additions for both odd and even fields are the same. What is more the number of lines used in each field is half of the total number of lines (291) allowing for interlacing. Each row is used twice in a TV frame, once in each odd and even field.

Each pixel voltage value stepped out of the CCD imager by the horizontal shift register contains two colour component values added together.

Correlated double sampling

All pixel samples coming out of the imager contain noise from a gating pulse and general electrical noise, as shown in Figure 10.35. Not all of the pixel sample period is wanted signal, apart from the reset pulse period the transfer period has no signal, just noise. The signal value also has noise and there is correlation between the transfer noise and the signal noise. The two noise values are related and this can be useful.

In the correlated double sampling circuit there are two gating pulses, t1 and t2; t1 samples noise, t2 samples signal and noise. The sample circuits are 'sample-and-hold'. In this case the sample is held to the output until the next sample, this effectively stretches the signal. Both outputs from the sample-and-hold circuits are supplied to an amplifier: continuous signal with noise is on the positive input and noise only is on the inverting input. Within the amplifier the noise is subtracted from the signal and in effect cancels. Therefore clean contiguous signal comes out of the CDS circuit. (shown in Figure 10.36). It is because the noise in the Tg period is correlated to the noise in the signal that cancellation occurs.

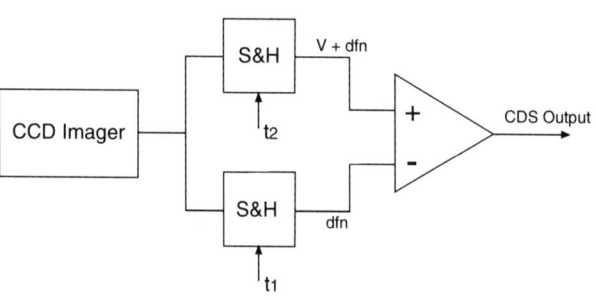

Figure 10.36 *Correlated double sampling circuit*

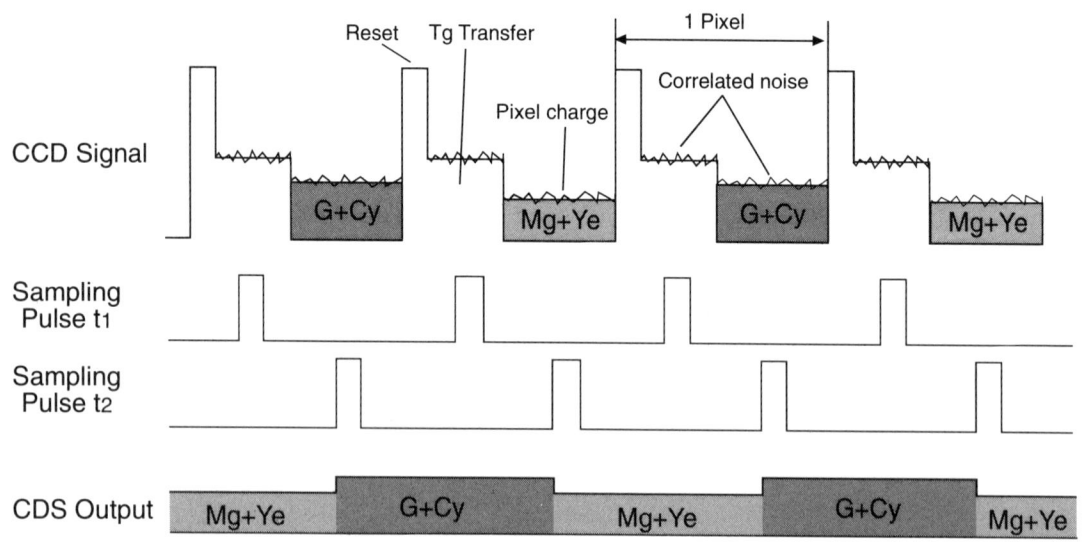

Figure 10.35 *Correlated double sampling waveforms*

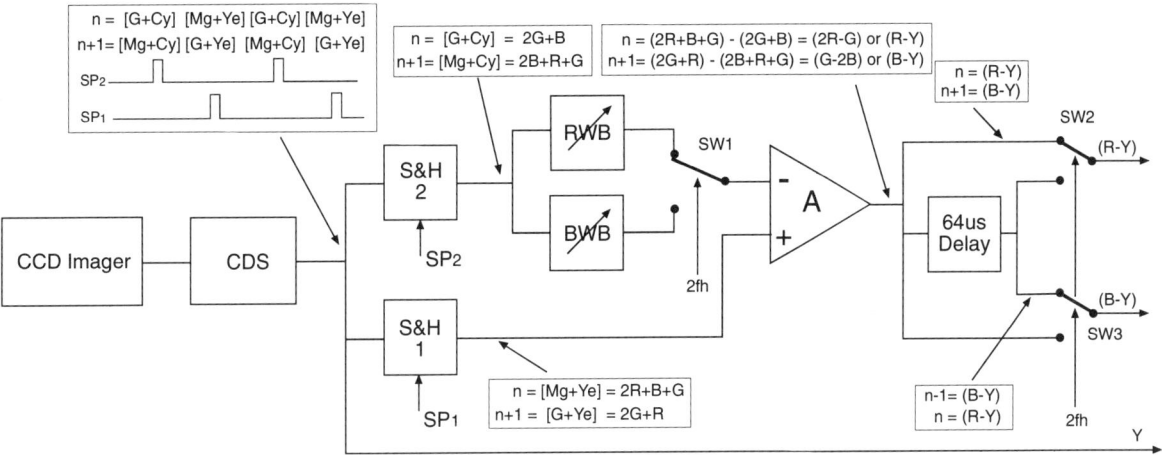

Figure 10.37 *CCD imager colour separation matrix*

Chrominance derivation

After the signal comes out of the correlated double sampling circuit, automatic gain control (AGC) is applied to the signal and black level clamping to stabilise the amplitude and DC level. Now the signal from the CCD has three formats: the Y signal, and the chrominance components (n) and (n+1). For each alternate TV line of 512 pixels, each pixel also alternates between two chroma components:

[G+Cy] [Mg+Ye] or [Mg+Cy] [G+Ye]

Line (n) = [G+Cy] [Mg+Ye] [G+Cy] [Mg+Ye] [G+Cy] [Mg+Ye] [G+Cy]

Line (n+1) = [Mg+Cy] [G+Ye] [Mg+Cy] [G+Ye] [Mg+Cy] [G+Ye] [Mg+Cy]

Now the signal arrives at two more sample-and-hold circuits with sample pulses SP1 and SP2. These sample pulses are at a high frequency, up to 12 MHz or more, i.e. half the pixel rate, and displaced by one pixel as shown in Figure 10.37.

Consider the first line (n). SP2 samples only the [G+Cy] pixel values, missing out the alternate [Mg+Ye] and leaving a gap. This gap is filled with [G+Cy] due to the action of the sample-and-hold circuit. SP1 pulse samples the [Mg+Ye] pixel values and holds the value between samples also.

Therefore for line (n) the output of S&H 2 is continuously the values of [G+Cy], and the output of S&H 1 is continuously [Mg+Ye]. For line (n+1) the outputs are [Mg+Cy] and [G+Ye] respectively.

From the colour mixing shown in Figure 10.38 the following can be deduced:

[G+CY] = (G+G+B) = (2G+B)
[Mg+Ye] = (R+B+R+G) = (2R+B+G)
[Mg+Cy] = (R+B+B+G) = (2B+R+G)
[G+Ye] = (G+R+G) = (2G+R)

In amplifier A, forming a subtraction matrix, the two alternating pixel values of each line are subtracted to obtain the difference values for line (n) we have [Mg+Ye] − [G+Cy].This expands to:

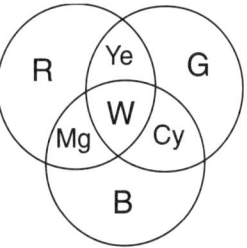

Figure 10.38 *Colour component diagram*

Video and Camcorder Servicing and Technology

$$(2R+B+G) - (2G+B) = (2R - G).$$

For line (n+1) we have [G+Ye] - [Mg+Cy]. This expands to:

$$(2G+R2) - (2B+R+G) = (G-2B)$$

This result in practical terms is that for line (n) the output from amplifier A is (2R-G) and for line (n+1) it is (G-2B).

If we take the formula Y = 0.3R + 0.59G + 0.11B, then the highest colour value by far is green. Therefore it can be assumed that G approximates to Y. It can be argued that (2R-G) is the (R-Y) component and that (G-2B) is (Y-B) or the (B-Y) component inverted.

At this point the CCD imager colour information is line alternate (R-Y)/(B-Y) and in order to obtain continuous signals the following Line Delay circuit is used.

Within amplifier A is a mixing matrix to convert (R-G) and (B-G) to (R-Y) and (G-Y), respectively, by applying the formulae:

$$(R-Y) = (R-G) + Kx (B-G)$$

and

$$(B-Y) = (B-G) + Ky (R-G)$$

Where Kx and Ky are fixed gain values for the mixing amplifiers; see Figure 10.39.

A half-line switching signal is applied to the matrix to select the two colour components on a line-by-line basis as follows. On the first line the signal is for example, (R–G). This is sampled by the sample-and-hold circuit (as was the previous (B–G)), and the value is maintained for the next two lines until it is refreshed by another sample. This means that at all times the two values of (R–G) and (B–G) are present at the summing circuits. As a result of this the calculated values of (R–Y) and (B–Y) are present at the inputs to the following gating circuit and are selected to the output by the half-line switching signal. They come out as line sequential (R–Y)/(B–Y) for separation in the next stage.

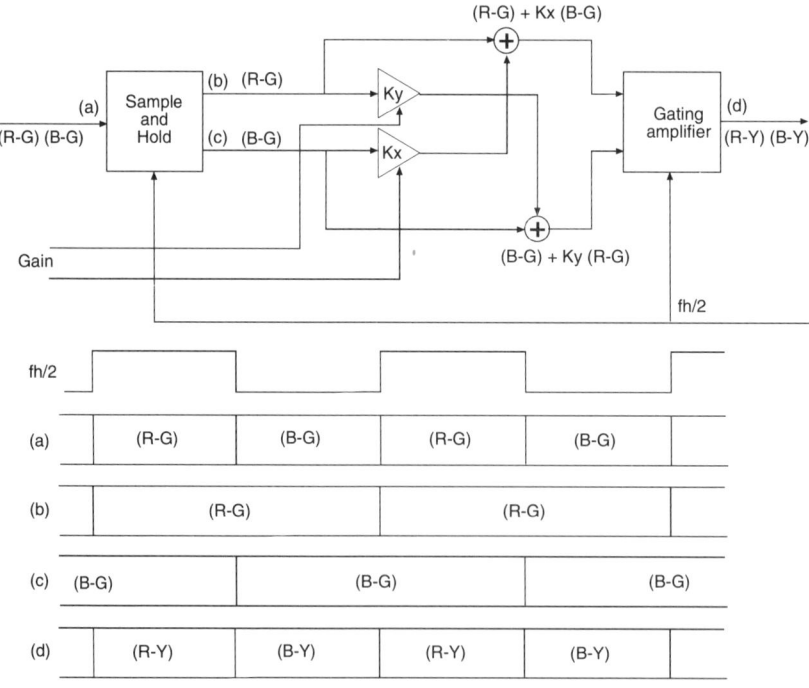

Figure 10.39 *(R - G)/(B - G) to (R - Y)/(B - Y) convertor*

Camcorders

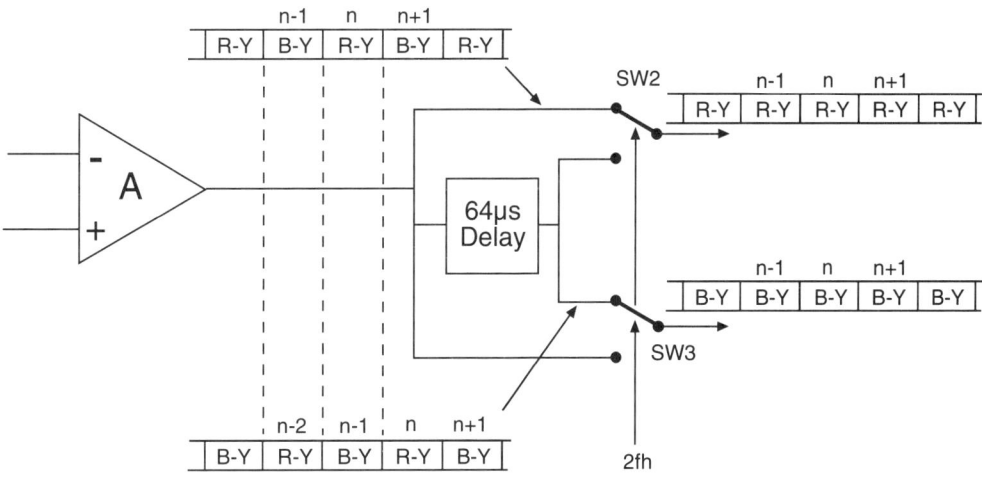

Figure 10.40 *(R-Y)/(B-Y) separation*

Delay line 'sample-and-hold'

Figure 10.40 illustrates the circuit and workings of the (R-Y)/(B-Y) separator. It works in a similar way to the previous sample-and-hold sections to provide continuous signals.

For line (n) the switches are in their upper positions and (R-Y) is on the direct path. Out of the delay line is the previous line (n-1) which is (B-Y).

For line (n+1) the switches are at their lower positions with the direct path now being (B-Y) and the delay line path being (R-Y).

When SW1, in Figure 10.37, is in the upper position the output of amplifier A is (R-Y); in the lower position the output is (B-Y). RWB therefore sets the (R-Y) level and BWB sets the (B-Y) level. Level control is software set and stored in an EEPROM.

At this point the (B-Y)/(R-Y) signal is amplitude and phase-modulated onto the colour carrier at 4.433619 MHz for recording, as described in the VCR colour recording section. Luminance and chrominance are individual components and there is no luma/chroma cross-talk. As with standard video recording the two signals are separate during the recording and replay process and are only combined at the E/E monitoring and replay video output terminals; unless the camcorder has S-VHS capabilities in which case they remain separate on the Y/C 'S-connector'.

Triple CCD imager block

In a triple CCD imager camera head the colour signal output is R, G, B from three separate imagers mounted and sealed on a prism block. All of the three imagers are of the same construction as a single CCD imager camcorder, i.e. they are one-third inch CCDs and they have separate RGB filters built in dichroic prisms. There are no individual filters over each photodiode; instead a single dichroic prism block is constructed bonding the prisms and the CCD imagers permanently together (see Figure 10.41). There are two distinct advantages in a triple CCD dichroic prism block arrangement: a greater colour dynamic range, and a higher contrast ratio when compared to a single imager.

A much greater colour range means that subtle colour differences can be observed, such as the shades of grass in a field or the different colours of crayons in a box where there are many shades of the same colour.

Horizontal resolution

A practice, common with triple CCD imager blocks, is employed to increase the horizontal resolution by a factor of 1.5.

You would expect the pixel arrays of all three imagers to be aligned in the vertical and horizontal planes of the prisms, such that the light falling on pixel dx/dy on the green imager will be the same light on the same pixel in both red and blue arrays. This is not so. Both the red and blue imagers are accurately aligned with each other but the green imager is horizontally displaced by half a pixel.

In Figure 10.42(a) the pixels are aligned, as one would expect and in (b) the pixel alignment of the green imager is as it is in practice. There are 542 pixels in a row that produce 542 green and 542 red and blue samples; when matrixed together for the Y signal = 0.3R + 0.59G + 0.11B the samples average out to approximately 750 pixels of horizontal resolution. It would always be possible

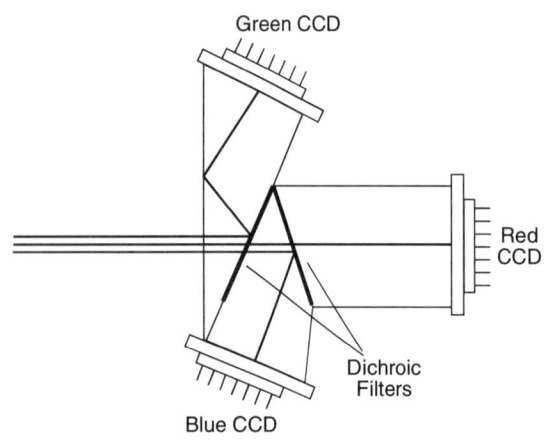

Figure 10.41 *Triple CCD imager prism block*

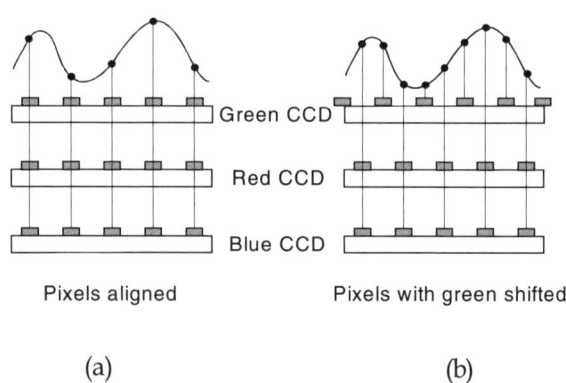

Figures 10.42 *CCD pixel alignment*

Camcorders

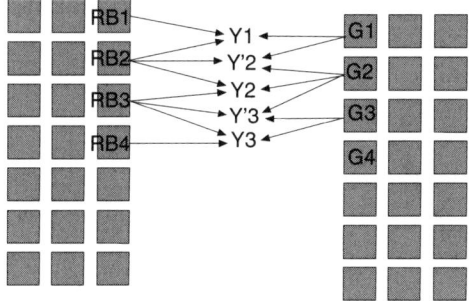

Figure 10.43 *Interleaved line scanning*

to increase the number of pixels on a given size of chip in order to increase the resolution, but as the photodiodes get smaller, so does the electrical charge. Smaller photodiodes give rise to a reduction in the signal-to-noise ratio and noise becomes much more significant at lower light levels, so this is a compromise.

All three imagers are accurately aligned in the vertical direction such that the light falling on any given row of pixels is the same row for all three. This maintains the colour registration in a similar way to a colour TV tube.

If digital zoom, digital imager stabilisation, and photo still are called for, then the vertical resolution is required to be higher and a two-field interlaced picture is developed from a single frame. To achieve this the timing pulses of the green imager (V1, V2, V3 and V4) are shifted by half a line and the green output is interleaved between the red and blue lines. This effectively increases the number of TV lines in each field Y and Y', as shown in Figure 10.43.

Y1 is the sum of: (RB1 + RB2)/2 + G1
Y'2 is the sum of: (G1 + G2)/2 + RB2
Y2 is the sum of: (RB2 + RB3)/2 + G2
Y'2 is the sum of: (G2 + G3)/2 + RB3
Y3 is the sum of: (RB3 + RB4)/2 + G3

and so on. Therefore, one field is made up of lines consisting of (RBx + RBy)/2 + Gx and the other field is made up of (Gx + Gy)/2 + RBy.

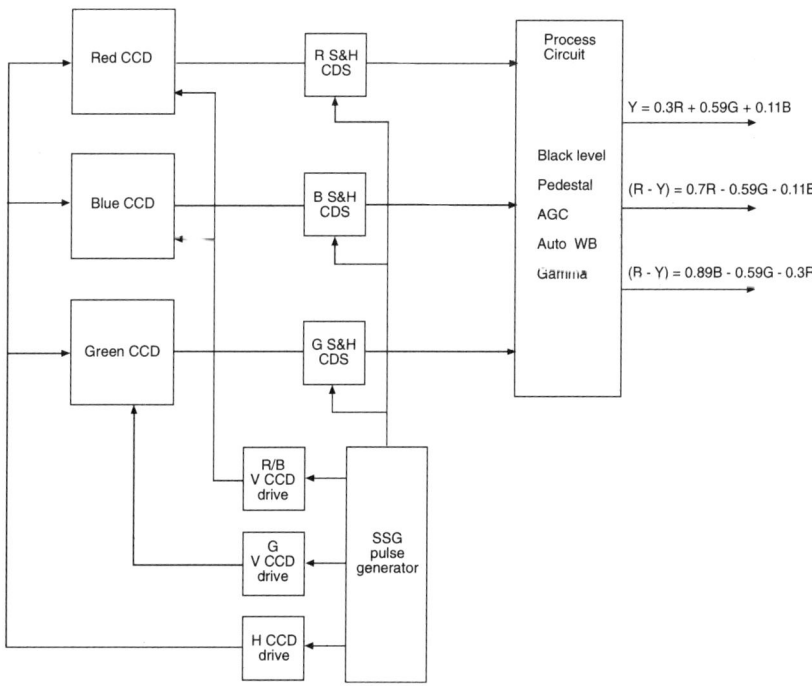

Figure 10.44 *Triple CCD imager timing and signal processing*

Analogue / digital camera processing

So far the luminance and chrominance camera signal processing has been analogue. By carrying out camera signal processing in the digital domain addition features can be incorporated that are not possible as analogue signals. Examples of these are: digital zoom, image stabilisation, and a range of effects such as wipe, mix, tints and solarisation; also the operation of other camera control systems benefit from digital signal processing. Figure 10.45 outlines the basic signal processing for a digital camera head within an analogue camcorder.

Colour separation is similar to that previously described and the mathematical calculations incorporated in this process are more suited for digital processing. A small difference is that the colour components stay in their complementary colours of yellow, magenta and cyan through most of the colour processing.

For white balance purposes the components are further processed to YL (Y signal at the lower chroma resolution), R/G and B/G as the white balance reference information is stored in the EEPROM in that format, see Figure 10.46.

Colour as seen by the human eye is as much interpretation as measurement. For example when comparing red and yellow paper, it is felt that yellow is brighter than red even though measurement may prove different.

Interpretation of colour can be in one of three ways: hue, saturation, and brightness.

Hue refers to the colour, for example red, green, blue, magenta etc. Saturation is the intensity of the colour: red may just be red or a vivid red; it is the level of the colour. Brightness is more subjective, two different colours will have different hues and the same saturation but one may look brighter than the other. A further aspect is that the human brain/eye combination can interpret colours from a source of heat and can also compensate for such changes. For example: if a lady is wearing a red coat and the weather is overcast, it looks red; if the sun comes out the coat does not change colour from a human perception point of view but a camera will 'see' a very different shade of red, tending towards magenta. This is due to the colour temperature changing, and the inclusion of infra-red content, and the addition of blue light from the sky.

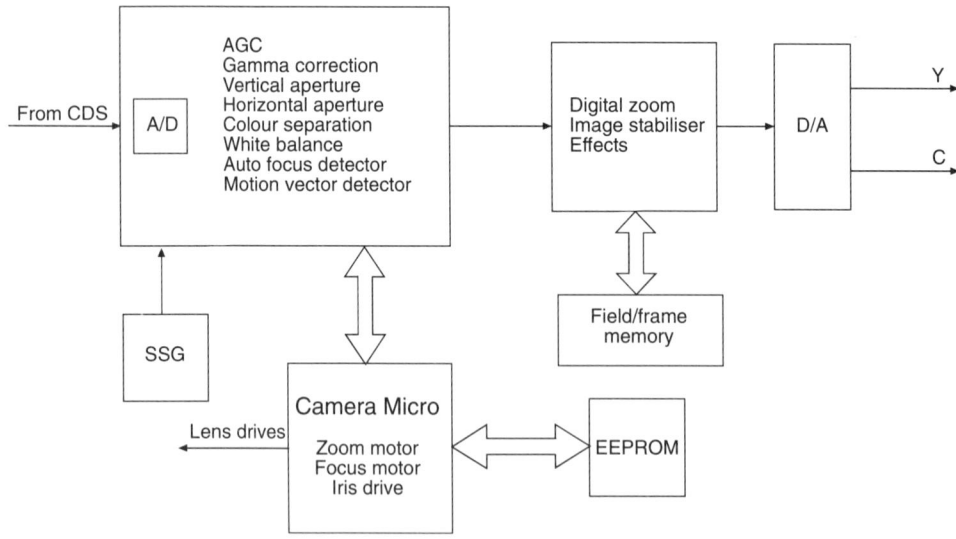

Figure 10.45 *Digital camera head signal processing*

Camcorders

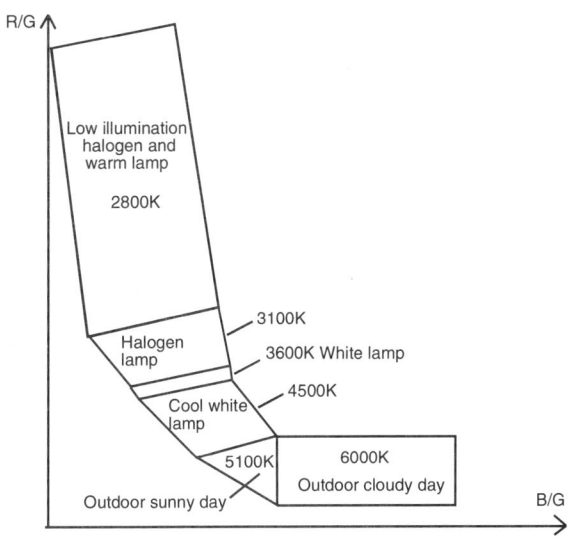

Figure 10.46 *White balance map in EEPROM*

Figure 10.46 is the colour temperature map in the EEPROM and Figure 10.47 is a chart of colour temperatures for various lighting conditions.

There are more aspects that affect the colour as seen by a camcorder. For example, the amount of light or brightness diminishes as the camera zooms in to telephoto, as it does when the auto iris level changes. All of these variables combine to make it as difficult as possible for a camera head to produce natural colours in all conditions. As seen in Figure 10.47 there are lower and upper limits between 2800 K and 6000 K wherein the auto white balance cannot function and manual control must be used.

Figure 10.48 is a chart of brightness levels measured in luminous intensity (lux). Between 10 and 250 lux additional lighting is required. At these levels both autofocus and white balance are unstable and may well be switched off. Below 10 lux colour may also be switched off to reduce chroma noise.

At high lux levels, in bright sunlight with blue skies, neutral density filters are necessary to reduce light levels to the lens to avoid white balance problems. In such conditions auto white balance is not stable and manual white balance must be used to ensure natural colours in the scenes during filming.

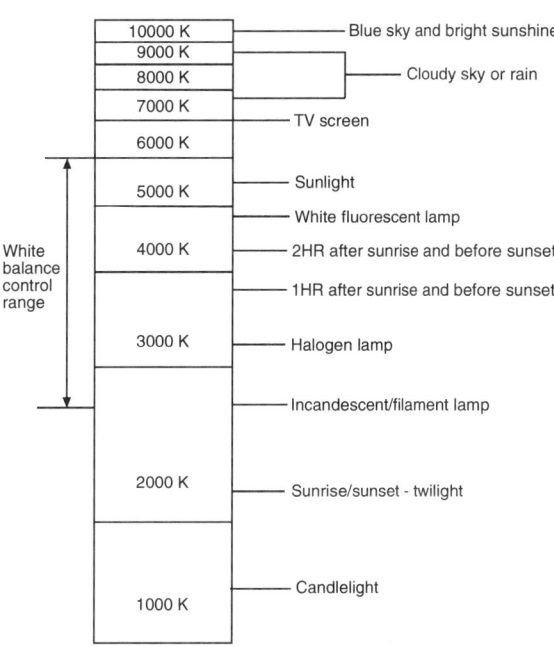

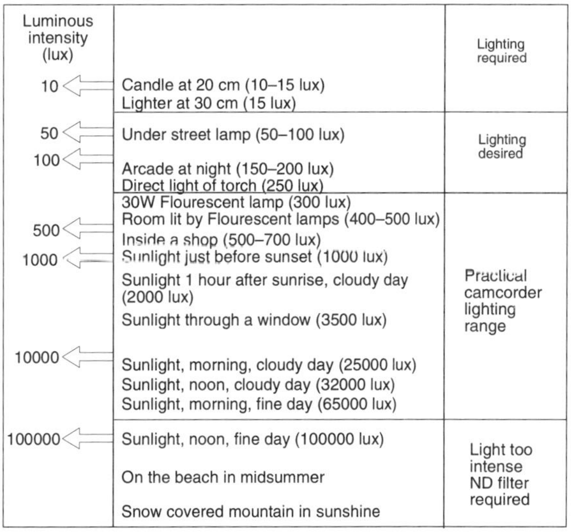

Figure 10.47 *Lighting conditions and respective colour temperatures.*

Figure 10.48 *Lighting conditions and respective Luminous intensity*

209

Video and Camcorder Servicing and Technology

Auto white balance

In general almost any colour can be made up from various quantities of red, blue and green, including of course white. By measuring the levels of red and green against blue and green from the imager, a section of the white balance map is selected from the EEPROM memory to set the parameters. The camera microcomputer also takes into account the zoom position, iris position and, if fitted, an infra-red sensor and/or white balance sensor. Then the white balance control comparator sets the (R - Y) and (B – Y) levels such that the output from the white balance amplifier is zero for each. It is then judged that the white balance for that scene is correct.

Setting up the white balance circuits is very specialised and time consuming. A 3200 °K light source is used to set the indoor levels to zero on a white surface. By using a blue filter, C10 or C12, outdoor conditions can be simulated and the outdoor white balance set. In between these are a multiple of various options for cloudy day, fluorescent incandescent lights, and many more. These are calculated by the PC software and set in the camera, if you are lucky that is!

Image stabiliser

It may be called the digital image stabiliser (DIS) or electronic image stabiliser (EIS) or indeed the super image stabiliser (SIS). Whatever the system is called there are two ways in which it can be achieved without resorting to mechanical means.

One way is to write the scene into a field memory and read it out with modified timing pulses according to motion vector calculations. The other is to make use of the CCD imager scanning timing pulses and modify these. Using the CCD is not very satisfactory for horizontal movement, as the photodiodes can introduce delays distorting the picture.

For the field memory image stabiliser, the image coming from the CCD is written into a memory, the size of which is larger than that of the image in both horizontal and vertical directions. It is an image within an area, called a partition, see Figure 10.49.

Any change in the framing of the image in either direction due to camera shake causes an output from the motion vector detector, as before, and after images are compared for motion vector calculations. An output from the motion vector

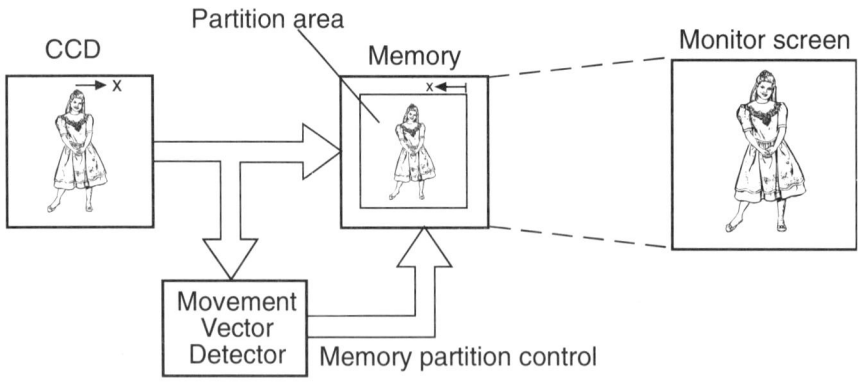

Figure 10.49 *Image stabiliser using memory partition*

detector then adjusts the position of the image partition within the allocated area. If, for example, an object within the scene moves to the right by 'x' amount, the partition is moved to the left by the same amount; the target object in the scene effectively remains in the same position on the monitor screen. This works in any direction as long as the partition remains within the memory area, if not then the target object moves.

Partition area CCD

In this example the CCD itself has a partition area, which is the size of the image area produced by the lens situated within an active scanning area that is larger, as shown in Figure 10.50. As described above, when the motion detector is active to correct for camera shake then the scanning pulse timing of the CCD is altered to move the partition to compensate. This method is cost effective as it does not require an extra memory, but it does result in image impairment in the horizontal direction and the picture size varies with the image stabiliser on or off.

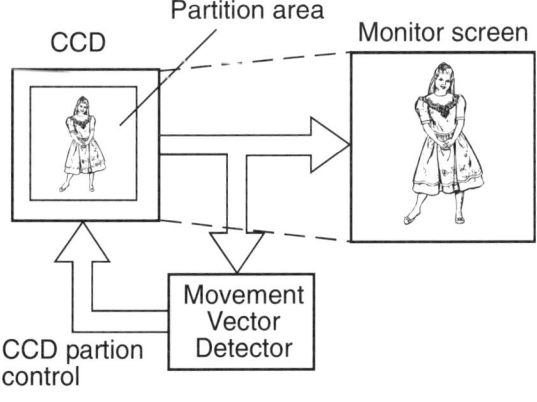

Figure 10.50 *Image stabiliser with CCD partition*

CCD and memory image stabiliser

A third type of image stabiliser has a high density CCD with a partition and a field memory. Scene displacement in the vertical direction is compensated for in the CCD and horizontal displacement in a field memory. Digital zoom is incorporated into this system and use the same memory so both cannot be used at the same time. Digital zoom is only possible when the image stabiliser is turned off.

High grade image stabiliser

In addition to the CCD having a partition there is a memory IC present, also with a partition area. This overcomes the horizontal distortion caused by a CCD partition; see Figure 10.51.

The active area of the CCD is 726 lines by 858 pixels and the partition is 576 lines in height. It is active in the vertical direction only, having a margin of ± 75 lines. For the memory, IC204, the active area is 858 pixels wide and the partition 720 pixels giving a working margin of ± 79 pixels. This results in a total image area of 576 lines by 720 pixels for a digital camcorder although it may be slightly less horizontal resolution (670 pixels) for an analogue model. IC102 processes the analogue CCD output for correlated double sampling, black level, AGC clamping etc., before the signal is converted to 8-bit digital format in IC201. IC202 controls the digital zoom function in conjunction with the field memory. For the purposes of image stabilisation, the microcomputer IC203 determines the vector motion in both the vertical and horizontal directions, sending vertical partition control to the CCD and horizontal partition control to IC204. IC202 also controls the flow of signal data from the CCD and to and from the memory, outputting the result to IC205, the D/A converter. There is no change in picture size when the stabiliser is switched on, nor is there any conflict with the digital zoom function; both can be used simultaneously.

Video and Camcorder Servicing and Technology

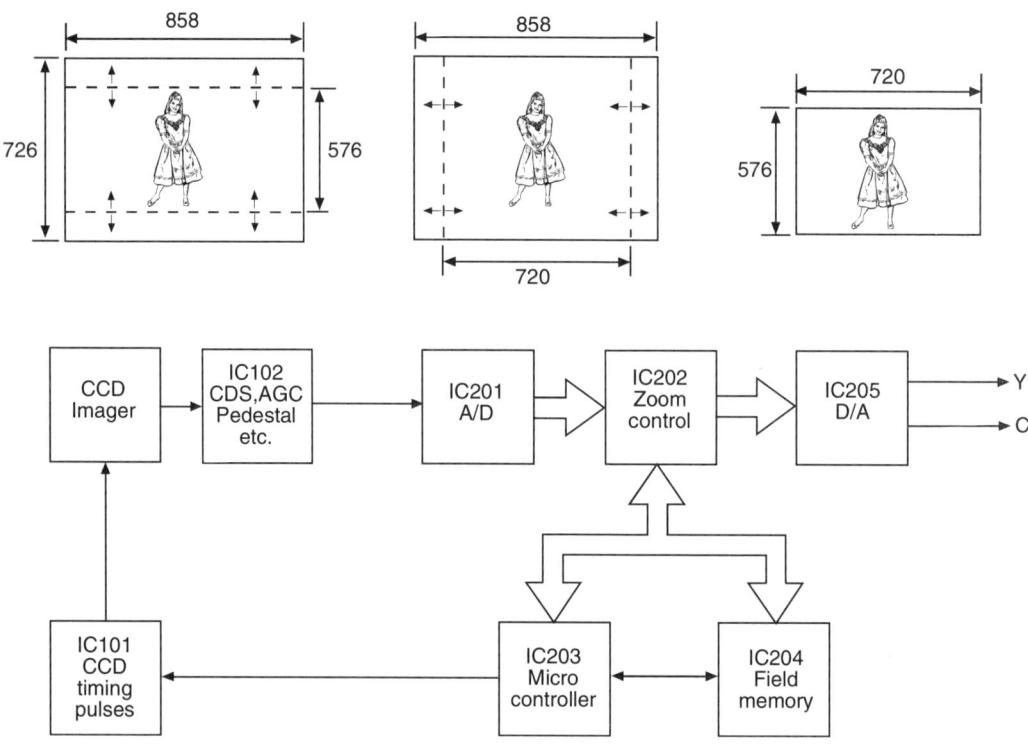

Figure 10.51 *Image stabiliser with CCD and memory IC paritions*

Digital zoom

All camcorders have the optical zoom provided by the lens. In addition to this a digital zoom function may be available and functions to 'magnify' the picture stored in the field memory rather than optical 'zooming'. Zooming is accomplished by taking a centre section of the scene stored in the memory chip and expanding it to fill the video output screen by gradually reducing the area that is expanded, as shown in Figure 10.52.

Obviously when a small section of the picture is enlarged both the line and pixel element structures become obvious and steps are taken to reduce this effect.

Interpolation and inter-compensation

Both of the terms interpolation and inter-compensation refer to the same process. Where the zoom ratio is large, extra lines and extra pixels are added by mixing adjacent components to form additional lines or pixels to increase the apparent vertical and horizontal resolution.

The digital zoom magnifies the picture by reading out the central area of the scene stored in memory. As the zoom increases the number of lines decrease and the picture may take on a 'venetian blind' effect. To fill the gaps between lines and make the picture smoother extra lines are added and these are constructed from adjacent lines. Similarly, in the horizontal direction extra

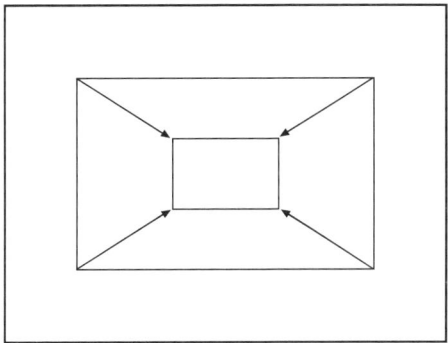

Figure 10.52 *Memory IC zoom area*

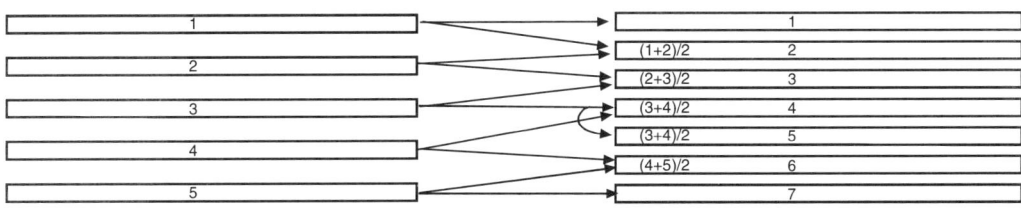

Figure 10.53 *Creating 7 lines from 5*

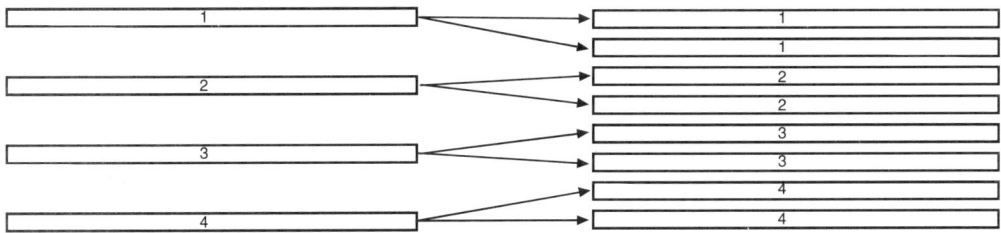

Figure 10.54 *50% zoom and all lines are duplicated*

pixels are constructed from adjacent pixels.

Referring to Figure 10.53 we can see how 7 lines can be created from 5. New line 1 is the same as line 1 and unaffected. Line 2 is constructed from lines 1 and 2 added together and then divided by 2 to equalise the levels. Adding lines 2 and 3 and dividing the result by 2 constructs new line 3. Other lines are constructed in the same way, excepting line 5 which is a duplicate of line 4, and line 7 which is unaffected.

This process of adding lines progresses as the zoom increases up to a point of 50% zoom (Figure 10.54), where all lines are duplicated. As the level of zoom increases the number of available lines decreases and then duplicated lines are interpolated to create extra lines in a similar method to that shown in Figure 10.53.

Pixels in the horizontal direction are treated in the same way, they are either duplicated or extra pixels are constructed by interpolation of adjacent ones.

Mixes and wipes

Both mixes and wipes can be added as special effects by using the scene stored in the field memory along with the current output from the CCD imager. After the user has filmed a scene rec/pause is selected and the last scene image is stored in the field memory. When record is again selected the stored scene appears and is gradually replaced by the new scene according to the special effect that has been chosen.

Digital strobe

This effect is produced by writing to the memory at extended intervals while reading out at normal field rate. This may be at a ratio of 20:1, where the scene is written only once in 20 fields missing the 19 in between and producing the strobing effect.

Digital gain up

It is possible to make the CCD imager more sensitive in low lighting levels by giving the photodiodes more time to charge up. This is achieved by reading out the CCD only once in eight fields, thus giving the CCD eight times longer to charge up in poor lighting conditions. Whiles this gives a good picture in low light it also gives rise to a slight strobe effect in scenes with rapid movement.

Negative function

This function is easy to obtain in the digital domain by inverting the data bit value. For example, a byte of 11110000 becomes 00001111.

Solarisation

The effect of solarisation is produced by dropping data bits from each byte and so reducing the number of luminance and chrominance levels. An 8-bit byte may become a 4-bit byte reducing the number of greyscale levels from 256 to 16. As there are now fewer levels the solarisation effect is evident where subtle changes in luminance and chrominance values are combined to a single value out of the D/A decoder.

Mechanical image stabiliser

There are various mechanical ways of compensating for camera shake; two examples are given here. One relies on piezoelectric velocity sensors and the other upon Hall effect gyroscopic measurement. Both devices sense movement in the vertical (pitch) and horizontal (yaw) directions.

A closed loop system is shown in Figure 10.55 for the variangle lens version. Each sensor produces a small voltage output when moved and the polarity of this voltage will change with

Camcorders

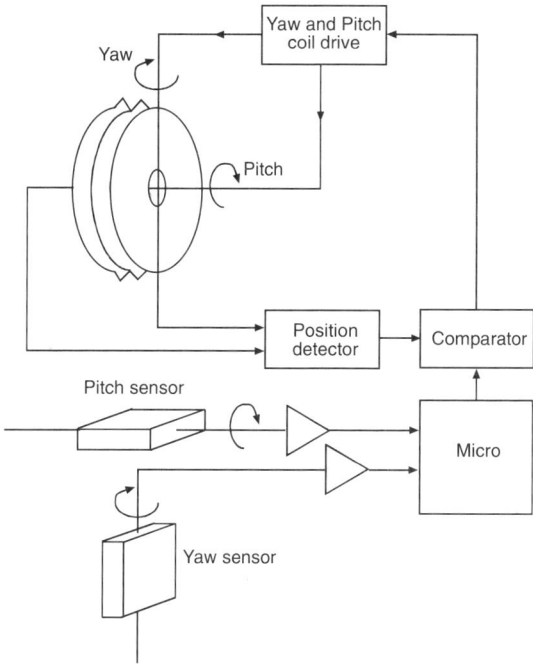

Figure 10.55 *Mechanical image stabiliser control system*

the direction of movement. It is amplified up to the logic level of a microcomputer that calculates the movement speed and vector before providing a correction output drive. In the example circuit, the drive is shown for the variangle yaw-and-pitch coil drive although it can be the lens drive in the second example.

Variangle lens stabiliser

The first compensation system relies on a new device known as the variangle lens that is constructed to operate as a prism; a cross-section is shown in Figure 10.56.

Two glass plates with reinforced metal outer rims are connected together by bellows. The space between is filled with a silicon oil that has a refractive index of 1.5, the same as glass. The variangle lens acts like a variable prism; by tilting the glass plates with respect to each other the optical path can be deflected. The variangle prism forms the compensation control element whilst

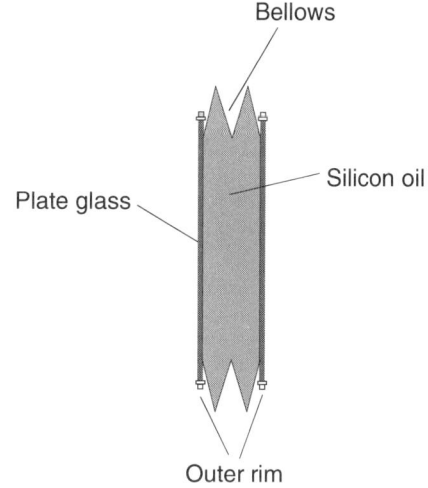

Figure 10.56 *Variangle prism construction*

215

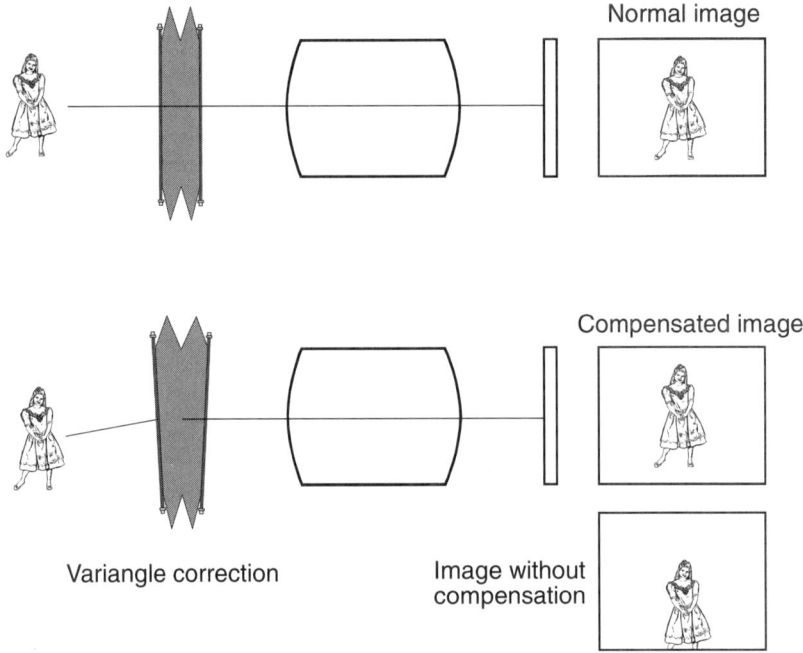

Figure 10.57 *Variangle lens operation*

two piezoelectric angular velocity sensors provide for the detection end.

A piezoelectric pitching sensor (pitch is vertical) and a yawing sensor (horizontal) detect the direction and amount of shake. The microcomputer calculates the result and drives two actuator coils to twist and tilt the variangle lens. The drive is in the opposite direction to the shake in order to compensate. For example, as in Figure 10.57, if the camera is slightly tilted upwards erroneously due to camera shake the wanted scene travels downward within the output picture area. The drive to the variangle lens then tilts the lens downwards in the opposite direction to compensate by bringing the target scene back up the picture. The amount of compensation is 1.5° for vertical and horizontal shake. It is more effective for high zoom values where shake is more noticeable.

Advantages for this system are that it is not dependent on sensing movement within the picture scene as gyroscopic physical movement is detected by inertia. The picture has full resolution as it is not artificially or electronically zoomed, and the full picture area is maintained.

Moving lens stabiliser

A second example of an image stabiliser uses a lens mounted upon a vertical slider unit. This in turn is mounted upon a horizontal slider to give the lens movement in both directions; see Figure 10.58. Each slider axis has a coil mounted at one end that is positioned within a 'U'-shaped magnet to a linear motor; current applied to the coil causes the lens unit to move up or down. A similar arrangement causes the whole assembly to move in the horizontal direction as the magnets are fixed to the case and the coils to the sliders.

Within the lens the image stabiliser is mounted between the zoom and focus lens assemblies and is independent of the zoom ratio or focus position. An image from the stabiliser is centred upon the CCD imager. In the case of camera shake this image can be moved in any direction to compensate. As previously shown a microcomputer receives inputs from the sensors, calculates the movement vector, and drives both linear motors to compensate for the image displacement.

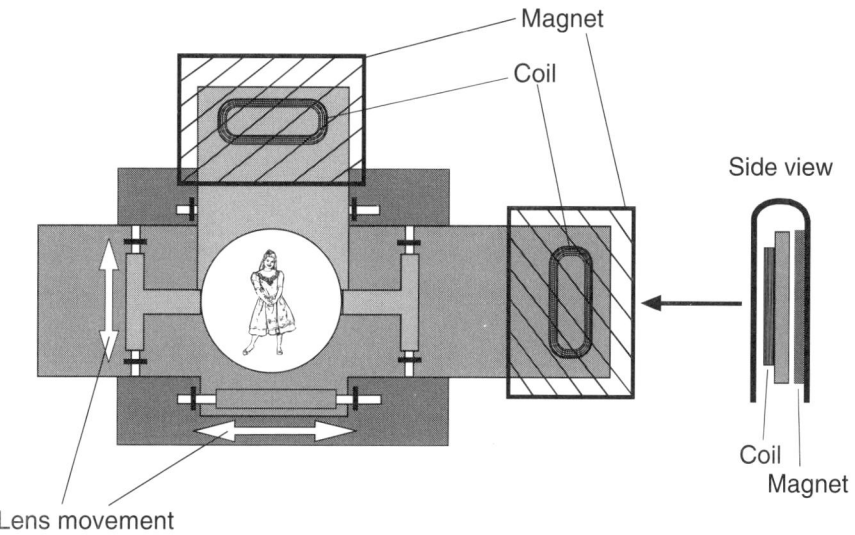
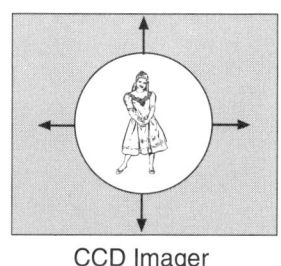
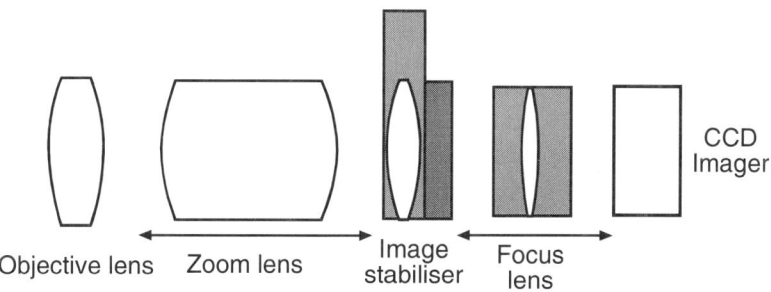

Figure 10.58 *Moving lens image stabiliser system. Stabiliser unit (top) with CCD imager faceplate (middle) and lens construction showing the position of the stabiliser (bottom)*

Automatic focusing

There are a number of quite different techniques for deriving an automatic focus (AF) system, ensuring that the image or subject is focused onto the saticon, or CCD, faceplate. Some are based on measuring the distance between the camera and the subject, others use the crispness of the image on the photo-sensitive pick-up tube or imager. There are advantages and disadvantages to all of the systems. A closed loop system identifying the maximum resolution or frequencies by calculation is the most effective.

Ultrasonic focusing

This is a very basic system using ultrasonic sound to measure the distance between the camera and the subject and uses techniques based upon marine ultrasonic detection. An ultrasonic beam of a frequency around 40 KHz is directed out from the camera to the target subject and bounces back to a receiver (Figure 10.59). A timer measures the time taken between transmission and reception of the beam in order to determine the distance of the subject and the focus is set accordingly.

It is not a closed loop servo system as the focus point is set indirectly by measurement, and initially the focusing system has to be calibrated, or set-up. If drift occurs between the measuring section and the focus drive then it cannot automatically compensate and has to be recalibrated.

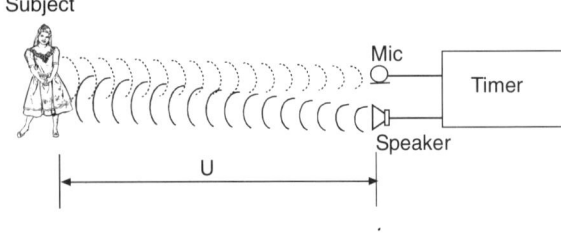

Figure 10.59 *Ultrasonic focusing system*

The main limitations include the fact that the ultrasonic system cannot work through a glass window, as the glass reflects the beam so that the camera focuses upon the glass and not the subject through it. Ultrasonic beams can bounce off adjacent objects and can be reflected off at an angle from inclined objects instead of back to the camera, making reliable focusing a problem on reflective surfaces.

Infra-red focusing

In an infra-red (IR) system the infra-red beam is emitted from a light-emitting diode and is introduced out of the lens system via a projecting lens mounted on top of, or adjacent to, the main lens. It travels out as a focused beam to the subject image. As the infra-red beam travels out of the camera it spreads, although the spread is not very wide, producing a circle about 80 mm diameter some 3 m from the camera. This is used as the focusing 'spot', hence the camera will focus upon any object the spot hits, providing that enough infra-red light is reflected.

As shown in Figure 10.60, the beam hits the subject and is picked up by a separate receiving lens, normally mounted beneath the main camera lens and focused upon a receiving IR photodiode. The distance 'W' is fixed between the IR projecting and receiving lenses so that U can be calculated by trigonometry. In practice the distance 'd' between the optical centres of the receiving lens and photodiode is used, as the photodiode is attached to a cam on the focus lens ring. When the focus ring rotates, driven by the motor, the AF control moves the photodiode assembly up and down increasing or reducing 'd', until maximum IR illumination of the photodiode is obtained. If the camera is off focus then the photodiode will receive less illumination, as the received beam will not be on the centre of the diode.

This autofocus system operates as a closed loop to maximise the infra-red beam onto the photodiode. Two photodiodes are actually used, to improve the control range and provide for more accurate focusing of wider lens settings by

Camcorders

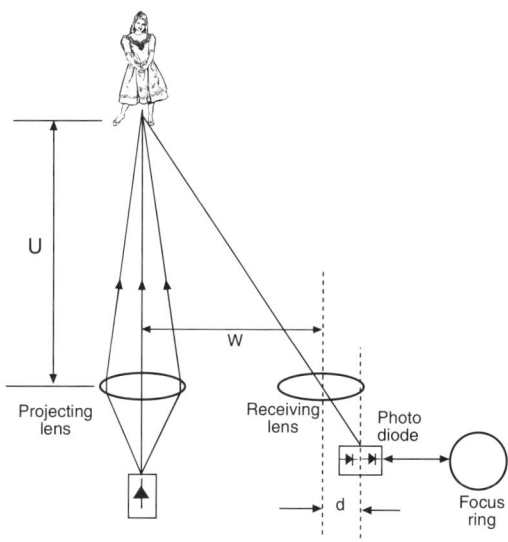

Figure 10.60. *Infra-red focussing system*

maintaining even illumination on both photodiodes. Operation is as follows: the infra-red beam is modulated and is switched off when not in use. As a check the beam is pulsed on and off at intervals. If focusing is incorrect then the received beam will not produce equal outputs from both photodiodes. The focus motor is operated to move the photodiodes to an equal illumination position. In doing so the focus ring is moved to obtain the correct focus.

The mechanical tie-up between the rotating focus ring and lateral movement of the diode receiver is critical for linear operation and must be correctly set up. Setting up is by adjusting the distance 'd' for best focus at three metres.

Limitations are similar to ultrasonic focusing, in that there are problems working through glass, as the camera will focus upon the glass. The system also suffers from parallax and reflection problems but not to the same extent as ultrasonic methods.

Frequency detection

As the image of a subject is focused on the pick-up imager it becomes crisper and sharper, and the high frequency content of the electrical signal will increase. Autofocus adjustment (Figure 10.61) is used to maximise the high frequency content of the video signal and is therefore a closed loop system. One drawback is that moving subjects create spatial frequencies, which are higher than when stationary, and can cause the autofocus to hunt backwards and forwards for an elusive mid-point.

Through camera lens

Referred to as the TCL system, it comes in two forms, both which are used in a camcorder. The TCL system avoids parallax errors and allows for a greater autofocus range between telephoto and wide angle than other types, while not being affected by glass.

The type of system first used is shown in Figure 10.62, where a portion (30%) of available light input is deflected off via a half prism, mirror and autofocus lens to the TCL sensor. As the autofocus prism cuts down the available light it also reduces the sensitivity of both the camera and the autofocus system. Accuracy is maintained

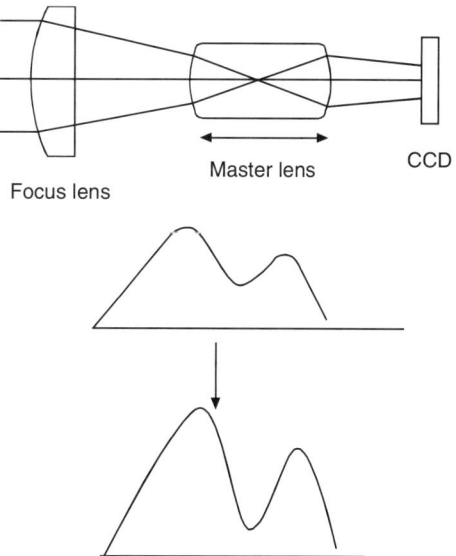

Figure 10.61. *Focusing by detecting high frequencies*

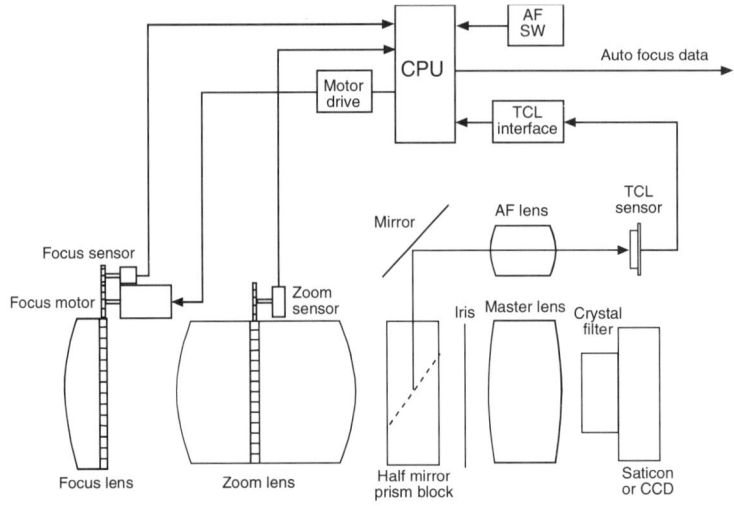

Figure 10.62 *TCL focusing with split prism*

by the fact that the light is split after the main lens, so that the TCL sensor obtains the same image as that on the imager.

A second version (Figure 10.63) has the autofocus lens attached to the main focus lens so that the two move in parallel. This allows for more light to reach the imager and the TCL sensor thus increasing the sensitivity of the camera, as well as allowing the autofocus to work at lower light levels. Accuracy in this case is maintained by tracking the autofocus lens and the focus lens mechanically.

In both types of autofocus lenses the special TCL sensor (a CCD detector) converts the light into an electrical signal sufficient for a microcomputer to calculate focusing errors and adjust to eliminate them. The principle is similar to that of the split prism used in photographic cameras using two derived images.

Figure 10.64 illustrates the split prism

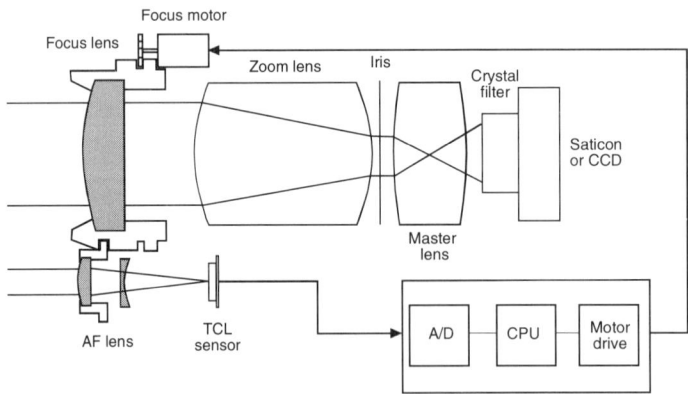

Figure 10.63 *TCL focussing by mechanical tracking*

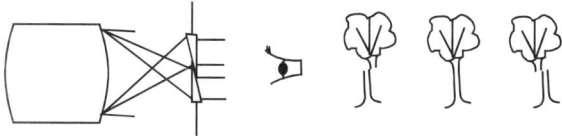

Figure 10.64 *Split image focus detection*

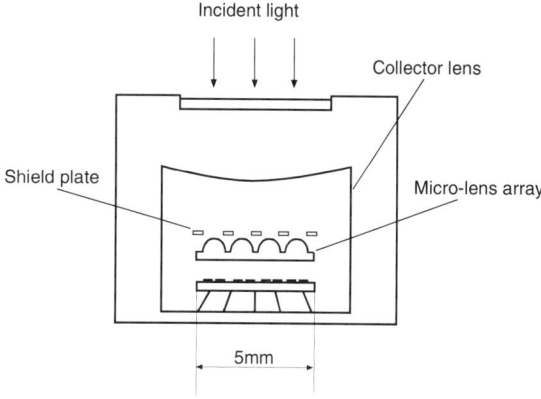

Figure 10.6 *TCL sensor assembly*

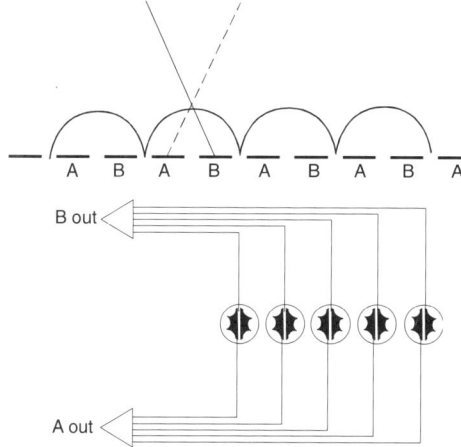

Figure 10.65 *Micro-lens and photodiode array*

photographic principle whereupon two images are produced within a central circle. They are displaced with respect to each other except when the lens system is in focus. A microcomputer can automatically gain the correct focus by sensing the displacement.

The two images are produced by two sets of photodiodes 'A' and 'B', mounted beneath a micro-lens array in a strip as illustrated in Figure 10.65. CCD registers are used to shift out the charges as two analogue sample signals. The complete CCD image sensor is mounted within an enclosed unit as shown in Figure 10.66, where incident light passes through a protective shield plate and collector lens onto a strip of micro-lenses covering the pairs of CCD photodiodes.

If the image is not focused correctly upon the micro-lens array then the two images A and B produced upon the CCD sensor will be displaced;

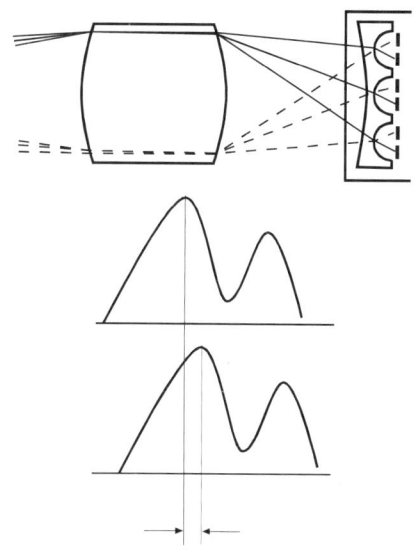

Figure 10.67 *TCL image shift when out of focus*

see Figure 10.67. Only when the lens is focused will equal amounts of light pass through any micro-lens and equally upon A and B photodiodes, and produce equal and simultaneous A and B outputs.

A more detailed illustration of the A and B images formed on the CCD sensor is shown in Figure 10.68. In Figure 10.68(b) the focal plane is formed on top of each micro-lens and so equal amounts of light fall upon the A and B

Video and Camcorder Servicing and Technology

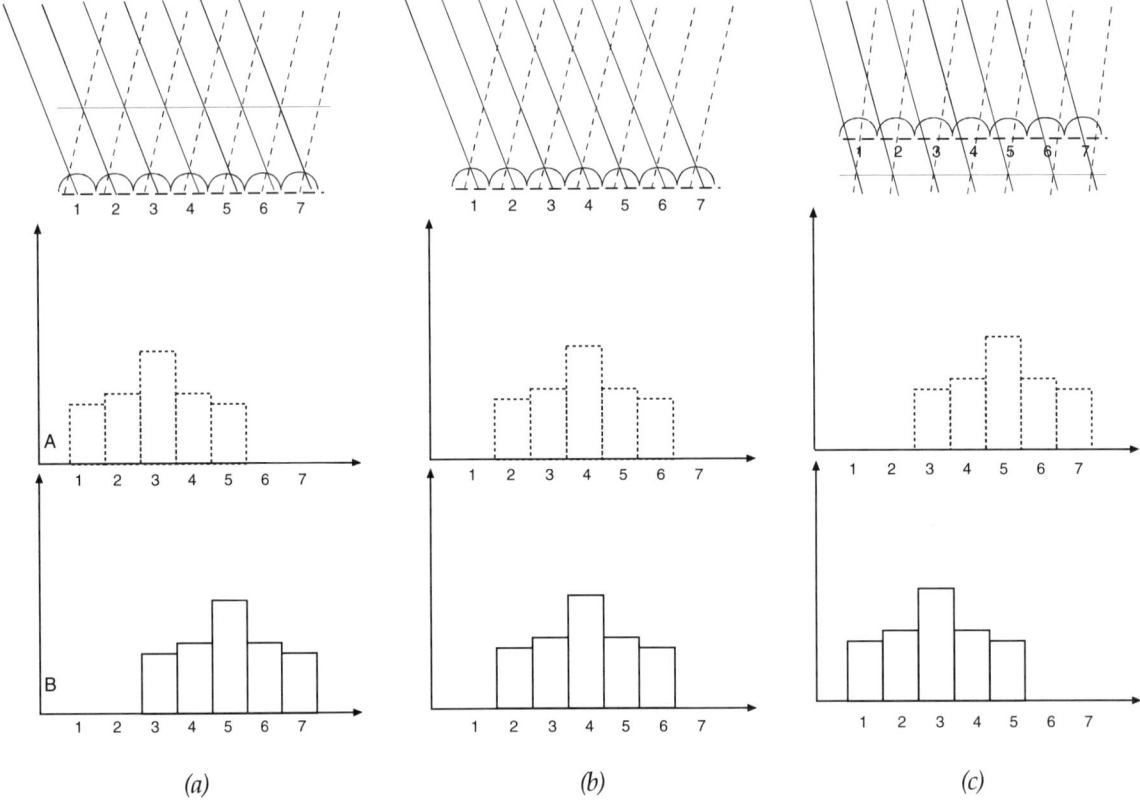

Figure 10.68. *Image shift when in and out of focus*

photodiodes. The two images are coincident and form a unity output.

In Figure 10.68(a) the lens system is in front of correct focus and the focal plane is well in front of the CCD chip. In cells 1 and 2 light only falls on the A photo diodes and so it puts the A image in front of the B image as the cells are shifted out by the CCD registers.

In Figure 10.68(c) the lens system is behind the focus and the focal plane falls to the rear of the CCD chip. This time, in cells 1 and 2 light only falls on the B photodiodes and the B image is consequently in front of the A image as the CCD elements are read out.

From this information the microcomputer can deduce front or rear focus and obtain the correct focus by coinciding the two images. Whilst this system works very well, limitations still exist. For example, the system cannot function in low light levels on plain surfaces or in conditions where the charge on the CCD cells is insufficient to allow for focus detection. Even if the subject is bright, such as white wall, there may be insufficient contrast to form an image and the A and B outputs will be a straight horizontal line Coincidence cannot, therefore, be confirmed and the autofocus system will shut off. Also an image with horizontal lines (say, venetian blinds) will cause the same problem as a white wall while an image with vertical lines (say, fence posts or trees) will confuse the coincidence determination by creating a number of A and B images the same, so that displacement cannot be determined.

Initially the sensor formed a sensing rectangle window in the centre of the picture area as used in some early camcorders. The later ones had a rectangular focus window mounted at 45°, which increased in size when zoomed in and reduced at wide angles; see Figure 10.69. This was achieved by the use of the moving AF lens, as illustrated in Figure 10.63. This improved the autofocus performance between wide angle and telephoto.

The autofocus system follows a microcomputer algorithm program.

1. The autofocus/manual switch is checked for auto position, and the AF system switched on.
2. Start CCD photodiodes charging and monitor the charge, setting the average charge to the centre of the A/D converter range.
3. Read out A and B images from CCD cells and convert to digital information.
4. Compare image deviations: if low contrast go back to start.
5. Convert results from split image deviation into focus deviation, i.e. distance from TCL sensor to focal plane. If zero then exact focus has been obtained. Return to start.
6. Convert focus deviation into distance from focal plane to camera subject.
7. Calculate focus ring rotation.
8. Switch on focus drive motor and rotate focus ring by calculated amount. Return to start and switch off motor.

Calculation of the amount of focus ring rotation required, and subsequent rotation is confirmed by pulse issued by the focus sensor pulse generator attached to the focus ring, or focus motor.

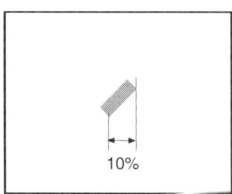

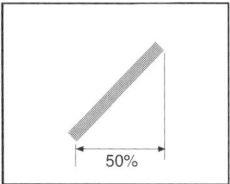

Figure 10.69 *Slanted autofocus window. The upper one is 10% when at wide angle, rising to 50% zoomed in to telephoto*

Frequency detection autofocus system

The full autofocus system is shown in Figure 10.70. Note that the component arrangement in the lens has been altered; the zoom assembly is now at the front and the focus assembly is at the rear of the lens. This type of lens is much smaller and lighter and designed for compact camcorders. Focusing is much quicker as the focus elements have a much shorter travel. With the advantages of a smaller lens with faster autofocus, low power and low noise, there has to be a downside and this is that optical zoom/focus tracking is no longer possible.

Larger lenses with the focus at the front and zoom at the rear had optical zoom/focus tracking. When an object was zoomed into to telephoto and the focus manually adjusted it would remain in focus as the lens was zoomed out to wide angle and back to telephoto. Now the tracking is not part of the lens construction and has a software solution, i.e. a zoom focus tracking curve held in a memory, and this is explained later.

Lens construction

At the front of the lens, just behind the front lens element there is a small optical zoom sensor. This is used to set up the zoom assembly reference position during initialisation. Zoom drive is a small electric motor with a long helical gear drive shaft, on to which the zoom assembly slots. As the motor turns clockwise or anticlockwise then the zoom lens moves backwards or forwards.

The focus motor is very different. It is a linear motor sitting on two parallel chrome runners, the component parts in Figure 10.71. On one side of the focus lens assembly is mounted a rectangular coil and on the other is a small long bar magnet. This bar magnet has many N/S segments impregnated upon it to generate a signal in the MR resistor matrix as the lens moves backwards or forwards. When assembled, the coil surrounds a U-shaped closed loop magnet. This is the linear drive to move the lens backwards and forwards

Video and Camcorder Servicing and Technology

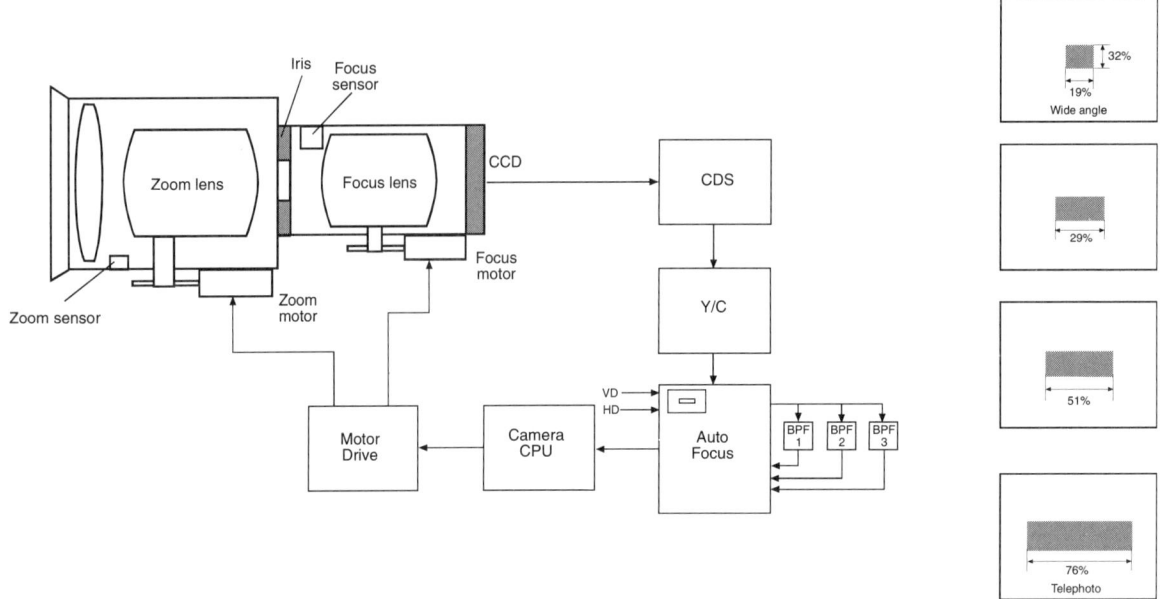

Figure 10.70 *Auto focus system*

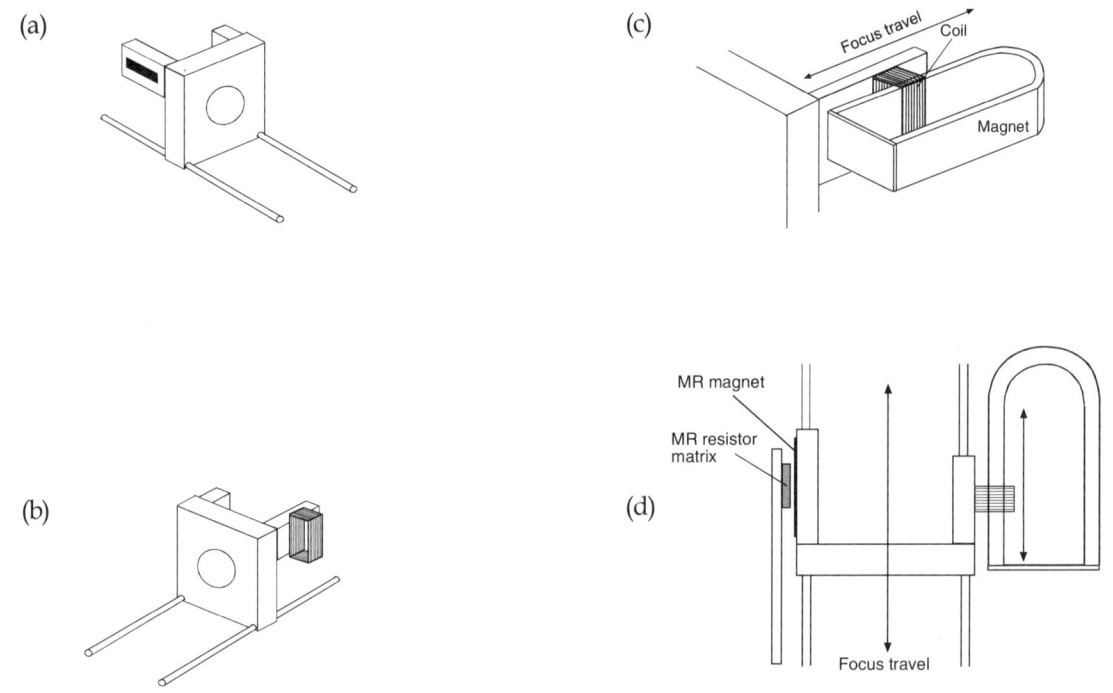

Figure 10.71 *Linear motor lens unit showing (a) MR bar magnet, and (b) drive coil. (c) View of linear motor coil mounted upon the fixed 'U' magnet. (d) Top view of the focus linear motor fully assembled.*

Camcorders

as the current flow changes direction in the coil.

Beween wide angle and telephoto the autofocus sensing area is enlarged. Most users are unaware that the autofocus only looks within a central area for the highest frequencies. This area is constructed by vertical and horizontal drive pulses (HD, VD) and counters within the autofocus IC.

MR sensor

Magnet resistors are used as the sensing elements in an the MR sensor. When a magnetic field passes through a magnetic resistor its resistance changes up or down according to the polarity of the field. This is converted to an electrical signal by applying a bias current through the resistor as part of a potential divider. By passing segmented magnetic fields across an array of magnetic resistors the current flow is modified and a small sinusoidal signal component is induced upon the bias current. A voltage component can be measured across a load resistor which forms part of the potential divider network.

As shown in Figure 10.72, the array is comprised of magnetic resistor elements A and B spaced apart by divisions of 200, 100, and 50 µm. Note also that the current flow is in opposition on each pair in series due to the way that they are wired. As the bar magnet has been magnetised with approximately 30 N/S segments, each 200 µm long, a sinusoidal waveform is produced at the MRA and MRB outputs when the magnet passes by the MR array. Due to the physical arrangement of the magnetic resistor elements these two waveforms are 90° apart and this allows the CPU to calculate direction and speed of the

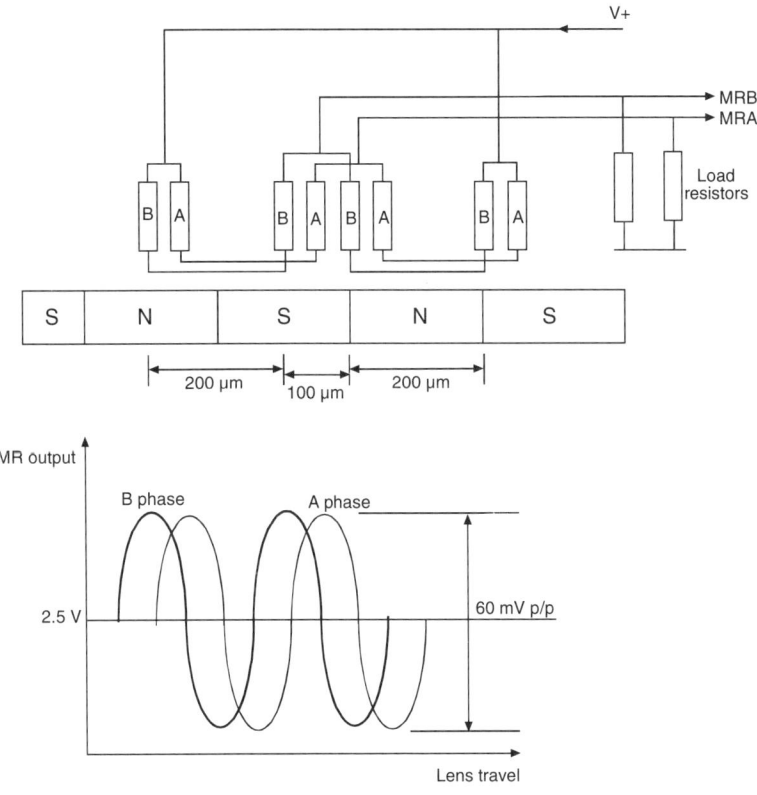

Figure 10.72 *Magnetic resistor array with the MRA and MRB waveforms*

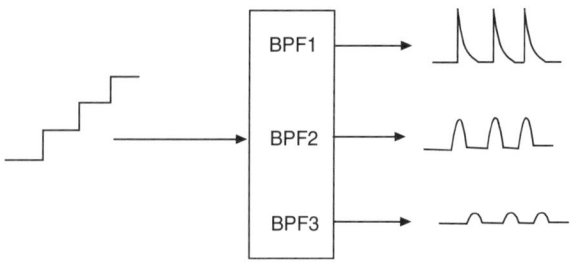

Figure 10.73 *Three-way bandpass filter*

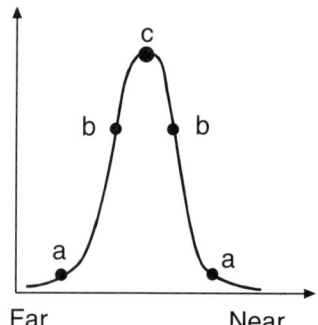

Figure 10.74 *Bandpass filter focus tracking curve.*

passing magnetic bar. Within the CPU pulses are used to determine the focus lens position and travel by counting the segments of the bar magnet from a fixed reset point during initialisation. The MR feedback also provides damping of the lens movement to stabilise operation.

Initialisation

When the camcorder is first powered up, the position of both the zoom and focus lenses are required to be measured.

Motor drive is applied to the zoom lens until its most forward position is detected by an optical sensor. During this period the drive clock pulses are counted, then the lens is reversed back to its original position using this count, and its position is logged.

Drive is also applied to the focus lens and it is reset then the CPU looks to achieve focus by maximum frequency, counting MRA/MRB pulses from the reset point. During initialisation the camera output is muted and may remain so until both auto focus and white balance are set. If either cannot be set it may be several seconds before the picture appears. This is a useful diagnostic aid in indicating a potential problem.

Autofocus detection

An output from the CCD is sent to the autofocus detect circuit. This may be an analogue or digital signal; the following is in analogue format. By the use of two or three band-pass filters the focus quality can be measured. If the lens is well out of focus then there will be no output from BPF1, very little from BPF2, and at least some from BPF3; see Figure 10.73.

If the focus is at either point (a) in Figure 10.74, then the autofocus system will need to search for an output from BPF2 at point (b) to know that it is on the right track. Once the point (c) is measured as the maximum output the focus will stop, but it will travel from (b) up to (c) and over the top of the curve before reversing back to point (c) to test for maximum output from BPF3.

Zoom/focus tracking curve

In principle the operation of a optical zoom/focus tracking lens is straight forward; zoom into an object in the distance, adjust the focus and then zoom out again to wide angle, while the target object stays in focus. The focus lens moves quite a lot in such a lens and it makes autofocus

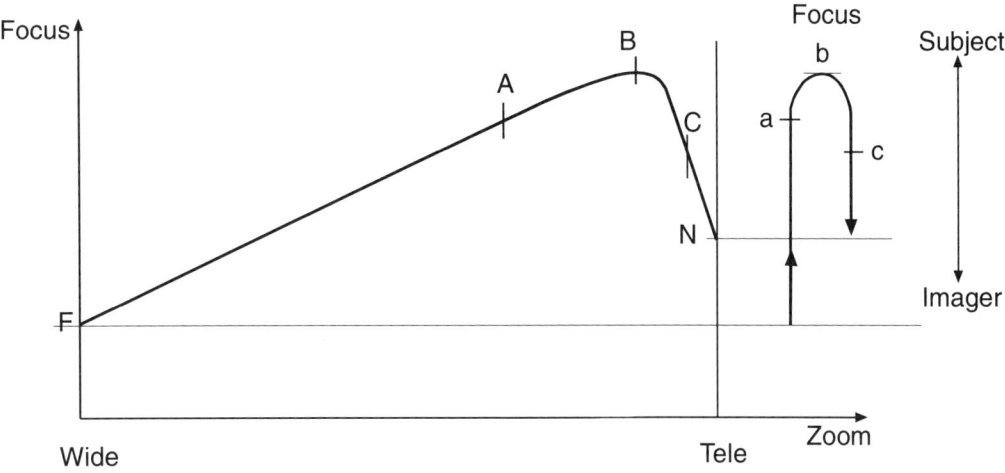

Figure 10.75 *Zoom/focus tracking curve as applied to the focus lens from zoom positional information*

quite sluggish. As camcorders got smaller so did the lenses and one way of making the autofocus operation quicker was to swap the positions of the zoom and focus elements of a lens over, so the focus was moved to the rear where the lens travel is much less. Now optical zoom focus tracking, once the proud boast of a good lens manufacturer, is not possible.

To overcome the problem of zoom/focus tracking the lens is fitted with either zoom and focus position sensors or, after initialisation, positional pulse counting identifies the lens element position. Whichever way is used the position of the zoom lens has to be monitored so that the focus lens can be set according to the tracking curve held in the auto focus RAM. The curve is shown in Figure 10.75.

As the zoom lens moves from wide angle (point F) to point A, in a linear section of the curve, the focus lens is moved forward, Both track up to points B and (b), where the focus lens reverses, and from B to N on the curve, the focus lens moves backwards towards the imager. As this is without the benefit of the autofocus, tracking has to be checked in manual focus by the same method as checking and optical tracking lens; i.e. zoom into a test chart at least 20 ft away or use a collimator lens, set the focus and slowly zoom out to wide angle. The target and it surroundings should stay in focus, if not, then the tracking curve needs to be set up. Although autofocus can cover up small deviations from the tracking curve it does make the autofocus erratic and slow to respond.

Early lenses were difficult to set up to get good zoom/focus tracking, requiring a lot of time and patience setting several software values and physically adjusting a focus encoder position. Other versions were semi-auto set up where the software produced a sample zoom/focus curve and tracked the lens's curve against it. Just a few points on the curve were set to match the lens to the sample, and this was not so difficult.

Later versions were set up by the software in the PC, so that one could sit sipping a cup of coffee while the lens was set up.

Auto iris control

A camcorder iris has to be able to regulate the amount of light falling upon the imager in all possible circumstances. The artificial intelligence controlling the auto iris cannot pick and choose to 'see' light or dark areas in a scene in the same way as the human eye. Where the user is shooting outdoors, care has to be taken to keep sky out of the top part of the picture or the iris will close down too much and objects on the ground will be too dark. In the most sophisticated auto iris systems the scene is split into segments, between 8 and 16, and the amount of light falling in each segment is measured with the result being the average amount of luminance level. Each area is judged by fuzzy logic, giving consideration to some segments above others. For example, less consideration is given to the top row of segmented areas if the light level is high as it is judged to be sky and so the iris will be slow to respond.

Figure 10.76 is a diagram of a typical auto iris system. A signal from the CCD imager is integrated in IC1, the auto iris central processor, or CPU. It is within this that the image is divided up into 16 segments and the scene judged form this. A DC value is sent to IC2 and damping feedback from the iris is added.

An iris motor is constructed as a moving coil in a similar way to that of an analogue test meter and has a bias spring to close the iris when power is removed. It requires damping to prevent rapid variations in iris response to changes that may cause the image to flicker. Auto iris level is set by software and stored in the EEPROM. The value is summed to IC2 along with the damping feedback. Also attached to this point is an iris open/close function for test purposes.

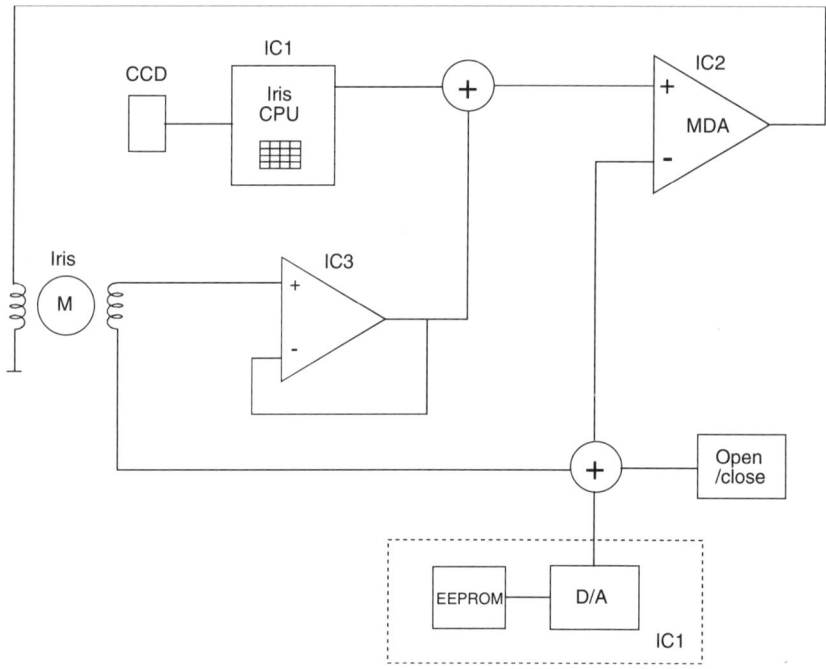

Figure 10.76 *Auto iris control system*

Auto white balance

There are five data sources for auto white balance deductions in order to determine the light source and colour temperature.

- Iris. The amount of light or intensity will determine the iris position; open due to low light – indoor, closed down due to bright light – outdoor.
- IR sensor. If infra-red is present then there is heat, the source is either sunlight or incandescent; otherwise it is fluorescent.

If a 50 Hz flicker is present the source is electrical, incandescent or fluorescent; otherwise sunlight.

- R/G and B/G colour difference.
- Zoom position. Colours in a scene are easily affected at full zoom, or telephoto; the response time of the auto white balance is extended and it is slower to respond to colour changes.
- Low contrast. This judges that the scene is of a single colour and again the auto white balance response time is extended.

The image from the CCD is split up into a number of segmented areas, either 8, 12 or 16 depending upon the model, and the (R-Y)/(B-Y) component for each area is analysed. In most cases the segment area where (R-Y)/(B-Y) are the least is judged as white, and a closed loop is formed with the colour gain amplifiers to reduce the values of (R-Y) and (B-Y) to zero. There is a limit to this function as there is a set reference level for each component above which it is judged that no white component exists in the scene and the closed loop white balance operation is inhibited.

Auto white operation

Figure 10.77 is a sample flow diagram for the auto white balance light source measurement to determine the prevailing lighting conditions.

First the iris is checked; closed indicates a strong light source, possibly sunlight. Then the infra-red sensor is checked for a heat source that will produce a DC level and if it has an AC component. Heat and AC means a halogen heat source, whereas heat and no AC indicates sunlight. Where there is no heat/DC level but there is an AC component, the light source is fluorescent. No heat level and no significant AC component suggests that there may be another light source and the colour map is checked against the RG/BG components to measure the colour temperature. Once the light source has been detected a decision is made as to when to check again, or whether to maintain the same colour correction values for a given number of fields. A field count starts ready for the next check, this could be as long as 5–10 seconds.

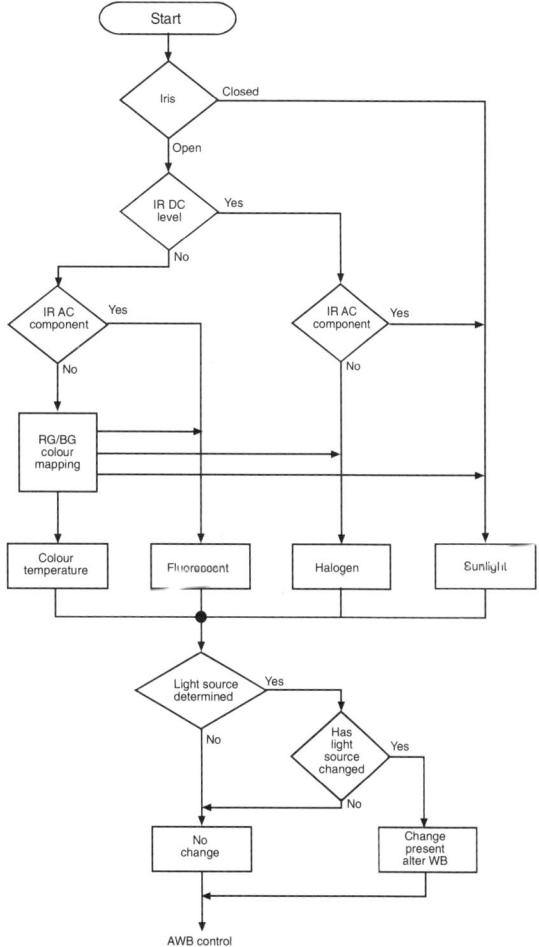

Figure 10.77 *Auto white balance flow chart*

11
Digital camcorders

The DVC tape ribbon is 6.35 mm wide and 7 μm thick and it is expected that the 60/90 min recording time will also be extended to 120 minutes in LP as thinner tapes are developed. Standard DV tape is metal evaporated (ME) as opposed to metal particle (MP). However, there is provision within the specification for metal particle tapes. Although is it of a much higher performance, metal particle tape is more abrasive than metal evaporated and drop out increases with continued usage. They are also less suitable for domestic use where the consumer expects to be able to re-use tapes many times.

A comparison between VHS and the DVC format is given in Figure 11.1.

Each tape cassette has four terminals at the rear right hand side for tape data and memory access. After the tape has been loaded the system control microcomputer within the VCR checks the cassette to see if a memory facility is present within the cassette; if there is then it is read. If no memory is found to be present then the terminals are read and then further measurement takes place to determine the type of tape used in terms of thickness, type (ME or MP) and grade (consumer or pro). This is called the BCID or basic cassette identification system.

Full details of the information contained within the cassette memory are not yet available. However, in the terms of pre-recorded material then it would be the titles, content and running times in a similar manner to the CD audio player and basic cassette information. Very few domestic camcorders will read/write to the IC memory.

The cassette construction

The DV cassette uses metal evaporated tape in order to sustain the high output and stability required for digital recording. Four contact

Video Format Comparison

	VHS	**S-VHS**	**DVC**
Recording System	Analogue	Analogue	Digital
Resolution	230 lines	400 lines	500 lines
Colour bandwidth	0.5 MHz	0.5 MHz	1.5 MHz
Video s/n	43 dB	48 dB	54 dB
Jitter	Fair	Fair	Exellent
Audio	FM/Linear	FM/Linear	PCM
Tape storage	Fair	Fair	Excellent
Tape speed	23.39 mm/s and	11.695 mm/s (LP)	18.831 mm/s (12.55 mm/s LP)
Tape width	12.65 mm	12.65 mm	6.35 mm
Cylinder speed	1500 rpm and	2250 rpm (C-format)	9000 rpm
Cylinder size	62 mm and	41 mm (C-format)	21.7 mm
Head azimuth	± 6°	±6°	± 20°
Track pitch	49 μm and 25 μm (LP)		10 μm

Figure 11.1 *Systems comparison*

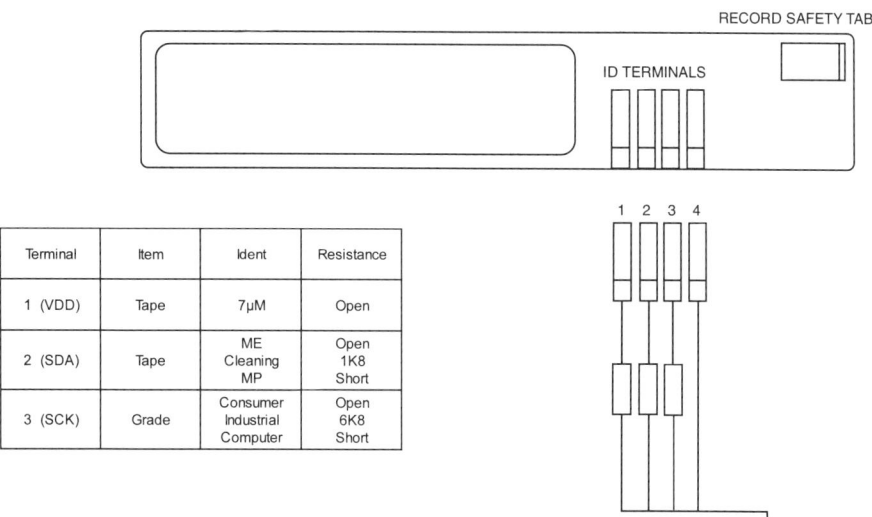

Figure 11.2 *Basic cassette identification system*

terminals are provided on the rear, and the resistance between these terminals and ground provides the information for the BCID to determine the tape thickness, type and cassette application. Where a more advanced cassette is used which has an internal memory IC, then the terminals change function. Pin 1 becomes a power line and pins 2 and 3 are for the I²C bus communications. When a cassette is inserted the systems control CPU first tries to read the BCID memory IC by the serial communication bus. If this fails then it reverts to reading the internal ID board resistance for basic data. When BCID is used the resistance to ground of the terminals 1,2 and 3 denote the characteristics as shown in the table in Figure 11.2.

Digitisation of a video signal

In order to digitise an analogue signal and maintain its integrity the sampling rate must be at least twice the maximum analogue signal frequency. For example, the CD audio sampling frequency is 44.1 kHz which is more than twice the 20 kHz maximum audio signal frequency. The reason for this is that the frequency spectrum of the sampling frequency (Fs) spreads from (Fs−Fmax) to (Fs+Fmax), where Fmax is the highest analogue frequency (see Figure 11.3). It is not good for the lower part of (Fs-Fmax) to mix with the upper part of the analogue signal spectrum, i.e. Fmax. If it does then beat harmonics are formed and they add to both frequency spectrums causing signal distortion and patterning. Therefore (Fs) must be greater than (2 x Fmax) in order to avoid negative fold back and aliasing in the lower part of the frequency spectrum. For a video signal with a bandwidth of DC to 6 MHz the sampling frequency must be greater than 12 MHz. A good choice is that of a frequency which is a factor of four times the colour sub-carrier frequency (4fsc), i.e. 13.5 MHz. An added advantage of this choice is that the digital sampling signal can be phase-locked to line syncs. This means that for any

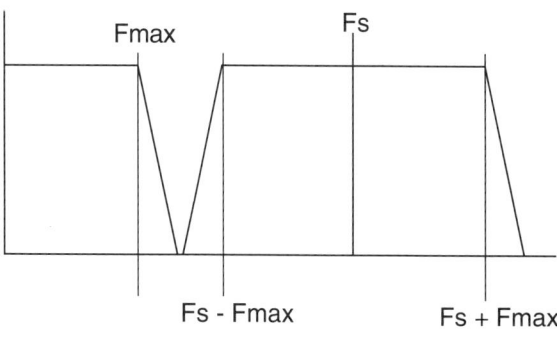

Figure 11.3 *A/D spectrum*

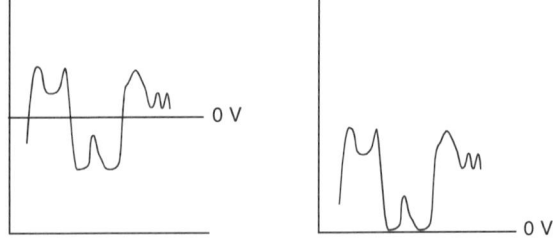

Figure 11.4 *Chroma clamping for A/D conversion*

given TV line the sampling takes place at regular intervals along the TV line and this symmetry is repeated for every successive line. A given frame is therefore made up of a symmetrical array of sample points in a fixed grid. In a camcorder this array is already present on the CCD imager as its pixel array. Phase-locked sampling is called orthogonal sampling.

Chroma signals are treated as two components, (R-Y) and (B-Y), and are sampled at half of the luminance signal at a sampling frequency of 6.75 MHz. Chroma signals are bi directional, being centred on 0 V and having both positive and negative values. To overcome this the A/D converter shifts the 0 V point to the most negative value by clamping, so that the signals are always positive and sampling is carried out from this new reference, see Figure 11.4.

When a video signal is digitised there are two important factors to consider. One is the sampling rate and the other is number of bits used in a sample binary word. In the two examples illustrated in Figure 11.5, there is a portion of analogue video signal. It is then sampled by 27 samples and each sample is given an 8-bit binary word to describe the amplitude of the sample.

After digital processing the signal is converted back to its analogue form in a D/A converter. The waveform is rebuilt using each sample. However, the level given by a sample word remains the same until the next sample describes a new level. Now the waveform is not as smooth as it originally was, but is made up of steps. This distortion of the signal is called quantisation noise.

Digitisation should ideally mirror the signal so that the resultant analogue signal looks as good as the original with no noticeable quantisation noise. This can be achieved by increasing the number of samples/s and by increasing the number of binary bits for each sample. As well as reducing the quantisation noise the resolution of the signal is improved and the resultant analogue signal is as good as the original. If only 2 bits are used then the signal can only have four quantisation or grey scale levels. Eight-bit words allow for 256 grey scale levels (2^8), and with the sampling rate of 13.5 MHz this is sufficient for

(a)
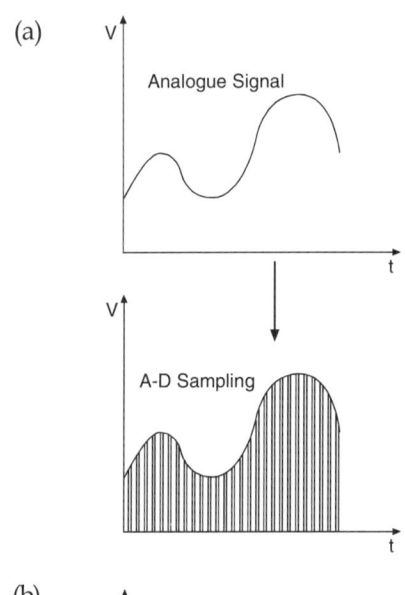

(b)
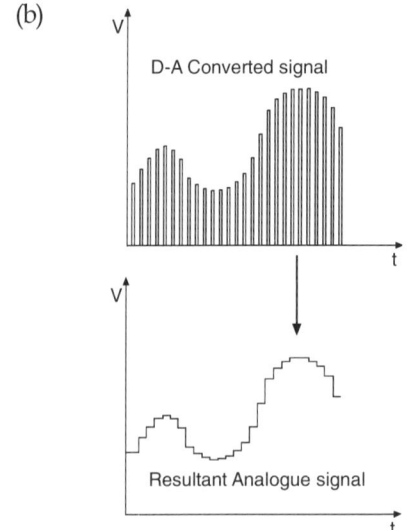

Figure 11.5 (a) *A/D conversion;* (b) *D/A conversion*

high quality video digitisation.

The full bandwidth of uncompressed video is about 240 Mbit/s. For domestic digital recording this is reduced to around 40 Mbit/s. MPEG2 reduces the bandwidth even further to 20 Mbit/s (high profile), 15 Mbit/s (main profile) or down to 4 Mbits/s (low profile). 15 Mbit/s is the bandwidth used for DTV and 9.8 Mbit/s maximum for DVD.

Digital camcorder block diagram

In Figure 11.6 video signals from the CCD imager are processed by the usual correlated double-sampling circuit to remove noise components, with black level clamping and blanking of the unwanted additional black areas of the imager. After this the signal is digitised at 9 or 10 bits, depending upon the manufacturer, before being converted to standard 8 bits for digital processing. Once in this digital format camera signal processing is carried out in a single IC, the digital signal processor (DSP). These processes include AGC, white balance, colour matrixing to (R-Y) and (B-Y) and Y levels, autofocus and zoom/focus tracking. Digital functions, such as image stabilisation, digital zoom and special effects, are also carried out in this IC by the use of the adjacent camera frame memory. As there is only one memory it is not possible to combine certain functions. Digital zoom and special effects both use the same memory IC and are not available at the same time; the customer can choose either.

All clock lines, sync and timing pulses are generated within the sync signal generator IC (SSG) using crystal oscillators which form interlinked phase-locked loops. These are used to time all signals from the CCD imager to the servo, deck functions, record and replay data, clock lines and data carried along the I²C bus linking all ICs.

Once processed in the camera section the digitised video image, at frame rate, is transferred to the shuffle memory. From this the E/E signals are passed on the D/A converter for monitoring. From the shuffle memory IC the digital recording video signals are shuffled to average data blocks and compressed by about 5:1. Digitised audio data is added at this point and both are stored in a second memory, the ECC memory (stands for error correction control). Error correction parity calculations are carried out at a high level and the parity checksums added. The data is then passed on to the DCI (digital control interface). Within the DCI integrated circuit, sub-code and tracking data are added to the video and audio data before being recorded on to the tape.

In playback the digital data is recovered using a phase-lock-loop. Tracking and sub code data are removed and passed onto the servo microcomputer. Sub-code, time and date, and frame count are sent to the D/A converter for 'on screen' display via the I²C bus. Video and audio data error correction is carried out in the ECC IC. In the case of the video data this is sent to the ECC memory, which serves as compressed data frame store, the data is stored in small blocks in the memory corresponding the position of the data in the original analogue picture. Any data blocks that are severely corrupted, such that the error correction is ineffective, are not written into the memory. Therefore previous data remains in that location. This is not too dissimilar to analogue replay drop out compensation. In the worst conditions the same still frame will be repeatedly read out of the ECC memory until the data off the tape is good enough to update the data in the ECC memory. Severe data errors will give rise to pixel blocks being displayed in playback, a common complaint due to clogged video (data) heads. Video data is then read out of the ECC memory and joins the audio data again at a point where both could be outputted via the IEEE1394 digital connector for digital copying or editing. The data rate from this DV connector is about 40 Mbits/s. Further on in the replay path the video signal is expanded and put into the shuffle memory from where it is read out to the D/A converter, where syncs and blanking are restored to give a composite analogue video signal. Within the shuffle circuits the video data is shuffled during record, in playback the shuffle memory operates as a video store for playback special effects; there is no 'shuffling' in playback.

Video and Camcorder Servicing and Technology

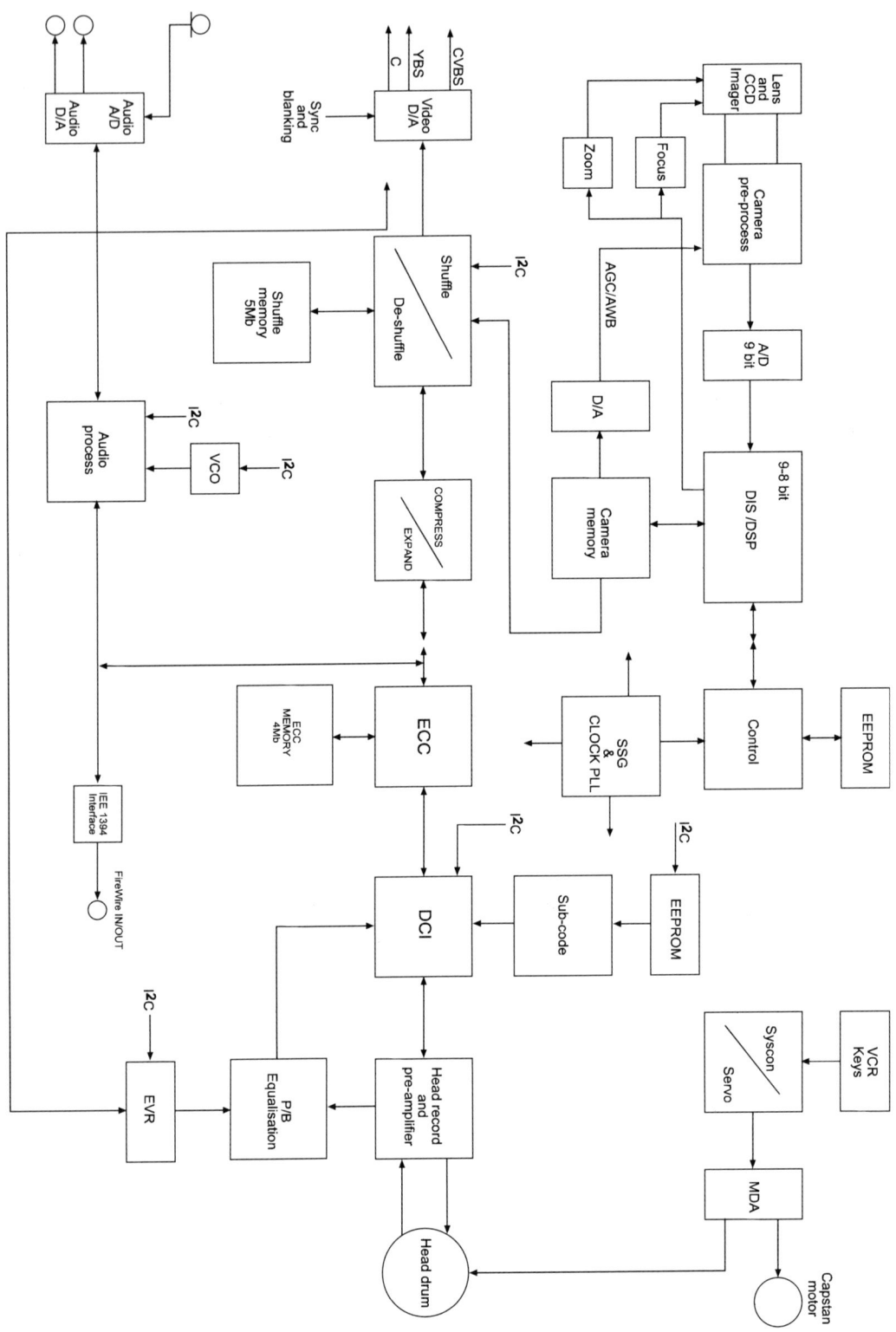

Figure 11.6 Digital camcorder basic block diagram

DVC tape format

DVC tapes are recorded as a helical track similar to VHS, except that 12 tracks are required for one frame compared to two for VHS; see Figure 11.7.

On the tape the recorded signal consists of the tracking data, audio data, video data and sub code data for each track. With the additional gaps (G1–G3), data 'run-in's and margins, the total amount of data allocated per track is about 139.5 kbits. For one TV frame consisting of 12 tracks this is a data total of 1.674 Mbits. At a TV rate of 25 frames/s the serial data stream on and off the video (data) heads is 41.85 Mbits/s. In replay a data locked PLL at 41.85 MHz recovers the data stream in a similar manner to a CD replaying.

Top and bottom edges of the tape are reserved for later use and designated optional tracks 1 and 2. We can forget about these for the time being.

As with all domestic videotape systems the head contacts the tape on the bottom edge, travels along and across and follows a long sloping path before exiting from the top edge. The angle of the digital video track path is 9.1668° and its effective length is about 33 mm. There is no audio/control track head nor is there a full erase head as these are not required. Erasure of the previously recorded data is accomplished by the recording data drive, which is high enough in level to fully record the new data and erase the old.

There are three tracking frequencies, f0, f1 and f2, and these are in digital format, not analogue as may be expected from video 8mm theory. Frequencies for the tracking data are given as: f0 is 0, f1 is 465 kHz, and f2 is 697.5 kHz. When processed in playback, these signals form a tri-state tracking signal for the tape servo.

The ITI (integrated tracking information) at the start of each track is a system data segment and contains information on track pitch and insert edit information. From this information the tape speed, SP or LP, can be determined by the CPU.

Post dub insert information is provided by the start sync block area (SSA) signal. It is used to indicate the playback position of the video data, audio data and sub-code data, by measuring a constant delay timing period after reading the previously recorded SSA. This is not too dissimilar to the way that a field sync pulse is used to indicate the start of an analogue picture in a standard VHS replay.

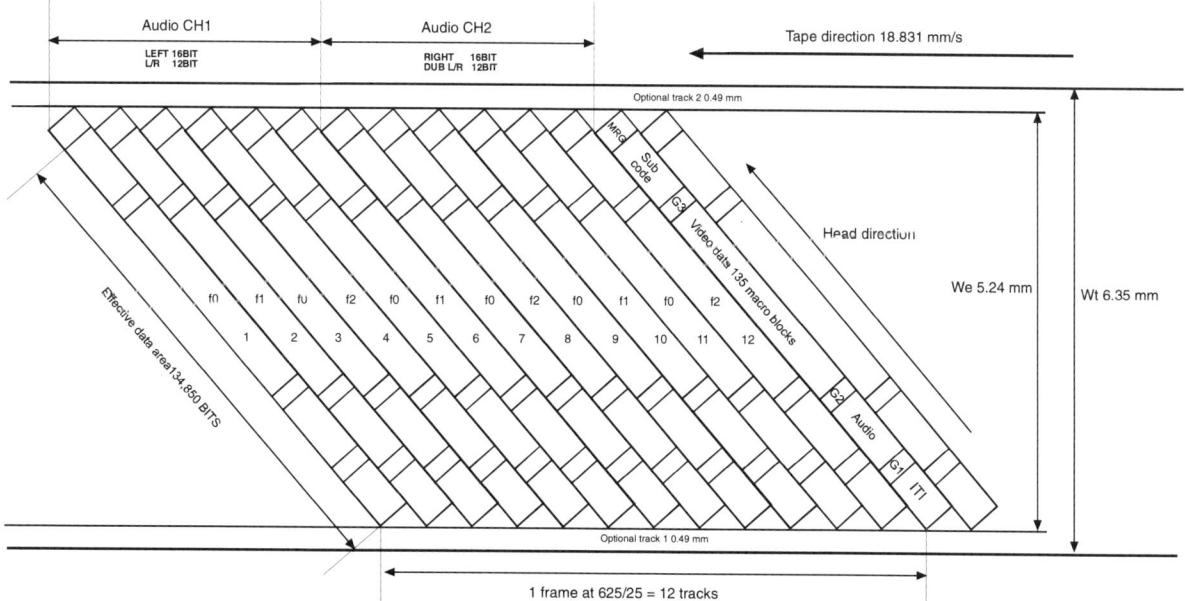

Figure 11.7 *Digital magnetic tracks on the tape*

G1 is a gap to separate the data, it is not blank because the replay PLL has to be kept running.

The audio section is next and there is an audio option that allows for post recording dubbing. Three stereo sampling frequencies can be selected; 48 kHz, 44.1 kHz (the same as CD and for playback only), or 32 kHz. The highest quality is 48 kHz with 16-bit linear sampling for two-channel working. Each channel is data compressed and spread over 12 tracks: Six for Ch.1 and then six for Ch.2 in twin-track stereo mode.

For four-channel operation, 32 kHz 12-bit non-linear sampling is used, tape tracks 1–6 are allocated to stereo tracks 1 and 2. Tracks 7–12 carry stereo tracks 3 and 4; these two tracks can be post dubbed. The audio section uses a total of 10.5 kbits per track.

After Gap 2 is the video data section. Data allocation is almost 112 kbits per track on which 135 macro blocks of data are recorded. Following the video and Gap 3 is the last section referred to as the sub-code area of 1.2 kbits. It contains the time and date information of the recording tape counter and the 'time code' i.e. min : sec : frame (as in 25 frames per second). Each video frame is time coded for accurate editing. Indexing information and still photo index are also stored in this section.

Following the sub-code data is an overlap margin area for continuity between the video heads for changeover switching.

The video heads

Each of the standard video (data) heads, shown in Figure 11.7, has a gap that is 12 µm in length to correspond to just over the width of the video track. In addition to this there is approximately 2 µm of leakage or fringing flux at each end of the head slot. This gives the video head tip an effective width of 16 µm, as shown in Figure 11.8. In playback the extra width is used to bridge the track being replayed (10 µm), in order to pick up the tracking pilot data from the adjacent tracks either side of the one being scanned for the

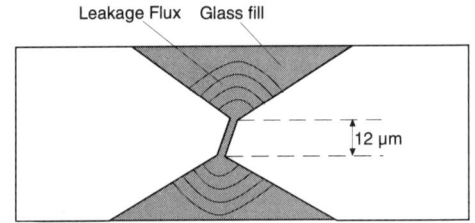

Figure 11.7 *Video (data) head construction*

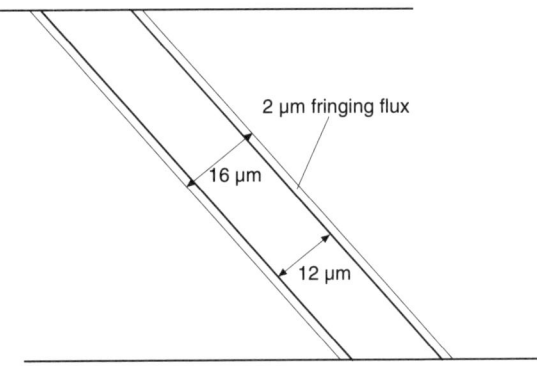

Figure 11.8 *Single track written by the head*

automatic tracking frequencies (ATF) tracking system.

In record the tape speed is designed so that the centre line of each subsequent track being recorded is displaced from the previous one by 10 µm. This means that the extra 2 µm width of the head over-records each previous track. Fringing flux of 2 µm reduces these tracks even further to an 8-µm wide magnetic track (see Figure 11.9). At the start of a recording the first track is laid down by a video head is 12 µm wide. The next track to be recorded is shifted along by 10 µm, determined by the speed of the tape so that the extra width originally recorded onto the tape is over-recorded. This lays down tracks on the tape that are effectively 10 µm wide, except for the fringing flux. Due to the fringing flux another 2 µm is over-recorded by each successive track, not fully, but enough to reduce the effective track width to around 8–8.5 µm. This does not cause any problems except that the replay error rate is probably higher than it could be if the full 10-µm width was available, but

Digital camcorders

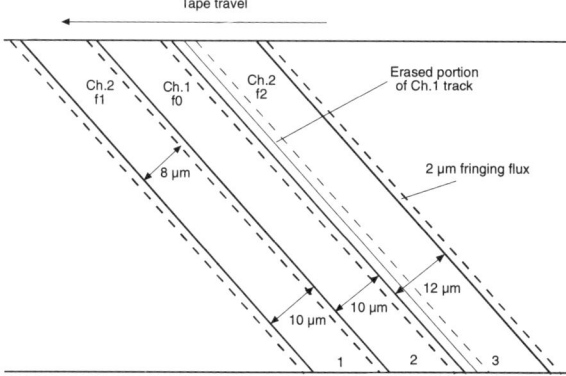

Figure 11.9 Track recording sequence showing how the track width is reduced by over-recording

it does increase the stability of the tracking system.

The tape in the track recording sequence diagrams travels from right to left, viewed from the virtual centre of the head drum.

The previous track in the sequence that is recorded is that of Ch.1/f0 and is initially recorded 12-μm wide excluding two edges of fringing flux 2-μm wide either side. By the time that Ch.2/f2 head passes across the tape to record the next track, the tape has moved on by 10 μm. Consequently this erases the trailing edge of the first track by 4 μm, which comprises the 2 μm of fringing flux and 2 μm of the track. As the new Ch.2 track also has 2 μm of fringing flux this also partially erases 2 μm of the Ch.1 track reducing its effective width to 8 μm.

DVC LP system

The benefits of having a 12-μm head with an additional 2 μm of fringing flux for the ATF to function are compromised in an SP/LP mechanism, where the head is of a later design that reduces the fringe flux to very low levels. However, if the fringing flux were left at 2 μm then it would cause many problems for LP working. In the LP mode the tape speed is reduced to two-thirds of the original, not by half as in analogue VCRs. Naturally this reduces the track width by two-thirds, from 10 to 6.67 μm. If the original head construction was used for LP with 2 μm of fringing flux, then the effective track width would be reduced to 4.67 μm by the fringing flux and this would reduce the level of the data so as to produce an unacceptable data error rate.

To overcome these limitations the video head construction (Figure 11.10) has been changed to a form that is suitable for both SP and LP working. Fringing flux is eliminated by manufacturing a head tip that is parallel in construction as opposed to the 'V' shape of the previous design. Due to the V-shaped head having large areas of supporting ferrite either side of the tip it was very robust. The new parallel head does not have the support of ferrite around the tip and so extra glass fill is incorporated to maintain durability and mechanical robustness of the heads.

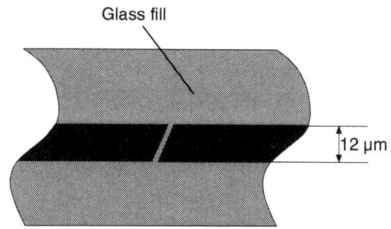

Figure 11.10 *LP video (data) head constructon*

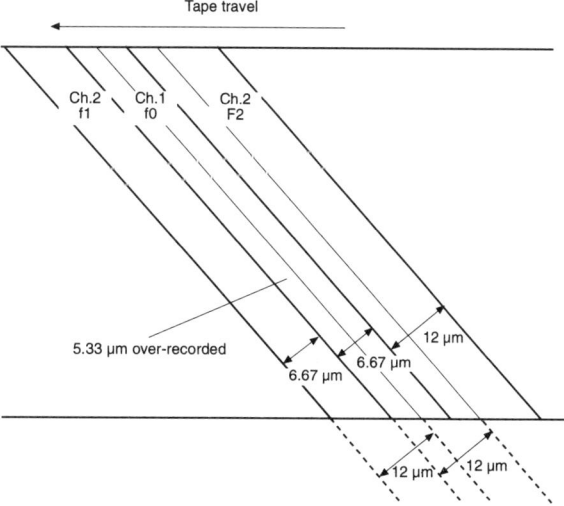

Figure 11.11 *LP head track recording sequence showing the degree of over-recording*

237

SP/LP head tracks

The first track laid down by this new head is 12 μm wide and subsequent tracks are recorded 10 μm apart in SP mode, each extra 2 μm is over-recorded leaving 10 μm wide tracks that are not affected by fringing flux. In SP playback there is less overlap to pick up the tracking signals from the tracks either side. However, this is compensated for in these SP/LP products by additional ATF pilot frequency amplification.

In the LP mode again the first track that is laid down is 12 μm wide; see Figure 11.11. As the tape speed now moves on by only 6.67 μm then 6.67-μm tracks are laid down. Each subsequent track that is laid down is 12 μm wide, and, as the tape is moving by 6.67 μm, then 5.33 μm of each previous track is over-recorded. As there is no fringing flux the tracks that are left recorded are not further reduced, thereby increasing the data pick up levels and reducing error rates.

It is worth pointing out that the newer fringe-free head cannot be fitted to previous SP only models as a replacement head. If it were, then playback would result in a reduced ATF signal and the ATF amplifier will not have enough gain resulting in poor and erratic tracking. Head drum assemblies differ between SP and SP/LP products and are not interchangeable. This, of course, does not affect interchangeability of the tapes recorded by either version in SP or LP mode. In LP replay the ATF levels are very high and so ATF gain is reduced accordingly.

ATF pilot signals (24/25 conversion)

There are three tracking signals: f0, f1 & f2. However f0 has no signal, and f1 and f2 are generated by division of the recording serial data rate so that there is correlation between the data and the pilot tracking signals making them transparent to the system. What does that mean? The recording data stream is 'modulated' by adding either '1's or '0's to each set of 24 bits, making 25 bits. Bear in mind that the data rate is 41.85 Mbits/s, so by adding so many '1's or '0's at varying rates the data stream can be 'frequency modulated'.

f1 is 465 kHz (record data rate of 41.85 Mbit/s divided by 90)
f2 is 697.5 kHz (41.85 Mbit/s divided by 60)

These pilot signals are not added to the data in the recording process as separate frequencies, they are integrated into the serial data stream by modifying the data. The 8-bit data is counted into groups of 24 and an extra bit is added so that there are 25 bits in all. The extra bit is either a '0' or a '1' and it is changed from '1's to '0's at the rate of the pilot signals. A product of doing this is that the DSV (digital sum variation) varies at the frequency of the pilot signal. As the DSV is the DC component of the data stream it can be used to add and extract the pilot signals which will appear as an AC component of the DSV; see Figure 11.12.

If the addition bit is a series of '1's then the average value of the DSV rises. If the bit is a series of '0's then the average value of the DSV falls. As the series of '1's and '0's change at the rate of a pilot frequency then during replay the DSV can be extracted and the tracking AC components can be filtered out.

In an analogue video recorder the main timing pulse is the drum flip-flop, it is the heartbeat of the system. A digital video recorder has many clock pulse lines from a single timing IC, and the

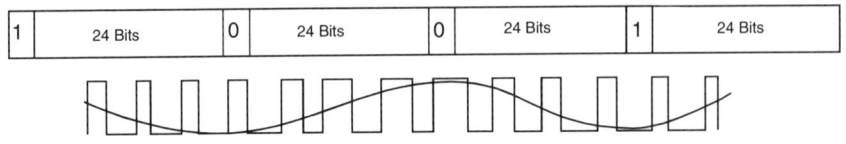

Figure 11.12 *DSV component of the data waveform*

Digital camcorders

clocks are phase synchronised. Digital video processing is dependent upon a frame rate pulse, the FRP, and this has to be synchronised to clock pulses and the head switching signal, HID. HID (head indication data) is high for Ch.1 head and low for Ch.2 head.

It is easy enough to work out that with two heads and six revolutions of the drum for a frame (12 tracks), that the head speed is 9000 rpm or 150 revs/s, giving an HID period of 6.666... ms (3.333 ms/track), or a frequency of 150 Hz compared to 25 Hz in a VHS recorder.

Tracking signal

Tracking signal control, pilot signal and tracking timing pulses are generated from the FRP and HID pulses in the DCI (digital channel coding IC) with an 18-MHz clock. From the FRP and HID pulses a switching signal called the TSR (tracking signal reference) is developed, and is almost identical to the HID signal. Pilot frequency and head relationship are controlled using the TSR as a head identification signal.

It is important for the servo system to replay the first track '0' at the rising edge of the FRP pulse (see timing diagram) to ensure correct 'framing' of the video data tracks. Two pilot frame pulses are available to determine the start of a replayed frame and the first track, only one is used in PAL, the PF0. Pilot frame PF0 corresponds to a start in tracking frequency sequence f0, f1. Three more sets of timing pulses are generated in the DCI IC, TRP0, TRP1 and TRP2. The combined binary logic of these and the pilot frequency signifies which number video data track (1–12) CH1 is playing back at any given time.

CH1 Track	TRP0	TRP1	TRP2
PF0 0	1	0	1
2	0	0	0
4	1	0	0
6	0	1	0
8	1	1	0
10	0	0	1

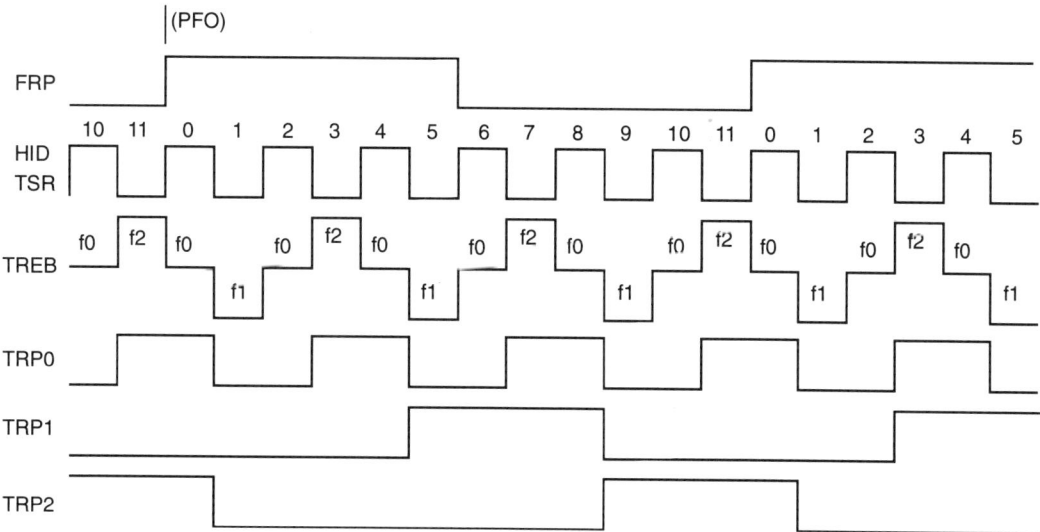

Figure 11.12 *Tracking waveform synchronisation*

Tracking error bit (TREB) signal

Another signal called TREB (tracking error bit) is used to control the phase of the capstan motor and maintain accurate automatic tracking. It is a tri-state waveform with three levels: 2.4 V, 1.5 V, and 0.6 V. It is developed by filtering the replayed data from the video heads using the digital sum variation calculation to recover the pilot signals f0, f1 and f2.

In the case of correct tracking, Figure 11.13(a), the TREB signal is high at 2.4 V from Ch.2 head playing back an f2 track and 0.6 V from Ch.2 head playing back an f1 track. As Ch.1 head is correctly centred on an f0 track then it picks up equal quantities of f1 and f2 from adjacent tracks as crosstalk which cancel each other out to the 1.5 V level. This produces a symmetrical TREB waveform. Note that the heads are wider than the tracks to accommodate the tracking system as we have seen previously.

In the case of the heads being in advance tracking phase then Ch.1 head is off-centre and picks up more of the f1 pilot than the f2 (Figure 11.14(b)). There is an imbalance in the f0 track crosstalk and the TREB signal for that track is low, almost 0.6 V. Ch.2 outputs of f2 and f1 remain the same.

In the third case, Figure 11.14(c), the heads are lagging in phase and the Ch.1 f0 imbalance is in favour of pilot frequency f2 and so the TREB signal at this time is almost 2.4 V. Again the Ch.2 outputs of f2 and f1 remain the same.

The average value of the TREB waveform is integrated to a varying phase error signal to correct the capstan motor. For continous control of the phase of the capstan motor the servo's aim is to maintain the TREB signal to 1.5 V for a Ch.1 (f0 track) replay. Capstan speed is set by the FG signal (897 Hz) to keep the speed constant and within the control range of the phase control system.

If Ch.1 head is dirty then the replay picture is blanked as the servo loses lock and mutes. If Ch.2 head is dirty then there will be alternate bands of pixel blocks or black lines on the replayed picture.

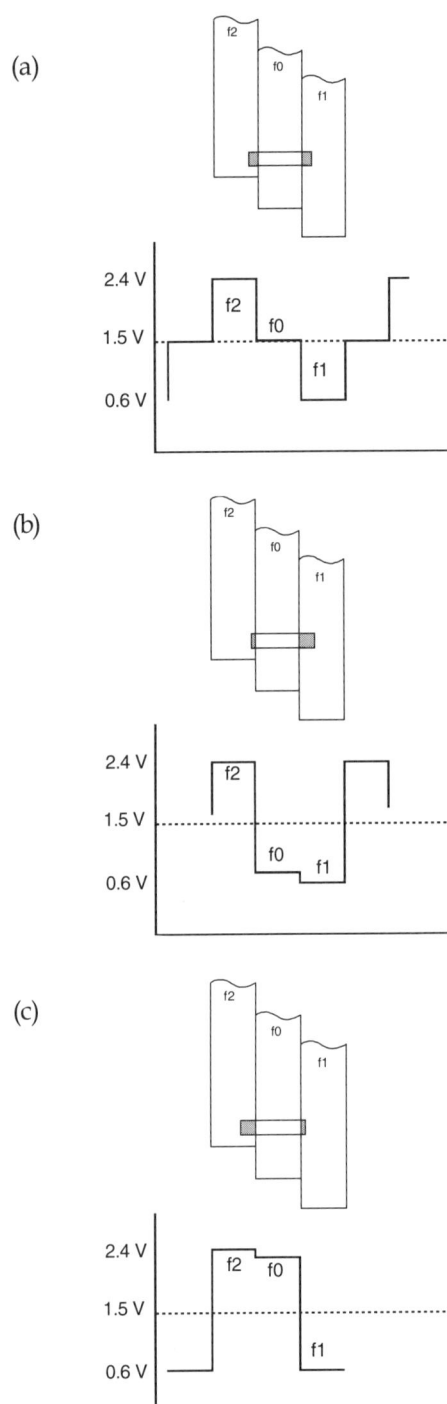

Figure 11.13 (a) Correct tracking; (b) head in advance; (c) head lagging

The DVC system

The digital video cassette camcorder is nothing like conventional VHS-C or 8mm, although the deck unit is similar to 8mm video in its overall operation and mechanical format, but very much smaller.

There are two rotary head tips with alternate azimuth of +20° and -20° mounted upon a 21.7 mm diameter head drum. The head tips are very thin and delicate, since the track pitch is 10 μm (6.67 μm LP), one fifth of VHS. The video head drum rotates at 9000 rpm producing a head switching signal of 150 Hz.

In broadcast component recording the system is 4:2:2, meaning that the luminance sampling is four samples to the two samples each of the components (B-Y) and (R-Y). DV NTSC is 4:1:1 where the luminance is sampled at 13.5 MHz and (R-Y)/(B-Y) at 3.375 MHz. PAL DV is termed 4:2:0, but because of the two-line sequential phase of PAL, the colour signal digitisation processing is line sequential and a more accurate description is 4:2:0/4:0:2. Y sampling is 13.5 MHz and colour component sampling is 6.75 MHz alternating a line of (R-Y) with a line of (B-Y), not unlike SECAM.

A digital camcorder imager has in excess of 600000 pixels but only 540000 are used. The remainder are masked off and some are utilised for black level reference.

For sampling, the active picture area from the CCD imager is set to 720 pixels horizontal and 576 lines vertical. This corresponds to the MPEG main level specification. Each pixel is digitised to an 8-bit word and there are 25 frames per second. Therefore the (Y) samples are: 720 x 576 x 8 x 25 = 82.944 Mbit/s. This approximates to a horizontal resolution of 700 lines.

For the colour components, only alternate horizontal pixels are digitised and so the colour samples are: 360 x 576 x 8 x 25 = 41.472 Mbit/s.

The total number of samples is around 124 Mbit/s. By complex digital video compression the recorded video signal is reduced to about 25 Mbits/s. This is not to be confused with the tape digital recorder data read/write rate of 41.85 Mbit/s as this figure includes audio and other data that are added into the stream.

The serial data is obtained from the active video area only. Syncs and blanking are removed and not added back until the A/D converter in the output stage.

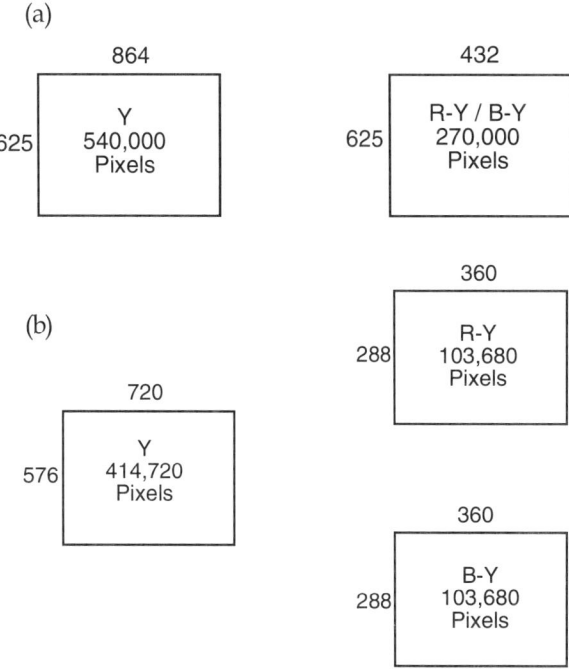

Figure 11.14(a) *Original imager picture area. (b) Pixels in the active picture area*

Camera processing

Pictures from the CCD imager are initially in analogue format as they are voltage amplitude representations of the light falling upon the pixels. These pixel amplitudes are matrixed forms of the green, magenta and yellow filters that are placed in front of the pixels in the CCD array. First, these pixel samples are matrixed into the recognisable signals of luminance, and the two colour components (R-Y) and (B-Y). There are no other video signal components, no syncs, no blanking, just raw video. The timing control of these signals is such that the syncs and blanking can be been inserted easily. This is due to the part played by the camera microcomputer and the SSG (sync signal generator) in generating the syncs for insertion at a later time, along with the timing signals for the imager scanning. Other digital timing signals are taken from phase-lock-loops all synchronised to an SSG master oscillator. All digital clock lines, HD, VD and framing pulses, come from the SSG circuit right up to the digital data going onto the tape; they are all synchronous.

Next for the camera picture, all noise and black level component signals are removed, leaving only those pixels that represent the active picture area. These are subjected to a degree of analogue processing. AGC is applied and the signal is clamped to a set black level.

For the luminance signal the array is 720 (H) by 576 (V) and for the colour components the array is 360 (H) by 288 (V). You will note that the colour component values are half that of the luminance. In the horizontal direction only alternate pixels are sampled, and in the vertical direction only alternate lines are sampled. This is as described for the CCD imager in the previous chapter. In order to maintain maximum colour resolution the horizontal lines are sampled alternately (R-Y) and (B-Y). For example if lines 1, 3, 5 etc were sampled (R-Y) then 2, 4, 6 etc would

be sampled (B-Y). This is also true for the pixels in the horizontal direction. It results in say odd pixel number samples for (R-Y) and even pixel number samples for (B-Y). The two colour component sample sets together sample the colour component of ALL of the pixels.

The bit rate for the data at this time is:
Y = 720 x 576 x 8 x 25 = 82.944 Mbit/s
Cr = 360 x 288 x 8 x 25 = 20.736 Mbit/s
Cb = 360 x 288 x 8 x 25 = 20.736 Mbit/s

The total is therefore 124.416 Mbit/s

A/D conversion: macro blocks

The pixels for digitising are assembled into blocks of 64 in an 8 x 8 matrix, that is 8 pixels horizontal by 8 lines in the vertical. So for our active picture area we have 90 blocks (H) by 72 (V) for the luminance signal, and 45 (H) by 36 (V) for the colour components (R-Y) and (B-Y). These are called DCT blocks as shown in Figure 11.16.

Next an (R-Y) and a (B-Y) DCT block are assembled with four luminance DCT blocks to form a macro block. This may seem strange at first, but don't forget that as the colour component samples are alternate pixels, then there are twice as many luminance pixels in both the horizontal

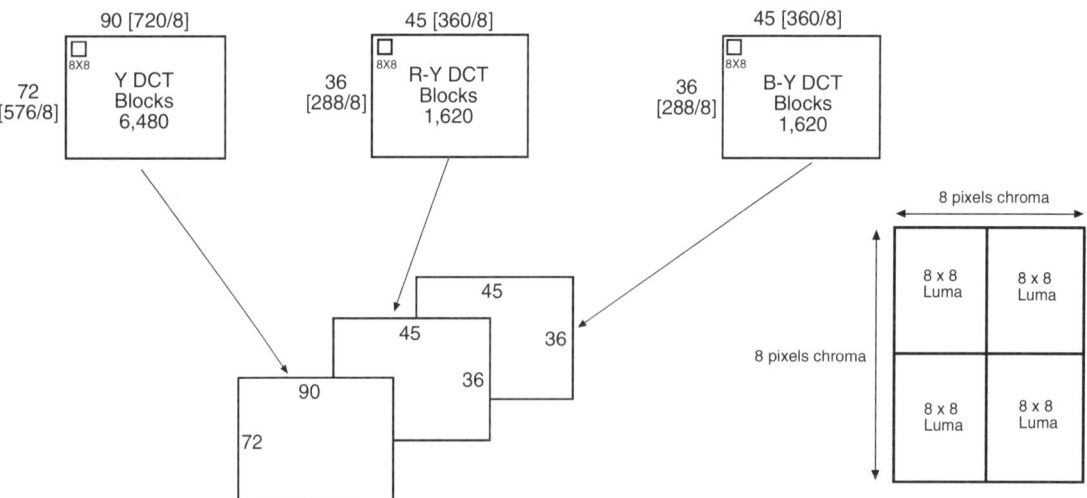

Figure 11.15 *Luminance and colour components assembled into DCT blocks in an 8 x 8 matrix*

Digital camcorders

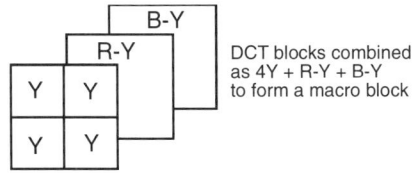

Figure 11.16 *Construction of a super block from 27 macro blocks*

and vertical directions for each colour component. Eight pixel samples by eight lines for a colour component covers the same picture area as 16 pixels by 16 lines of luminance samples, so they have to be kept together as a 4+1+1 to cover the same area of video picture information!

Super blocks

After combining each R-Y and B-Y DCT blocks with 4Y DCT blocks into a macro block, then an array of 27 macro blocks are assembled into a form 9(H) by 3(V) called a super block; Figure 11.17.

As shown in Figure 11.18 the whole picture can now be arranged as a frame in an array of 5 super blocks (H) by 12 super blocks (V). Each row of 5 super blocks is equivalent to one of 12 video tracks once they have been digitised and data compressed. This is how the video picture once compressed is recorded onto the tape as a frame comprising 12 tracks (each track has five super blocks).

Before data compression takes place the combined luminance and chrominance macro blocks are combined into super blocks. Each super block containing a total of 27 macro blocks. This divides the picture frame up into a 5 x 12 matrix.

To sum up, the TV frame has 12 vertical sections that correspond to the 12 video tracks of the magnetic recording. Each of these 12 tracks contains five super blocks that in turn contain 27 macro blocks. That is 5 x 27, giving 135 macro blocks per video data section. In cue and review modes the picture break up can be identified as super blocks, with some macro blocks where the picture breaks up into smaller vertical rectangles.

Super blocks are arranged in a matrix 5 wide, i.e. five super blocks across and 12 vertically, corresponding to the 12 video tracks recorded upon the tape per frame.

There are now 135 macro blocks per track, each macro block has a data volume that is close to but must not exceed 77 bytes. Macro blocks are read out of the memory as serial data, and error correction code and tracking data are added. Video data is placed between gaps G2 and G3 on the tape in the video data area.

Video data compression

One of the problems of digitising a video picture in 8 x 8 DCT blocks is that different parts of the picture contain different values of data. In some

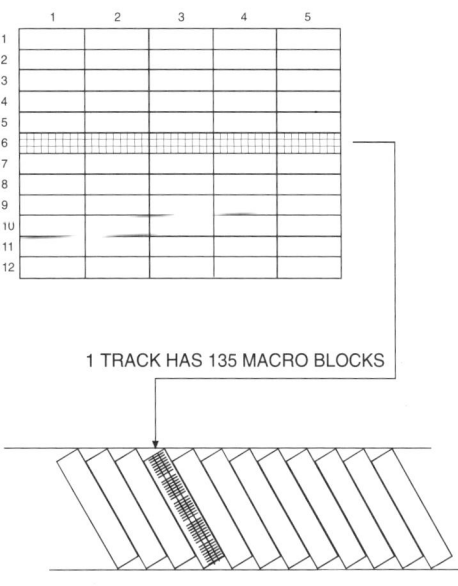

Figure 11.18 *Super blocks arranged on the recorded track in groups of five*

Video and Camcorder Servicing and Technology

parts of the picture the data value is high whereas in other parts the value may be low.

Let us imagine a seaside scene with a speedboat whizzing along on the sea. At the top of the picture there is nothing but blue sky and at the bottom large areas of yellow sand, but in the middle is the speedboat moving rapidly. Now there is no point in digitising the data for large numbers of DCT blocks of blue sky. Data can be reduced by saying "here is a block of blue sky and the next 20 blocks are the same". In other words digitise the first block and as there is no change in the rest do not digitise them, just say they are they same. In the centre of the picture a speed boat is moving rapidly. Here the DCT blocks are changing consistently, therefore each block has to be digitised, hence the data volume is very high for this part of the picture. As the amount of data space is fixed per track (approximately 83 kbits) then large amounts of data volume can overload the tape with resultant loss. On the other hand too little data is a waste of tape space. The aim is to obtain a consistent recording data stream, i.e. not too much and not too little.

In order to spread the data volume out over the whole frame, video segments are made up of macro blocks from different parts of the picture. They are digitised and then compressed; this is called shuffling and shown in Figure 11.18.

Video segments are made up of five macro blocks from different parts of the picture where some macro blocks will have small amounts of data and others large amounts. The aim is to optimise the video segments to average data quantities, in order to compress the video segment to meet a fixed criteria of data volume.

The area of tape that is allocated to video data can contain a maximum of 83 kbits (83160 bits) and as this is allocated to 135 macro blocks then it works out as 616 bits per macro block or 77 bytes (616/8 =77), Recorded data is limited to 77 bytes per macro block maximum, which is limited by the amount of data that be recorded onto the area of tape. Therefore a video segment of five macro blocks is limited to a maximum of (5 x 77= 385 bytes) after compression.

Within the shuffle IC and its associated 5 Mb of memory the stored video frame is broken down

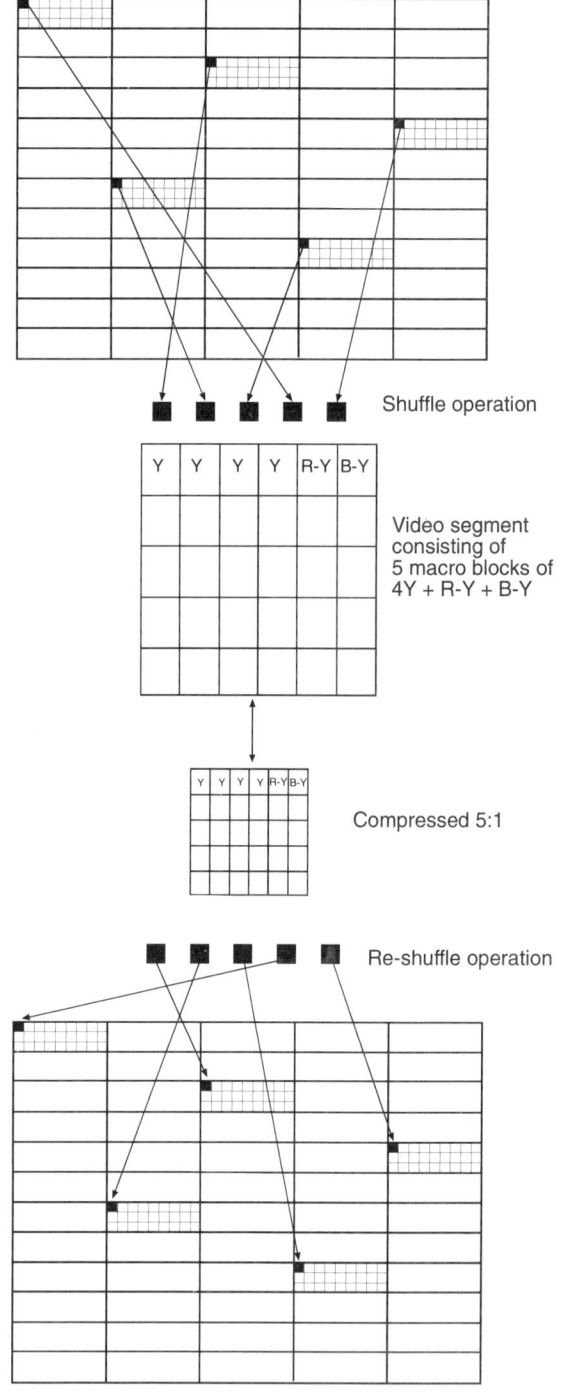

Figure 11.18 *Shuffle method of breaking down the picture, compressing and re assembling again*

Digital camcorders

by a fixed routine which builds video segments by a selection of five macro blocks from 5 different super blocks scattered about the frame.

Compression comparison

The complete uncompressed TV frame has 5 Mb of data stored in the shuffle memory. There are 1620 macro blocks in a frame which with simple division gives us a macro block data consisting of 385 bytes (or 3080 bits).

Before compression:
1 Macro block = 385 bytes
and a Video segment= 1925 bytes.

After compression:
1 Video segment is 1925/5 = 385 bytes
and a 1 Macro block = 77 bytes.

Under expected circumstances from CD technology these 324 sets of video segments would be recorded onto tape in their 'random' format, such as the cross interleaved reed solomon error correction (CIRC) used on an audio CD. This is fine for normal record and playback but not much good for picture search. This is because the 'random' data being replayed would exceed the error correction and reshuffling capabilities and would render the picture search impossible.

After compression it is required that the macro blocks in the video segments are reshuffled back to their original positions so that the data is recorded onto the tape as a contiguous digital video signal in just the same way as analogue video recording.

In replay the data is read off the tape into a memory (ECC memory). Then it is expanded, by the reversal of the compression process, and put into the shuffle memory. Video data is NOT shuffled or reshuffled in replay, as it is not necessary. From the shuffle memory the video data is read out into a D/A converter where the field sequence is re-established, and syncs and blanking are added back.

Video tape data map

A data map of the tape data is shown in Figure 11.19,. In order to understand the structure it is read across and down in the same way as reading a page of text. Each macro block is preceded by sync data and ident data to label the macro block. It is followed by inner parity data for error checking the macro block for bit errors. Outer parity for data correction is over 11 sync blocks and is a full checksum for bursts of data errors exceeding more than one macro block.

Error correction data is added in such a way that the 8-bit code still remains 8 bits. The intra-code of the video data is (85,77) for the inner code and (149,138) for the outer code. It means that the total code length for the inner code is 85 bytes (video + parity) and the data is 77 bytes. The outer code is 149 sync blocks (made up of 2 video Aux, 135 video data, 1 video aux 1, and 11 parity) and 138 data sync blocks (total of 149 less the 11

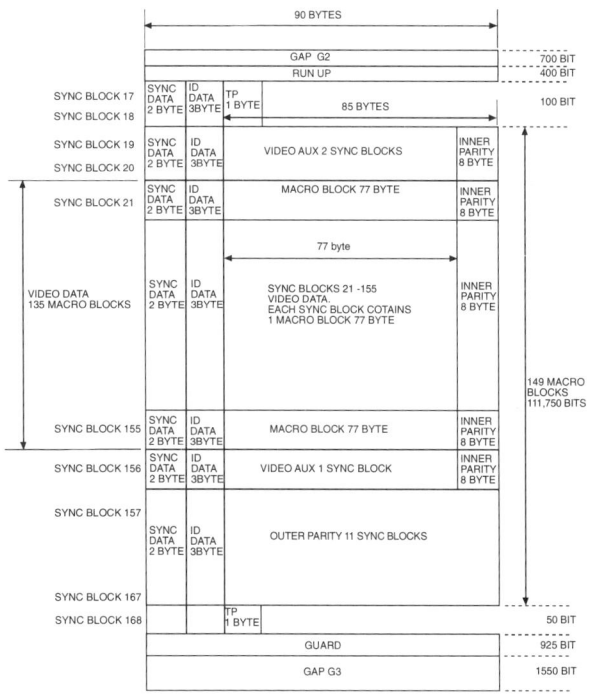

Figure 11.19 *Tape data map*

parity). These blocks are labelled on the track illustrated in the diagram.

The total data volume for the video data section of the track is calculated as follows:

140 sync blocks x 90 bytes x 8 bits x 25/24
= 111750 bits.

(An explanation of the 25/24 term is given in the tracking pilot signal description on p.238).

Error correction is also applied to the audio section but not to the sub-code section. The reason being is that the sub-code contains time, date and frame count. This data does not change over the 12 tracks, therefore if the data is lost on one track it can be read on the next one. For the time and date or time code, any single track read during a frame is sufficient.

Compression techniques: discrete cosine transformation (DCT)

This is the most complex part of the DVC system and although it can be split into several sections it has to be understood that a number of procedures take place simultaneously.

Discrete cosine transformation is a complex mathematical process which separates the DC and AC components of an N x N block matrix, in the case of digital video N=8, and places the cosine functions in the DCT block in ascending frequency order.

The formula for the Fourier transform is reproduced in Figure 11.20, and it can be explained a bit more simply.

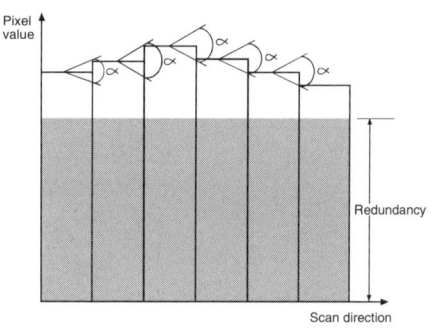

Figure 11.21 *Average DC level and AC component*

DCT is the first part of the compression system although the DCT process itself does not contribute to data compression. The simplest explanation of a complex mathematical transform calculation that it splits the video data into a DC component and an AC component by applying a Fourier transformation.

If we consider a very small selection of video samples in Figure 11.21, it can be seen that there is a great deal of redundancy as the average DC level is the same for each sample. It is not necessary to encode and 'transmit' the DC component for each sample, it can be done by defining the DC average for a small set just once and then say it is the same for the rest. In that way the amount of data can be reduced without any loss of information.

The amplitude changes between each sample, above the average DC level, are then transformed by the DCT process into frequency coefficients by calculating the cosine and defining just the coefficients. In this way the frequency coefficients represent the change in amplitude and the rate of change as a multiple of the average DC value. The figure shows the DC level redundancy and the

$$F(u,v) = \frac{2}{N} k(u) k(v) \sum_{x=0}^{N-1} \sum_{y=0}^{N-1} f(x,y) \cos \frac{(2x+1)u\pi}{2N} \cos \frac{(2y+1)v\pi}{2N}$$

Figure 11.20 *Fourier transform formula*

Digital camcorders

cosine angles of the AC component.

It is important to note here that the frequency coefficients represent the frequency by value and sign. These are frequency domain values and are treated as such by the DCT process.

For example: if we take a small sample of the picture, such as a small section of an 8 x 8 block (Figure 11.22) and pair up the pixels as P1 and P2 in both the horizontal (pixel) direction and vertical (line) direction, it can be seen that there is very little change in value between P1 and P2 because they are next to each other. We can plot the levels of P1 and P2 (Figure 11.23) and show that the amplitude changes little, except where there is a sudden change, then this is an AC component part of the samples.

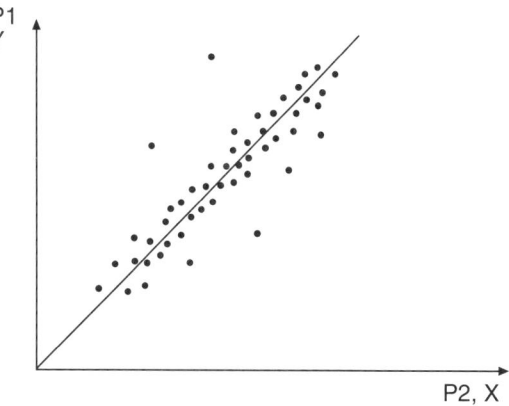

Figure 11.24 *Adjacent pixels P1 and P2 plotted against each other*

DCT frequency distribution

DCT takes the assumption that P1 and P2 have the same level a step further.

If we plot sample pairs of P1 against P2 on a graph, such as in Figure 11.24, where P1 is on the Y axis and P2 is on the X axis, the result is that the points are congregated along a 45° slope. This shows that there is strong amplitude correlation between adjacent pairs. Strong amplitude correlation means that the amplitude of adjacent pixels in both the horizontal and vertical directions are similar.

If we now rotate the graph by 45° so that the slope becomes a new axis X' and at 90° to this is the Y' axis, there is a different viewpoint. The energy concentrated along the X' axis is a near constant value, that is the DC level. On the Y' axis the energy is concentrated at a point and this is a representation of the AC component and so the two components AC and DC are separated on the X' and Y' axes, as shown in Figure 11.25.

This is the Fourier transform mathematical function that is carried out on the DCT blocks. Figure 11.25 shows a two-dimensional graph, in the horizontal direction. In reality it is a three-dimensional function on the 8 x 8 matrix consisting of the horizontal, vertical and amplitude

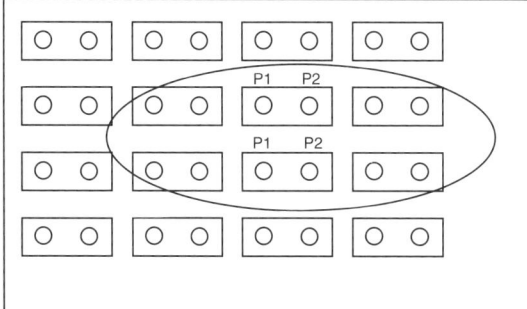

Figure 11.22 *section of the imager pixel array*

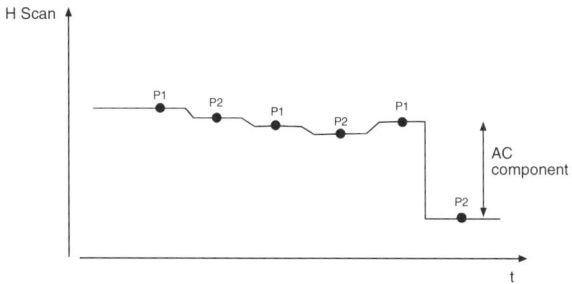

Figure 11.23 *Illustration of adjacent pixel levels*

247

Video and Camcorder Servicing and Technology

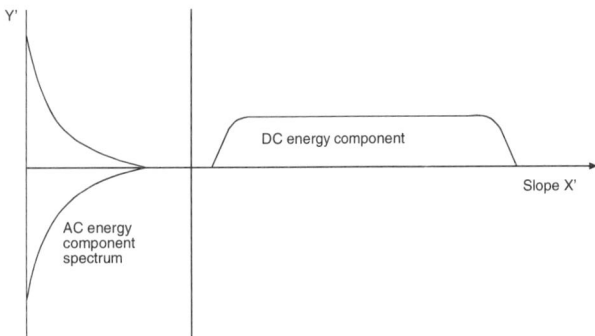

Figure 11.25 *X/Y graph of the DC and AC components*

axis, as shown in the weighting diagram of Figure 11.26.

Once the two components are separated they are reassembled into a DCT block in an 8 × 8 matrix of 64 pixels. In the upper left-hand corner is a single data sample representing the average DC level of the whole block of original video samples (Figure 11.27).

In both the horizontal and vertical directions samples are set in ascending frequency coefficient order, in such a way that the bottom right-hand corner has the highest frequency coefficient. At the same time frequency weighting is applied as shown in the three-dimensional weighting diagram, Figure 11.26.

Note: In the DCT Fourier transform, the analogue signal samples are transformed into a series of cosine harmonic coefficients, representing only the AC components of the block; these are in phase with the signal. These coefficients represent the Fourier transform in a single value. For example, if all of the cells in the 8 × 8 block have the same level because it is a sample of a scene containing sky, then all of the DCT coefficients are zero and only the DC block in the top left-hand corner has a value. The other 63 squares have no values in them, as there are no AC variables.

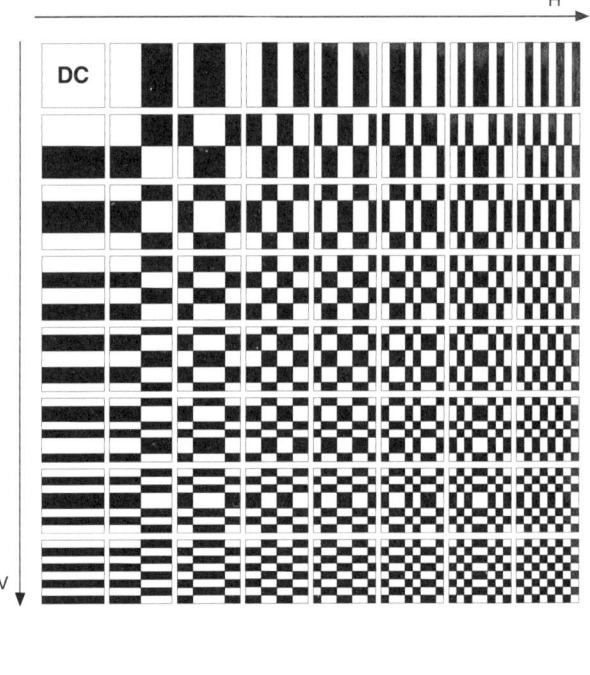

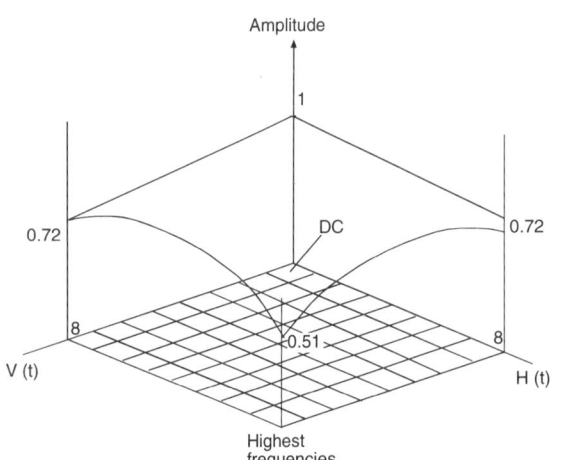

Figure 11.26 *Weighting diagram*

Figure 11.27 *A DCT block of 8 × 8 showing the DC and frequency distribution*

DCT processing

In the first example in Figure 11.28 we have a DCT block that contains the 8 x 8 matrix in which the values are analogue. Black is 0 and white is 100: it is a small sample of the picture

In the second stage, Figure 11.29, of the process the example is of the DCT block after digitisation and after descrete cosine transformation has been applied. A value of 230 has been given for the DC level in the top left-hand corner square and represents the average DC level of the other 63 squares. All of the values in the other 63 squares are the AC coefficients calculated by the Fourier transform. In reality they represent the change in level between the sample and its neighbour and the frequency of the original signal.

In the method of video data compression there is such a thing as 'what the eye doesn't see, the heart doesn't grieve over'. Certain frequencies and details within a picture that the imager picks up can be made redundant, as we cannot see them. Take for example a fast moving racing car with lots of small logos along the side. If the car passes by very fast the eye cannot easily pick out the details. These details are fast moving, low level high frequencies and so they can be abandoned.

The next stage in the process is to do just that. Areas of the DCT block are selected according to the laws governing human eye responses and allocated a division ratio. Mathematical division in a weighting process will reduce the AC coefficients within these areas, usually to 0.

H →							
60	60	60	61	61	61	61	62
60	60	60	60	61	62	62	62
60	60	60	61	61	62	63	64
60	60	60	62	62	63	63	64
60	60	60	61	62	63	63	64
60	60	61	62	62	63	63	64
60	60	61	62	63	63	64	64
60	60	61	62	63	63	64	64

Figure 11.28 *An 8 x 8 DCT block of 64 samples in analogue format where peak white is 100*

Adaptive quantisation

After descrete cosine transformation, the DC sample and the frequency coefficients are still in an 8 x 8 matrix and is mathematically reversible; therefore no data loss has taken place so far. Only frequency weighting has been applied. In an average picture the higher frequency coefficients are much smaller in amplitude and decrease with increasing frequency. The eye is not responsive to low level high frequencies and during the weighting process amplitudes that are lower than a fixed threshold are discarded.

The next stage is adaptive quantisation. In this section three processes are carried out in one move.

1. To reduce the data volume by dividing the data according to its frequency and placement within the grid block.

2. Vary the quantisation, i.e. the number of

H →							
230	-37	-6	-1	-2	-3	-2	0
-52	-2	2	1	1	0	1	1
-7	8	1	-1	1	0	0	0
-5	0	1	-2	1	1	1	0
0	-1	-1	0	0	1	0	0
0	-1	1	1	0	1	1	0
0	0	0	1	0	0	1	1
0	0	0	0	1	0	-1	0

Figure 11.29 *DCT block after A/D encoding. The DC value is 230, a light grey, peak white is 255. All other values are frequency coefficients in ascending order from top left to bottom right.*

DC	2	2	2	4	4	4	8
2	2	2	4	4	4	8	8
2	2	4	4	4	8	8	8
2	4	4	4	8	8	8	16
4	4	4	8	8	8	16	16
4	4	8	8	8	16	16	16
4	8	8	8	16	16	16	16
8	8	8	16	16	16	16	16

Figure 11.30 *DCT block showing the data division ratio according to the location of the frequency coefficients*

230	-37	-5	0	-1	0	0	0
-52	-2	1	0	0	0	0	0
-7	6	1	0	0	0	0	0
-5	0	-1	-1	0	0	0	0
0	-1	0	0	0	0	0	0
0	0	0	0	0	0	0	0
0	0	0	0	0	0	0	0
0	0	0	0	0	0	0	0

Figure 11.31 *The DCT block after frequency division has taken place. Note that most of the small high frequency coefficients towards the bottom right hand corner have been made zero*

'steps' according to the data classification and frequency component.

3. Vary either of the two above according to the data volume out of the adaptive quantisation. That is, if there is too high a data volume then the quantisation value can be reduced to a coarse quantisation and therefore reduce the number of samples and hence the data volume.

After the data has been transformed by DCT into a DCT block the first step of data reduction takes place. In order to understand the next step you have to be sure of the arrangement of a DCT block.

Within the block in our example the division of the data is according to frequency and it is this matrix in Figure 11.30 that is applied to the DCT block. DC in the top left-hand corner is largely unaffected and the division is taken as 1; no change there then. The data division values given for lower frequencies is 2, or 4 and 8 for mid frequencies, and the highest frequencies, 16.

Example

$00000111\ /1\ (2^0)$ = 00000111 = 111
no change.

$00000111/2\ (2^1)$ = 00000011 = 011
$00000111/4\ (2^2)$ = 00000001 = 001
$00000111/8\ (2^3)$ = 00000000 = 0
$00000111/16\ (2^4)$ = 00000000 = 0

This division prepares the data for the next process by decreasing the coefficients of the higher frequencies. Because of this process the low level high frequencies below a fixed threshold are discarded as they result in zero, as shown in Figure 11.31.

When the DCT block has been through the adaptive quantisation process the amount of data has been reduced, mostly at the low level high frequencies, and is not reversible resulting in data loss.

It is important to point out that the adaptive quantisation is not linear and is designed to reduce the quantisation value of frequencies that are not subjective to the human eye. In the range of 1–2MHz quantisation errors are very noticeable; therefore the quantisation value is higher in this region. This overall process has been devised with the sensitivity of the human eye in mind and is call physco visual masking

The final step is to read the data out in serial form for the next stage. A zig-zag scan is used to read out the data from the DCT block; see Figure 11.32. The DC level and the important lower frequencies are read out first, the higher and less important frequencies are last. This gives the adaptive quantisation circuit the option to discard more high frequencies if the maximum allowable data volume of the macro block is exceeded. For example if there is a lot of movement in the picture in that sample block, then the loss of data at the higher frequencies is not noticeable to the eye. For example, if a bird flies across the picture, the eye is not responsive to the fine detail of the feathers, just that a bird is flying past.

Variable length coding

VLC is based on an Huffman algorithm that works on the probability of occurrence of any given symbol and those that occur the most are given a low bit count whereas the least occurring are given a higher bit count.

For example: if you take random pages of a book and plot the occurrence of the symbols (letters), in all probability certain letters (A, E, I, O, U, S, T,) will occur more than others (Q, X, V, Z). If the book were to be digitised data volume can be reduced by giving a low number of bits (or quantisation) to the higher probability symbols and the largest number bits to the least occurring symbols.

It has been determined, by experimentation of average pictures, which symbols occur the most and those which occur the least, and this information has been stored in a 'look up' table.

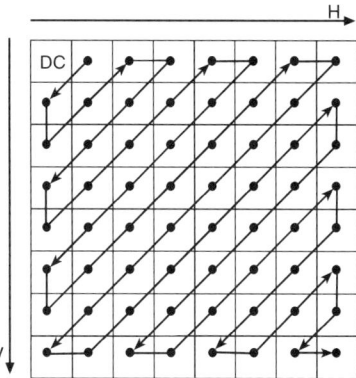

Figure 11.32 *Zig-zag read-out of the DCT block*

Data is read out of the DCT block in serial form into the VLC circuit and from our previous example looks like this:

230, -37, -52, -7, -2, -5, 0, 1, 6, -5, 0, 0, 1, 0, -1, 0, 0, 0, -1, -1, 0, 0, 0, 0, -1, 0, 0, 0, ...0

The VLC system is to reduce the data further by converting the DCT read-out into a format (run, amp) or (a, b,) where (b) is the data value or frequency coefficient and (a) is number of preceding zeros. This is determined from a 'look up' table.

Our example is then converted into the following sequence taken from Figure 11.32 as a zig-zag read-out.

(0, 230), (0, -37), (0, -52), (0, -7), (0, -2), (0, -5), (1, 1), (0, 6), (0, -5), (2, 1), (1, -1), (3, -1), (0, -1), (4, -1), + EOB (end of block).

From the table in Figure 11.33 the number of bits are:

(15) + (15) + (15) + (6) + (3) + (5) + (4) + (5) + (5) + (5) + (4) + (6) + (2) + (6) + (4)
    =100 bits.

By using the table shown, the length of the data word to describe the VLC data is selected from fixed values. This will also result in data reduction and the loss of information and is not

Video and Camcorder Servicing and Technology

fully transparent.

Refer to Figure 11.33 which is the 'look up' table for the VCL values. Take (0, 230) as the DC value. The amp value is 230 and the run is 0. Then from the top row of the table a 15-bit data word is selected.

In another example, (0, -7) results in a 6-bit data length, and (3, -1) in a 6-bit data length, and so on.

If the data length is worked out for all of the block VLC values and the 4-bit EOB is added then the total data is 100 bits. As the original DCT block had 64 x 8 bits = 512 bits, this represents a data reduction of 5:1.

All of this is heading towards the maximum bit limitation set by the tape capacity. The 8 x 8 block is digitised and compressed to 100 bits. A macro block which is 4Y + (R-Y) + (B-Y) totals 600 bits and a video segment is 5 Macro blocks = 3000 bits which is within the permitted maximum of 3080 bits for a video segment.

As the block of sky has only 8 bits or so, it is arranged into a video segment of the speedboat, a fast moving image where the block may be greater than 100, perhaps 400, and the maximum limit criteria can still be attained.

From this point on the compressed video data has audio, sub-code, and ITI data added to it before it is recorded onto tape.

AMP	0	1	2	3	4	5	6	7	8	9	10	11	12	13	14	15	16	17	18	19	20	21	22	23	.....	255
0	11	2	3	4	4	5	5	6	6	7	7	7	8	8	8	8	8	8	9	9	9	9	9	15		15
1	11	4	5	7	7	8	8	8	9	10	10	10	11	11	11	12	12	12								
2	12	5	7	8	9	9	10	12	12	12	12	12														
3	12	6	8	9	10	10	11	12																		
4	12	6	8	9	11	12																				
5	12	7	9	10																						
6	13	7	9	11																						
7	13	8	12	12																						
8	13	8	12	12																						
9	13	8	12																							
10	13	8	12																							
11	13	9																								
12	13	9																								
13	13	9																								
14	13	9																								
15	13																									
:																										
61																										

RUN

NOTE: Sign and parity bits not included
EOB: End of Block 4 bits
AMP: Absolute value
RUN: Preceeding 0s

Figure 11.33 *Variable length coding 'look up' table.*

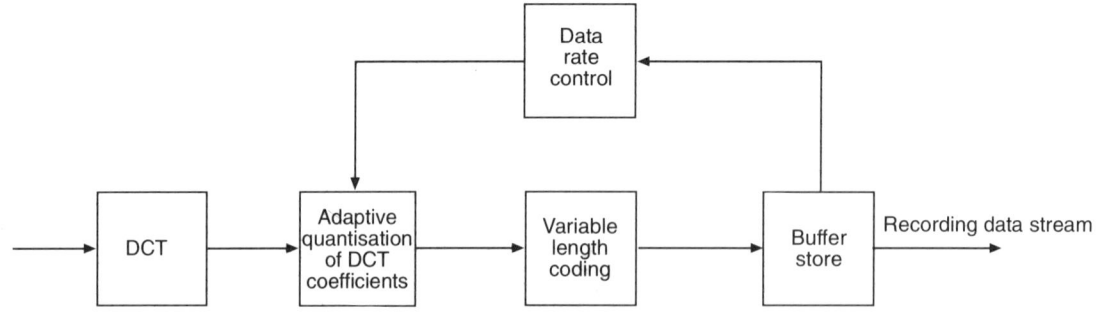

Figure 11.34 *Overall recording data flow control to regulate the data volume on tape*

Digital camcorders

DVC audio

DVC audio is recorded in the PCM (pulse code modulation) format, there is the option of two Channel (stereo) or four Channel (two stereo channels) sound tracks, these are:

Two Channel, 16 bit, 48 kHz sampling
Four Channel, 12 bit, 32 kHz sampling

There is a further option of playback only of two channel 16 bit at 44.1 kHz sampling, similar to the CD format, possibly for replay of music tapes.

The tape layout is different for the two recording options and is designed to keep the data volume the same on each audio section of the digital track whatever the option. For two-channel, 16 bit the first six tracks in the frame are the 'left-hand channel' and the following six tracks the 'right-hand channel'. In this format, Figure 11.36, there is no option for dubbing additional sound.

In the four-channel mode, Figure 11.37, the first six tracks in the frame are allocated to both the left and right channels of the normal stereo sound signal from the microphone whilst recording the original pictures. The next six digital tracks in the frame are allocated to the dubbed audio L' and R'. If audio dubbing takes place in the four-channel mode the original audio tracks remain intact and the dub sections are replaced without affecting the ITI or video data either side. In playback the menu options are: sound track 1,

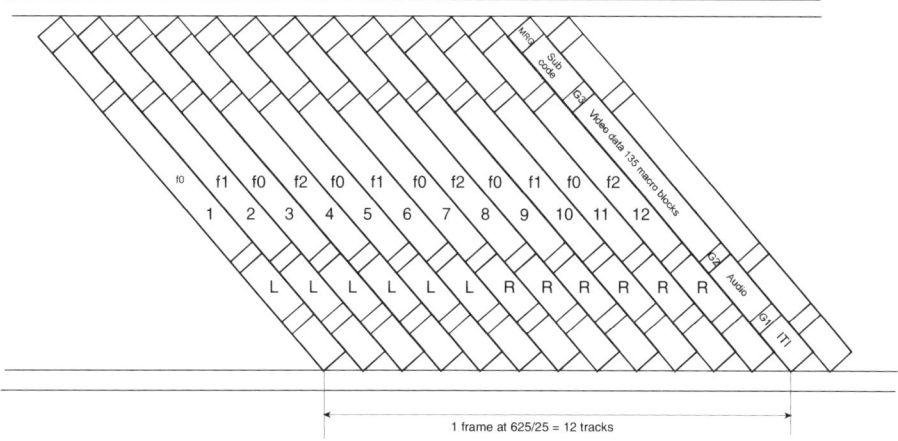

Figure 11.35 *Audio data 16 bit, 48 kHz*

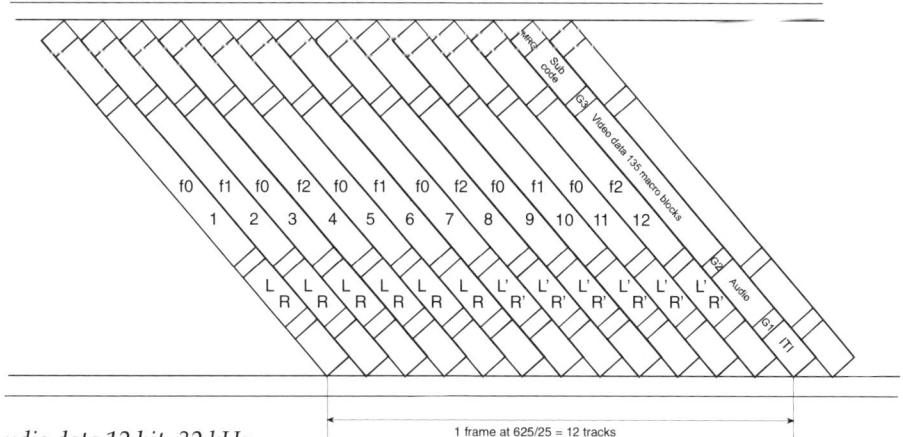

Figure 11.36 *Audio data 12 bit, 32 kHz*

253

sound track 2 and mixed replay of both. Customers are often confused by this and there are complaints of playback audio failure. The customer has set the recording menu option to 12 bit and the replay menu to sound track 2, which is blank unless dubbed!

The data volume for two-channel, 16 bit, 48 kHz sampling is:

2 x 48 kHz x 16 = 1,536 kbits/s
(768 kbits/s per channel)

The data volume for two channels of the four-channel 12 bit, 32 kHz sampling is :

2 x 32 kHz x 12 = 768 kbits/s for both channels.

This shows that 12 bit/32 kHz audio data fits two sound channels in the place of one channel of 16 bit/48 kHz.

768 kbits/s ÷ 25 frames ÷ 6 tracks
= 5120 bits /track

The audio data allocation for one had scan track is just over 5 kbits.

Audio error correction

The audio data is segmented and scattered in a fixed way on the recorded tape using a similar fashion to the CD, so if there is data loss then it does not affect a contiguous audio data stream. cross interleave reed solomon code (CIRC) uses interpolation to correct burst errors and then reed solomon error correction for single bits.

In the example in Figure 11.37, the audio data is contiguous and it is stored in the memory area. It is then read out in an interleaved format where the data bits are scattered over the data stream. In the case where some data is lost due to drop out or damaged tape then the data bits 37, 2, 6, 10, and 14 are lost in replay. After de-interleaving the data is restored to contiguous form and the lost data appears as random bits that can be corrected by the Reed Solomon error correction process.

Parity error correction

A simple example of error correction is illustrated in the drawing of Figure 11.38.

Figure 11.37 *Audio error correction by cross interleave reed soloman code*

Digital camcorders

The original data is multiplied by a series of prime numbers (3, 5, 7, 11) and the sum of this multiplication (152) is added to the data as a parity checksum.

In playback the data is again multiplied by the same series of prime numbers (3, 5, 7, 11). One bit is in error and this time the checksum is different (141). Now the replay calculation is checked against the parity checksum (152-141=11).

The difference between the two sums indicates which data bit has the error, in this case the one with 11 as the multiple. As the checksum is higher then the error is positive, a '1' is added to the data (7) in order to equate the two sums. Had the checksum been a negative value, indicating a higher value error (say 9), then the correction would be by subtraction.

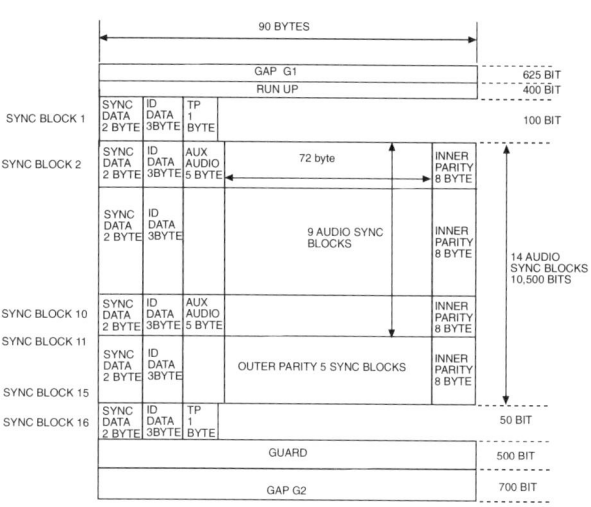

Figure 11.39 *Audio data track layout*

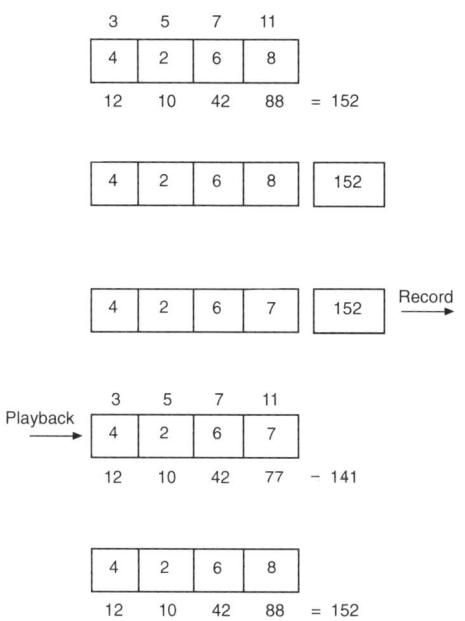

Figure 11.38 *Example of parity error data correction*

Audio tape track layout

The layout of the audio data on the tape is similar to that of the video data and is read like a page of text, across and down. It is arranged into sync blocks starting with sync data, ID data, audio auxiliary data, audio data and parity data. There are 14 sync blocks in total: nine audio data sync blocks and five outer parity code sync blocks.

The audio auxiliary data section contains the audio information, sampling frequency and bit data.

The main audio data section consists of 14 sync blocks x 90 bytes x 8 bit x 25/24 = 10,500 bits, and this includes the inner parity checksum data and the outer parity checksum data. Inner parity corrects a sync block row data and outer parity corrects for larger errors over a number of sync block rows. This is carried out prior to the cross interleave correction.

The audio data is allocated nine sync blocks; 9 blocks x 72 bytes x 8 bit = 5184 bits/track. This is sufficient space for the calculated 5 kbits of audio recording data, which is not compressed. Note that the value 25/24 is the ATF data 24/25 modulation.

255

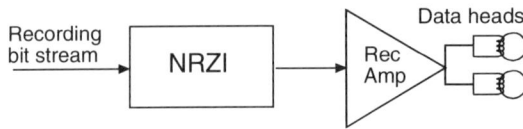

Figure 11.40 *Data recording block diagram*

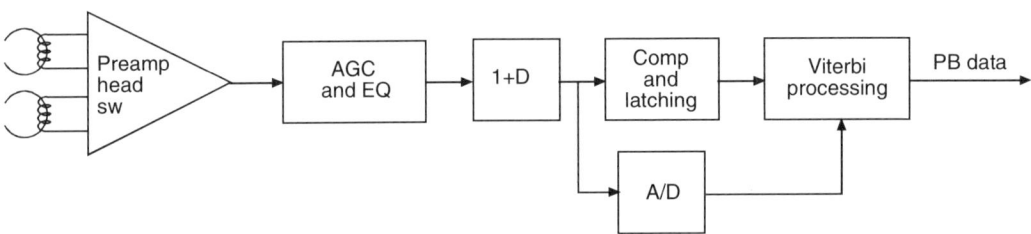

Figure 11.41 *Data playback block diagram*

DV data recording and playback

In a similar manner to that used in PCs to record data onto a hard drive disc, the data has to be encoded in record and decoded in playback using a non-return to zero (NRZ) encoding system. This is to avoid a DC component forming in the data stream when large quantities of 0s or 1s occur within the data and to avoid loss of replay PLL synchronisation. In a digital video camcorder a modified version of NRZ, non-return to zero inversion (NRZI) is used just prior to the recording amplifier; see Figure 11.40.

In order to recover the original data, two systems are employed: one is the NRZI decoder and the other is the Viterbi correction system for data affected by noise. I should point out that Viterbi is not a nomenclature, it is the engineer's name within NASA who developed the theory and circuit in order to recover noisy data from distant spacecraft.

The video data signal is replayed by the two heads in the same way as any video recorder, amplified and selected by the video (data) head switching circuits (Figure 11.41). AGC and equalisation are applied to compensate for HF losses before NRZI (1+D) decoding, after which the signal is split. One path is via an A/D decoder for Viterbi analysis and the main path is via a compensation and latching circuit to recover the original square wave bi-state format before Viterbi correction.

The replayed signal is in a tri-state format up to the (1+D) and Viterbi A/D blocks. This is in effect an AC waveform centred on '0', with an upper level of 1 and a lower level of -1. The bi-state signal is the normal 0 and 1.

A magnetic field is only produced across the gap of a recording head when the current in the winding is changing. A stable current (DC) does not affect the characteristic of the tape and records nothing. When a square wave, such as a bit stream is applied to a recording head only the transitions from low to high and high to low are recorded onto the tape. In other words the recording signal is differentiated as shown in Figure 11.42.

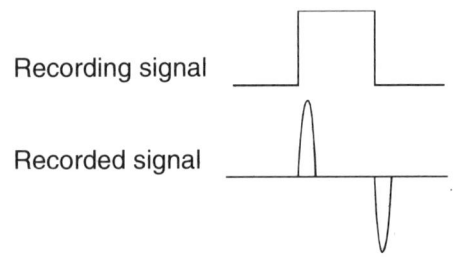

Figure 11.42 *Square-wave data differentiated by the process of recording onto tape*

Interleaved NRZI (non-return to zero inversion)

Within the digital video recording data stream there may be long strings of continuous 0s or 1s and this is not good for keeping the playback phase-lock loop synchronised. Also, it will cause the data stream to be lifted or lowered upon a DC level, formed in a similar way to the DSV recovery of pilot tracking signals. As these long strings would not have enough changes to keep the playback PLL circuit in sync with the replay data then data may be lost or corrupted. An NRZI encoder ensures that these long strings of 0s and 1s do not happen by 'chopping' them up to maintain clocking pulses for the replay phase-lock loop clock. A scrambled NRZI circuit is utilised. It consists of an exclusive OR gate and two delay lines, each of a period equivalent to a data bit, see Figure 11.43 and the waveforms in Figure 11.44.

(A) is the input data, (B) is the output data (C) delayed by two bit periods, and (C) is the recording output of the exclusive OR gate. In this configuration the exclusive OR gate is used as a serial multiplier, where

$$C = A + (C \times D \times D) \text{ or } A + CD^2$$

From $C = A + CD^2$ we can re-arrange to the form:

$$A = C - CD^2$$

Expanding this produces:

$$A = C(1-D^2) \text{ or } A = C(1-D)^2$$

From $A = C(1-D)^2$ we get an expression for the output of the NRZI circuit:

$$C = \frac{A}{(1-D)^2}$$

Now by expanding this for the recording signal we get an equivalent formula:

$$C = \frac{A}{(1+D)(1-D)}$$

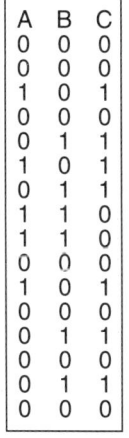

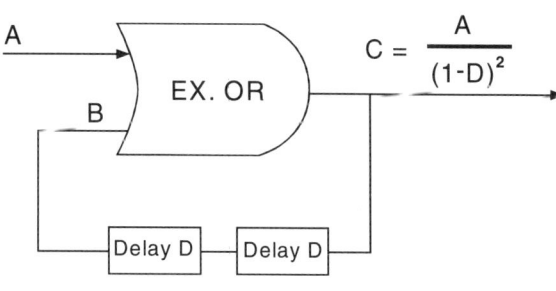

Figure 11.43 *Recording exclusive OR gate and logic diagram*

Video and Camcorder Servicing and Technology

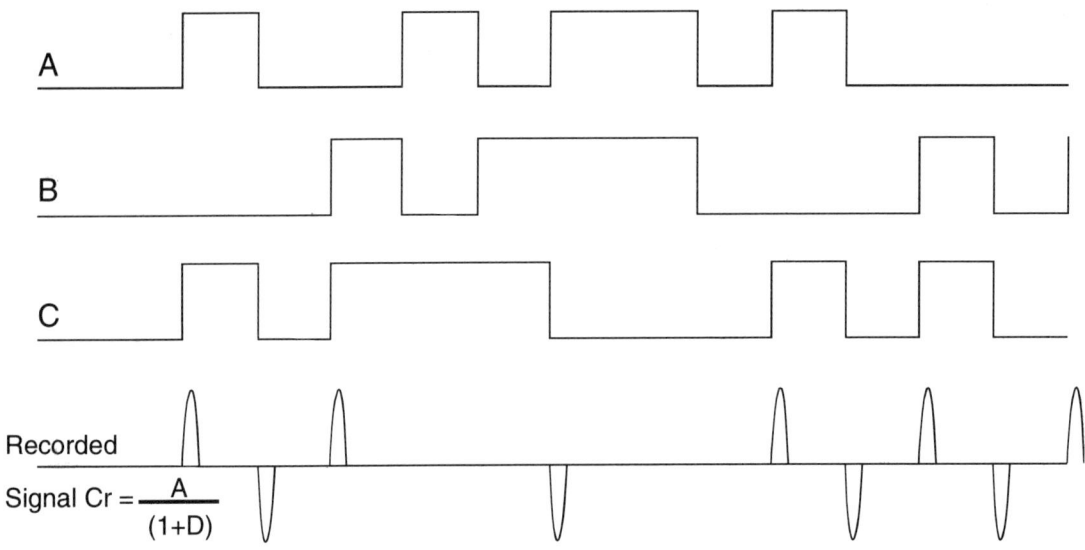

Figure 11.44 *Waveform diagram of the exclusive OR NRZI recording circuit*

In the recording process the data signal is differentiated by the video (data) heads as it is placed onto the tape and this is equivalent to multiplying the data by (1-D).

$$C = \frac{A}{(1+D)(1-D)} \times (1-D)$$

Therefore

$$Cr = \frac{A}{(1+D)}$$

Integration of the square waveform (A) into positive and negative pulses is the equivalent of multiplying (A) by (1-D) as shown in the diagram. Cr is the recorded magnetic signal upon the tape and is equivalent to A/ (1+D).

The two illustrations in Figure 11.45, show how the differentiation of the recording signal onto tape is the same as multiplying it by (1 - D) and that the recorded data is tri-state; having a 1, a 0, and a -1 state. The waveform diagram illustrates the NRZI process from (A) to $\frac{A}{(1+D)}$

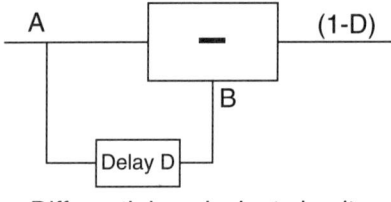

Differential equivalent circuit

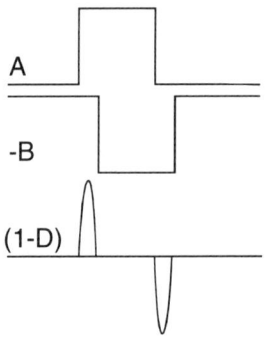

Figure 11.45 *Equivalent circuit of tape recording differentiation*

258

NRZI replay

The replay signal off the tape (P) is equivalent to Cr, so (P) is:

$$P = \frac{A}{(1+D)}$$

In the playback process the tri-state signal of $P = A / (1+D)$ is in either a 1, 0 or -1 state and is used in this configuration by the Viterbi error correction circuits for any tape drop out and noise affecting tri-state levels. Within the playback equalisation circuits, and part of the equalisation, is a (1+D) multiplier that converts the signal back into its bi-state format restoring it to its original format (A) (Figure 11.44). Negative (-1) pulses on the (Q) input are converted to positive pulses in the summing network of (P+Q).

Referring the playback waveform diagram (Figure 11.4) it can be seen that where pulses are of opposing polarity they are cancelled. Where either (P) or (Q) have a pulse, be it positive or negative, then there is a positive output for R=P(1+D). Following on is a clocked latching circuit that will produce a square wave for the period of the clocking pulses initiated by a output of the (1+D). Where two or more outputs occur in sequence then the latch will produce a longer square wave.

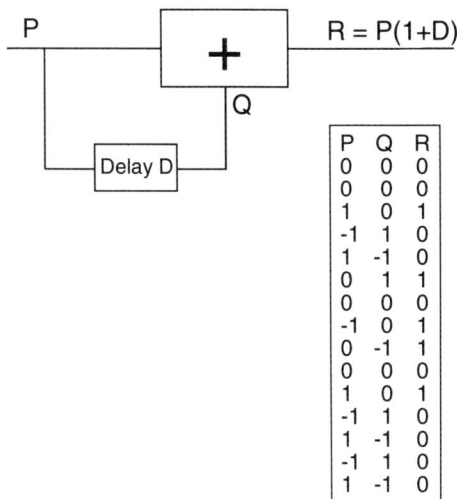

Figure 11.46 *Playback (1 + D) equalisation*

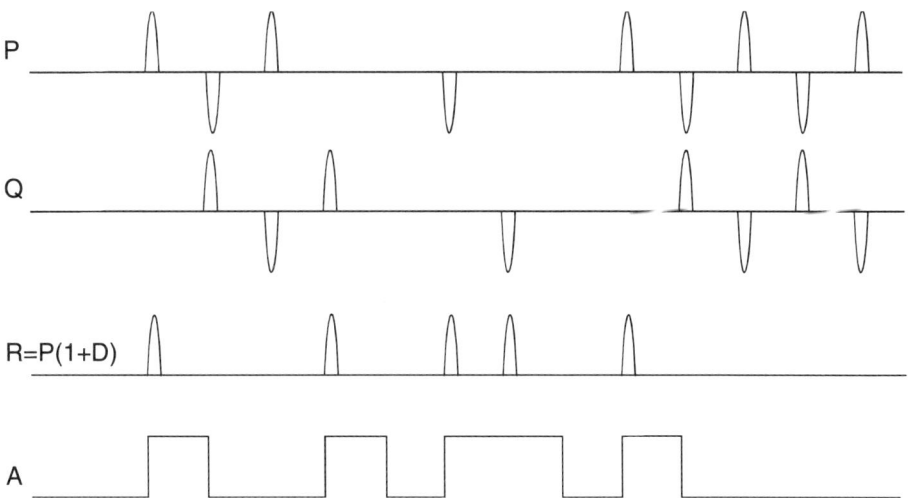

Figure 11.47 *Playback (1 + D) waveforms*

NRZI operation for a long string

Where a long string of 0s or 1s occurs the NRZI processing system comes into its own by 'chopping up' the data stream to ensure that there are as many clock pulses as possible to maintain the replay PLL clock synchronisation.

Figure 11.48 is an example where the input signal (A) has a long string of 1s. Due to the logic of the exclusive OR gate there are additional pulses recorded (shown in line P) during the extended high period. Also if (A) were to be left in the continuous high state it would tend to integrate into an unwanted DC level and the replayed data stream would not centre upon 0 V, it would lift up on the DC level causing missed data within the data slicing circuits that followed.

In replay the decoding of the (1+D) circuit (R) and the subsequent compensation and latching circuits restore the data back to the original (A). As an exercise check this waveform diagram against the logic diagram chart for the exclusive OR NRZI circuit to get the result (C) from the original signal (A).

Also confirm the result R=P(1+D) from the logic diagram and chart for a multiplication of (1+D) in the playback process; and that the clocked latching circuit then produces a long series of 1s from the string of (R) pulses.

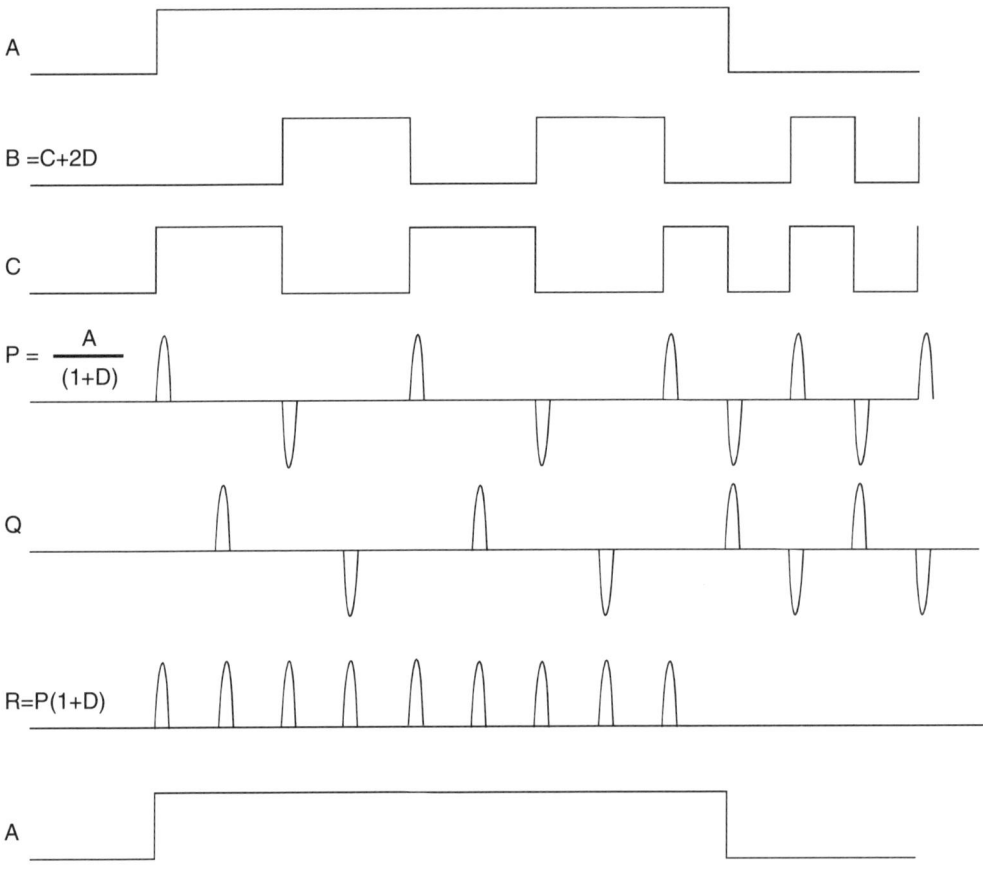

Figure 11.48 *Illustration of how the NRZI prevents long pulses from being recorded*

Viterbi error correction

In LP playback the data error rate of the tape is much higher than that in SP due to the reduced track width and increased noise, and subsequently requires additional error correction.

Binary data record drive to a tape head is a square wave whereas the replayed signal is not. This is due to the differential action of the inductive coupling of the video heads and rotary transformer. Replayed binary data is a tri-state signal comprising a positive pulse (1), a negative pulse (-1) and the zero level (0).

Recovery of the binary signal into its bi-state square wave is accomplished by the (1+D) multiplier and a latching circuit. This works fine as long as the signal-to-noise ratio is high and that the reed solomon parity error correction can accommodate the errors. If the signal is too distorted by noise as is the case in LP recording and poor SP playback then a more comprehensive signal recovery system is required.

Originally the problem was recovery of digital telemetry data from satellites and space where the data was severely corrupted by noise. As the noise was too great for data recovery by a variable slicing level system, a software solution using a complex algorithm was proposed by Dr. Andrew J.Viterbi. The algorithm analyses the data waveform and by storing sections, compares and examines the changes, decides the errors and provides corrections. Reductions in error rates of a factor of 100 can be obtained.

Firstly the replayed data waveform is A/D converted into a 6-bit word representing the voltage level of the samples taken from the signal that is waveform (P) in Figure 11.47. An explanation of this is that the A/D conversion is to measure the voltage level and polarity of the playback waveform (P) and the resultant voltage level is then given a 6-bit word to describe it. Viterbi error correction is processed in digital form by storing the values of the replayed signal pulses as a 6-bit word in order to compare strings of data.

Decision 1. In Figure 11.49 the levels are shown for samples (a) to (f). When the software examines these samples a decision is made as to the most likely binary level of the tri-state signal, with choices being 1, 0 or -1.

For samples (a), (b) and (f) the logic level is fairly certain in that they are 1, -1 and -1. Samples (c), (d) and (e) are uncertain, excepting that they are not negative, and so they pass on to the next decision stage.

Decision 2. This is performed upon a data train sequence. Within the magnetic coupling medium when a square wave is differentiated the 0s and 1s become tri-state (0. 1 and -1). It is certain that from the original square wave a -1 must follow a 1 if it is not 0. Therefore 0s are ignored as the data string is checked for alternating 1s and -1s.

Sample (d) is the highest level, it is positive and it follows the -1 at (b), therefore it must be a 1. Intermediate sample (c) must be a 0, because it falls between a -1 and a 1. Also by the same logic it follows that (e) is a 0 because it falls between a 1 and a -1.

If a number of samples cannot be decided upon then the sequence is stored in a memory to suspend detection while the next following sequence is processed. Viterbi looks for common sequences within a set number of bits and completes correction of the suspended data bits from information found during processing of the

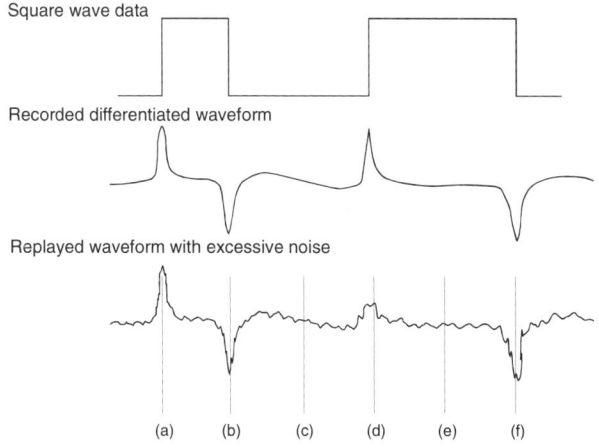

Figure 11.49 *Viterbi data recovery from noise.*

subsequent data, all of which contributes to the decision-making process. The number of bits stored can vary from manufacturer to manufacturer and the more bits that are used for comparison then better the error correction. The average is 30–40 bit samples stored.

In the circuitry the data bits are taken from the (1+D) stage for checking and then the Viterbi circuit applies correction to the recovered data outputted from the latching circuit in its two state binary form, as shown in Figure 11.42.

Digital 8mm

Sony's Digital8 system is a link between 8mm analogue camcorders and DVC camcorder. It is unique in that the mechanism is compatible for both analogue and digital formats allowing replay of analogue tapes as well as recording and replaying digital format tapes.

By retaining the 8mm tape width and a 40-mm diameter drum with a 221° wrap around and by increasing the speed to lengthen the track, it has been possible to record twice the amount of DV data on a single magnetic track. An increase in track pitch from 10 to 16.34 µm reduces the data errors and allows the recordings to take place on Hi8 tapes as opposed to producing a special ME or MP tape formula.

Digital8 can replay 8mm recordings, Hi8 recordings and Digital8 recordings either as analogue composite video signals or as a DV output to IEEE1394, i-link specifications. Most useful for PC editing purposes is that 8mm and Hi8 replays are converted into the digital domain and are available via the i-link output.

In most respects the digital recording and replay technology is the same as DVC and the data signal processing is similar, only the track writing and reading are different.

Figure 11.50 shows the difference between Hi8, Digital8 and the DVC formats. Because of the signal processing in the digital domain there is little difference between Digital8 and DVC specification. Both systems have a resolution of 500 lines, a colour bandwidth of 1.5 MHz, 16 bit or 12 PCM and audio sub-codes for the time and date.

Digital video and 8mm systems comparison

	Hi8	Digital8	DVC
Recording system	Analogue	Digital	Digital
Resolution	400 lines	500 lines	500 lines
Colour bandwidth	0.5 MHz	1.5 MHz	1.5 MHz
Video s/n	48 dB	54 dB	54 dB
Jitter	Fair	Fair	Exellent
Audio	FM/PCM	PCM	PCM
Tape storage	Fair	Excellent	Excellent
Tape speed	20.051 mm/s	28.695 mm/s	18.831 mm/s (12.55 -LP)
Tape width	8mm	8mm	6.35 mm
Cylinder speed	1500 rpm	4500 rpm	9000 rpm
Cylinder size	40 mm	40 mm	21.7 mm
Track pitch	34.4 µm	16.34 µm	10 µm

Figure 11.50 *Hi8, Digital8 and DVC format comparisons*

Digital camcorders

Track data distribution

Figure 11.51 shows the data distribution over six tracks in Digital8 as opposed to 12 tracks for DVC. Only a single margin area occurs at the end of each track for head crossover purposes. Frequency distribution for tracking purposes is over the six tracks and the tracking frequencies cover for two video track segments on a single path, see Figure 11.52.

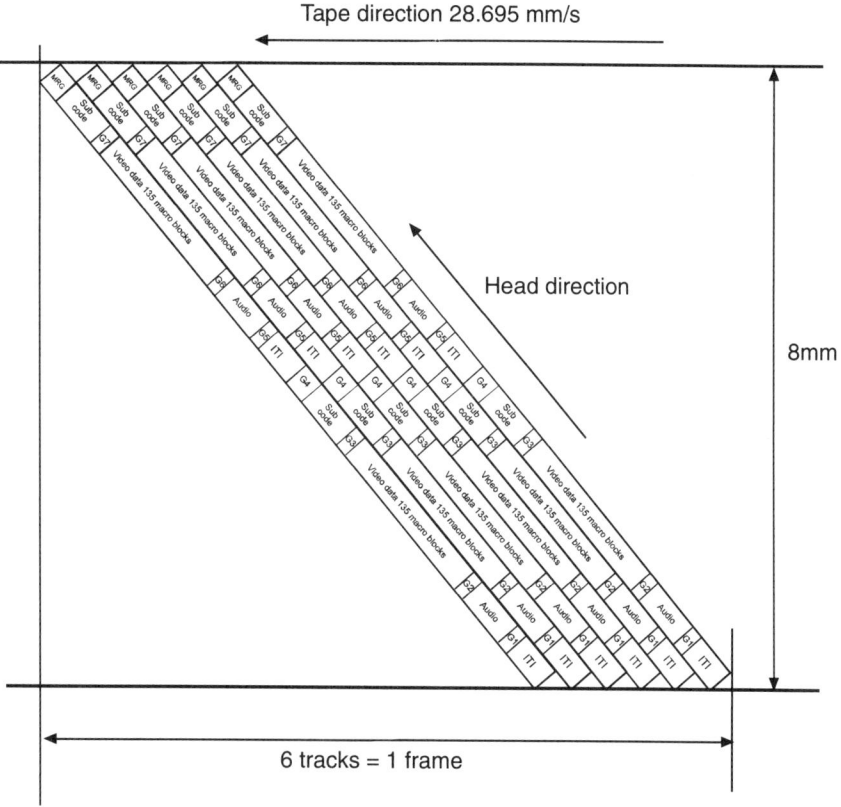

Figure 11.51 *Digital8 track layout*

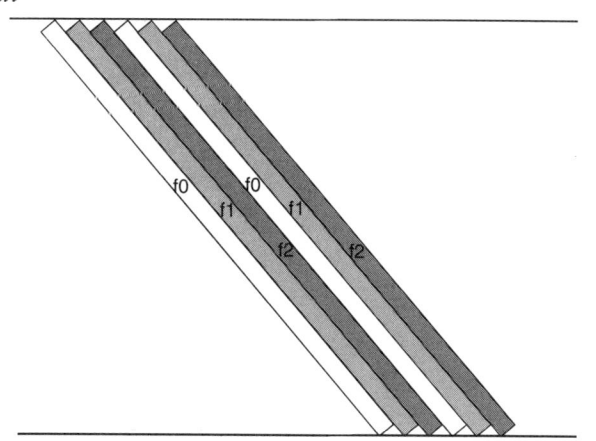

Figure 11.52 *Digital8 Tracking frequency distribution*

Audio

PCM audio is used in Digital8 with a 16-bit, 48-kHz sampling frequency for high quality and a 12-bit, 32-kHz sampling frequency for audio dubbing.

In the 16-bit mode the audio segments are similar to the DVC the first three tracks of six segments, as shown in Figure 11.53, contain left-audio data and the following three tracks contain six segments of right-hand audio.

In the 12-bit mode, as shown in Figure 11.54 the first three tracks are L/R audio of original audio recording and the following three tracks can be audio dubbed. Information of track segment timing for this purpose is stored in the ITI area as referred to earlier in this chapter.

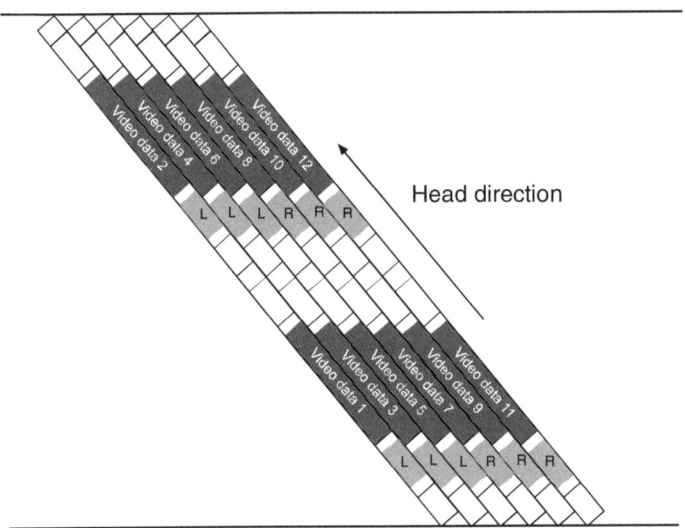

Figure 11.53 *Digital8 audio segments, 16-bit mode*

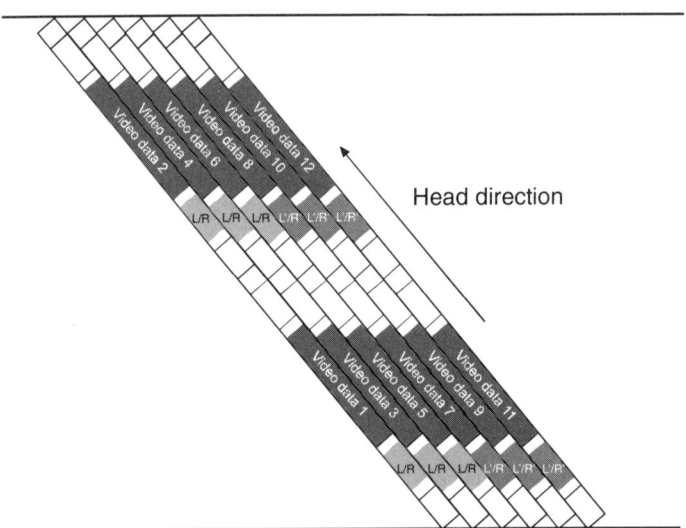

Figure 11.54 *Digital8 audio segments, 12-bit mode*

Digital still picture camera

A CCD imager, digital or analogue, provides a complete 625-line TV picture built up by scanning two interlace fields of 312½ lines each. If a still image is to be stored on a PC there are two problems. One is that the aspect ratio is not compatible, and the other is that the vertical resolution is that of one field; 312½ lines.

The aspect ratio of the TV picture is 4:3. In digital terms this is MPEG main level 720H x 576V pixels. A computer has two common aspect ratios: 640H x 480V pixels or 1024H x 768V pixels.

Many digital camcorders are found to have digital still camera functions inbuilt with picture storage on a card or memory stick. Aspect ratio compatibilty problems have to be overcome, with the addition of a compression system to reduce stored file size; standard JPEG is chosen for this.

Non-progressive scan uses a standard CCD imager scanned at field rate. The still picture is obtained by reading out the first (even) field into a memory then rapidly closing the iris and reading the second (odd) field as an interlaced field into the memory. Closing the iris prevents any movement distortion between fields. It is only possible to obtain a vertical resolution of 300 lines in the active picture area due to the way the imager is scanned for colour matrixing.

As previously described for the CCD imager on p.201, Figure 10.34, the first even field is constructed from rows 1 and 2, and the second (odd) field is made up from rows 2 and 3. It is because one row is used for both fields that the vertical resolution is approximately half of the 576 rows, that is around 300 lines.

Both the vertical resolution and aspect ratio problems are solved by using a development of the CCD imager to provide for frame still and by scanning the imager in either TV or PC aspect ratios.

Progressive scan CCD imager

In a progressive scan CCD imager all 576 lines are scanned to give a full resolution by scanning two fields simultaneously and then sorting them as required for various modes in a memory. Full motion recording in progressive scan mode will show a strobing effect as the recordings are 25 frames per second. This mode is for recordings where still pictures may be required for storage from the subsequent replay. Automatic systems may only enter the progressive scan mode when the photo shot function is selected.

Figure 11.56 shows the basic construction of a progressive scan CCD imager system. Consider that for video recordings the aspect ratio is MPEG main level, 720H x 576V pixels.

All 576 lines are scanned during the field period of 20 ms so that the scanning rate is 50 frames/s as opposed to the normal scan rate of 50 fields/s. The CCD imager has two horizontal output shift registers, HCCD1 (A channel), and HCCD2 (B channel) and during the scan alternate pixel rows are sent to each output. A channel outputs the odd fields and B channel outputs the even fields simultaneously.

Normal video is obtained by selecting alternate odd and even fields from these frames, such as the odd field from frame A and the even field from frame B, this sequence maintains real time motion and provides a standard interlaced TV signal.

Progressive scan for full motion recording takes the odd and even fields from frame A. Frame B is skipped and then the odd and even fields from frame C are recorded. As both fields are scanned simultaneously there are no motion vectors between the fields and this gives the visual strobing effect. To achieve this one field is taken direct while the other is stored temporarily in order to keep the odd/even sequence.

In photo shot mode the odd and even fields of a given frame are alternately recorded for the photo recording period which is about 7s.

It is not possible to use the same colour pixel arrangement for a progressive scan imager that is found in the standard CCD imager of Figure 10.34 where yellow, cyan, magenta and green pixels are mixed over 2 lines and the luminance content has a reduced vertical resolution.

A different arrangement has been developed with coloured filter pixels that will produce a full

Video and Camcorder Servicing and Technology

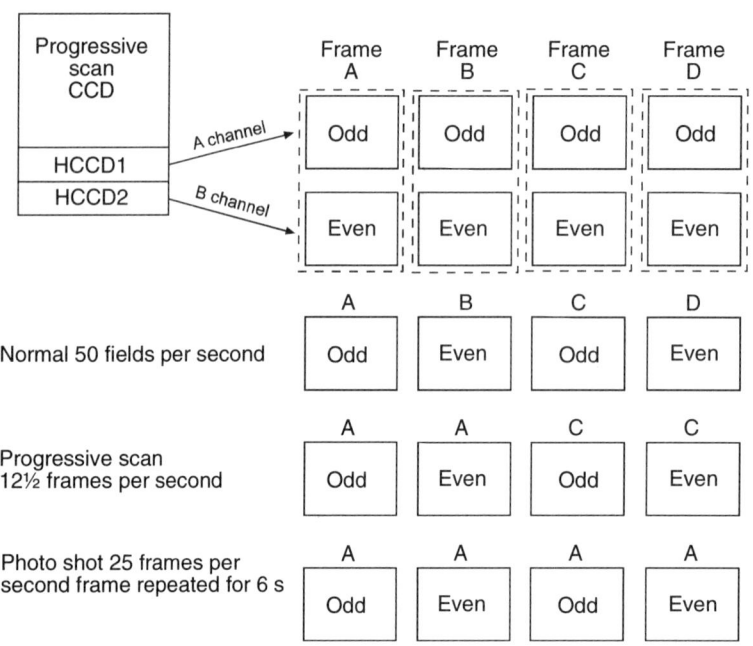

Figure 11.55 *Progressive scan imager structure*

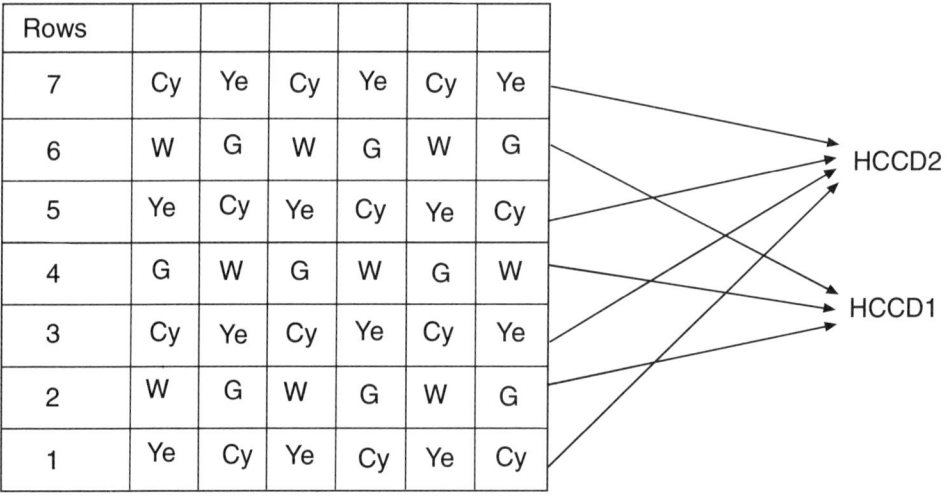

Figure 11.56 *Progressive scan imager colour pixel array*

luminance component in each row. This is shown in Figure 11.56 where the rows are directed to either output channel. It can be seen from this diagram that rows containing white and green are sent to HCCD1 (A channel) and those containing yellow and cyan are sent to HCCD2 (B channel). At first glance this may not seem to provide correct colours but colour matrixing does take place to produce suitable chrominance and luminance signal components.

Progressive scan components

Now where fs (13.5 MHz) is the horizontal scanning frequency as applied to the HCCD1 and HCCD2 registers, the following can be applied.

$$S_n = Y_n + \frac{C_n \cdot \sin 2\pi \cdot f_s \cdot t}{2}$$

The following colour components apply:

White   W = (R+B+G)
Green   G = G
Cyan   C = (G+B)
Yellow Y = (R+G)

An interesting result is obtained if we inspect the colours of the two A and B channel outputs:

(W+G) = (R+B+G) + G = R+B+2G
(C+Y) = (G+B) + (R+G) = R+B+2G

Each gives an R/B component and a green (luminance) component. The output of channel A is:

$$S_a = (W+G) + \frac{(k_w \cdot W - k_g \cdot G) \sin 2\pi \cdot f_s \cdot t}{2}$$

And the output of channel B is:

$$S_b = (C+Y) + \frac{(k_c \cdot C - k_y \cdot Y) \sin 2\pi \cdot f_s \cdot t}{2}$$

For R/B separation:

(W+Y) − (G+C) = (2R+2G+B) − (2G+B) = 2R
(W+C) − (G+Y) = (R+2G+2B) − (R+2G) = 2B

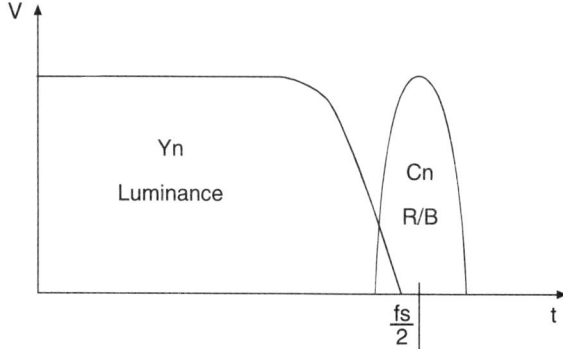

Figure 11.57 *Progressive scan imager spectrum*

Both A and B channel shift register outputs contain luminance in the form of green plus the colour component R/B, therefore full vertical resolution is obtained. This is similar to the system originaly used in the saticon colour tube described on p.191.

Digital still imager

Depending upon the make and manufacture, the progressive scan CCD imager can have up to 1000000 pixels, or more. This array does not make much difference to the normal DV recordings as these are fixed to the MPEG main specification of 720H x576V for a 4:3 aspect ratio; see Figure 11.58.

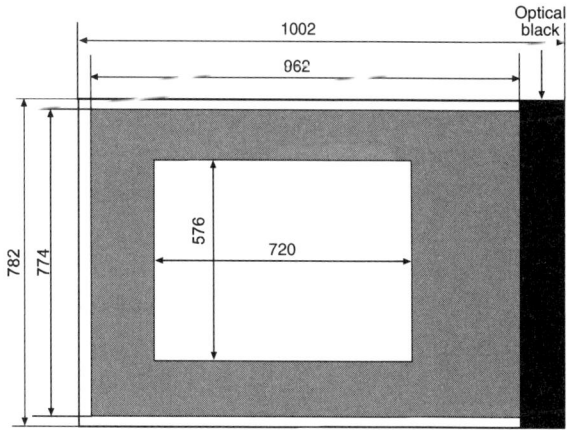

Figure 11.58 *Still camera progressive scan imager*

Digital still cameras (DSC) are incorporated into many digital camcorders and the extra pixel array is used to achieve PC aspect ratios: VGA at 640H x 480V, or XGA at 1024H x 768V, or even 1280H x 1024V. Where the pixel array may be a bit short of pixels for the PC resolution then extra pixels can be added by interpolation using the same method as described for digital zoom.

Digital still camera

Figure 11.59 illustrates a typical digital camcorder with an inbuilt digital still camera (DSC). The heart of the system is the camera digital signal processing (DSP) section. This contains the main timing and clock generation, CCD imager drives, zoom and focus control, and digital signal processing, including the picture effects. Within the CCD imager drive timing control the camera DSP can alter the aspect ratio for still picture capture and carry out any image stabilisation, digital zoom functions and special effects.

When the digital still camera functions are operational data is fed to and from the digital still camera processing circuits. Here a 16-Mb SDRAM memory is utilised for frame storage and field selection. Still picture data is written into the removable storage device, which can be either a multimedia card (MMC) or memory stick. The picture is compressed to a JPEG file format for compatibility with a computer. Early models used software compression but later models use a hardware IC.

There may also be an interface to a photo printer for direct printing of still camera pictures,

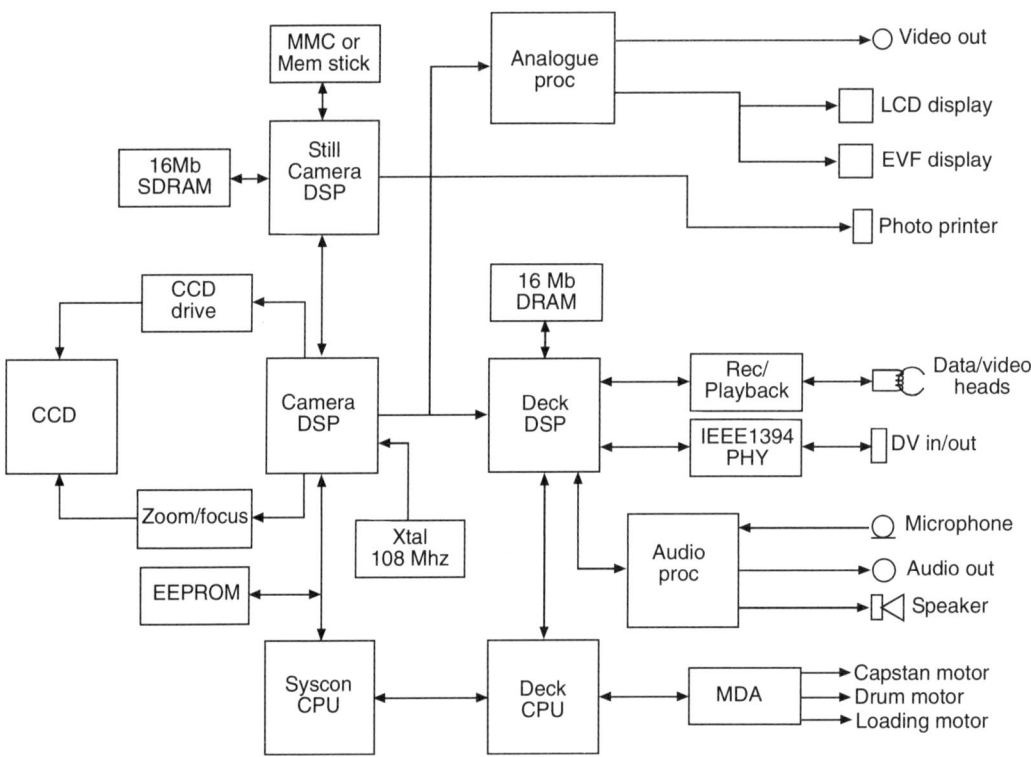

Figure 11.59 *Digital camcorder with digital still camera*

either from the CCD imager, storage device or tape replay.

Digital data signals are fed to the deck DSP for record/playback routing and DV communication data control, and to the analogue processing section for output and display.

Multimedia memory card

A standard format exists for the storage of JPEG picture files on multimedia cards within digital still cameras; this is the design rule for camera file system (DCF). It provides a standard for directory, and file structure for storage and management of image files and associated data on removable memories, such as multimedia cards.

There are three layers to the directory structure. The main directory is the DCF image directory and is called digital camera images or DCIM. Further to this is the DCF directory and then the image files in JPEG file format; see Figure 11.60.

Each DCF directory has an eight character directory designation where the first three characters are three digits numbered from 100 to 999, providing for 899 directories. The following five digits are arbitrary capital letters for naming the directory.

Within each directory are the image files with names consisting of eight characters; the first four characters are arbitrary capital letters or numbers. The following four characters are designated as numbers from 0001 to 9999 with the extension JPG for the file type. This allows for 9999 image files to be stored in each directory.

In Figure 11.60 the DCIM file has directories 100MYPIC and image files PICT0001 –0003. The still picture index within the digital still camera for image 2 will be shown as 100-PICT0002.

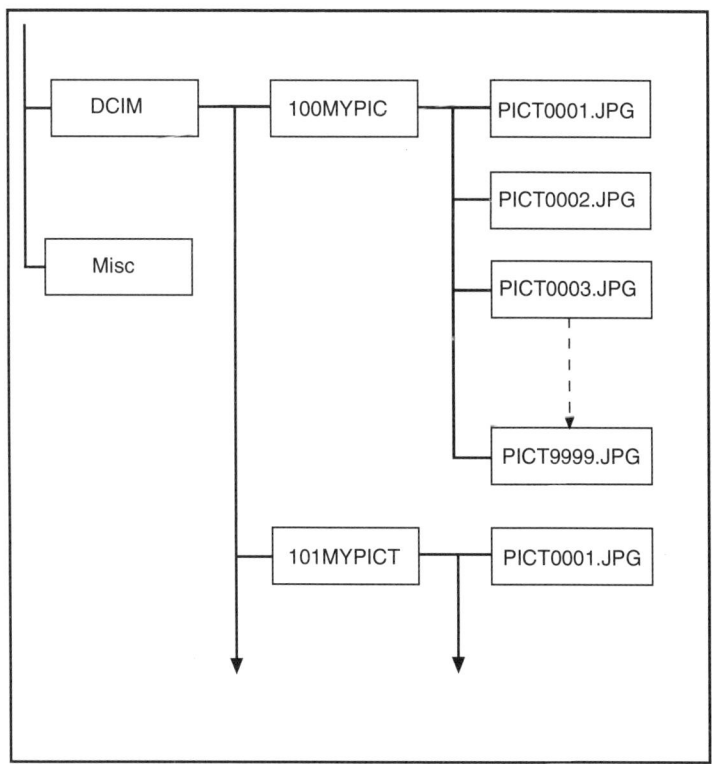

Figure 11.60 *Multimedia memory card directory layout*

		Fine	Standard	Economy
VGA 640 x 480	4 Mb	25	40	65
	8 Mb	50	80	130
	16 Mb	100	160	260
XGA 1024 x 768	4 Mb	10	16	32
	8 Mb	20	32	64
	16 Mb	40	64	128

Figure 11.61 *Comparison of MMC memory and image files*

Where there are 9999 images in a directory or if an image is numbered 9999, then a new directory is started and identified as 101MYPIC for a further 9999 image files, subject to sufficient memory.

Figure 11.61 shows the the comparison between MMC memory, resolution, and the number of image files stored. It is expected that storage capacity will rise.

Digital print order format

Along with the image files stored on the multimedia card there is an additional printing information file placed in the Misc folder. It allows the user to specify frames and the number of prints in each frame to be printed out.

A header section contains the version number of the file, the product name that stored the file and the time and date. Customer information, name, address and telephone number, is stored if the user inputs the data.

The user marks the prints whist checking stored images by DSC playback of what is contained on the multimedia card. This information is called the digital print order format (DPOF) and is stored in the Misc folder as a file; this file name is the 'autprint.mrk' file. A personal computer, printer or professional photo print service can read the autprint.mrk file and the specified images within the multimedia card can be printed out with captions.

12
MPEG and D-VHS

MPEG is a system defined by the Motion Picture Expert Group and contains a number of compression options of various data rates and picture quality.

Chapter 11 discussed the compression of video data by use of spatial redundancy where the amount of data in an 8x8 DCT block of a scene that has no detail, such as a portion of blue sky or a sandy beach, can be reduced. This was achieved by the picture content being defined by a macro block consisting of four luminance DCT blocks and two chrominance blocks and the only data in each DCT block is that of the DC level in the top left-hand corner. Where any detail is present then it is defined by the descrete cosine fourier Transformation as a set of frequency coefficients. MPEG-2 compression takes this a step further by introducing temporal redundancy. Originally the Motion Picture Experts Group developed MPEG-1 for moving video onto a CD at a data rate of 1.5 Mbits/s and low resolution video transmissions. MPEG-2 is an open system aimed at providing high quality pictures at data rates between 2 and 15 Mbits/s.

Temporal redundancy

Temporal redundancy is the concept of reducing the amount of data by taking a group of TV frames and comparing the amount and direction of any movement taking place between each frame, and then sending data to describe only that movement rather than the total data for each frame. It is not perfect, as motion artefacts can be seen in fast moving parts of the picture and any temporal errors that may occur can be seen over several frames.

Each group consists of 12 frames made up of three different types of frame: I frame, P frames and B frames.

Main profiles and levels

For good picture quality there are three main profiles that can be applied to each of four levels.

Simple profile
This profile can be used to simplify encoder and decoder circuits as it does not use bi-directional prediction B frames. The bit rate is much higher and the compression is lower because of this.

Main profile
This is the most common for broadcast and DVD and uses all three frame types: I, P and B. Encoders and decoders are complex and require a number of frame store memories.

High profile
This is intended for high definition digital broadcast.

Levels

		Max-data
High level	1920H x 1152V	80 Mbit/s
High 1440	1440H x 1152V	60 Mbit/s
Main level	720H x 576V	15 Mbit/s
Low level	352H x 288V	4 Mbit/s

For consumer broadcasting, digital recording, PC editing and DVD the main profile at main level is used giving an aspect ratio of 720H x 576V and a transport data stream with all three frame types.

Video and Camcorder Servicing and Technology

I frames

I frames are compressed using intra-frame compression of data identical to that of each DVC frame described in chapter 11. Intra-frame means that each frame is individually compressed without reference to preceding or following fames.

P frames

These are predicted frames that are constructed using a previous I frame or P frame as a reference, and need several stored frames. P frames are used to construct other P frames and B frames. An error in a P frames may be carried over to following P or B frames.

B frames

These are bi-directional predicted frames and are formed from bi-directional interpolated motion prediction between I and P frames. B frames are not used as a source for other frames as they are the result of motion interpolation calculations between I and P frames.

Motion prediction

By comparison of macro blocks between subsequent frames any motion can be analysed and from the direction of travel the resulting position in a following frame can be calculated and predicted.

The overall sequence of a group of 12 frames is shown in Figure 12.1(a) where the relationship between I, P and B frames. In Figure 12.1(b) two B frames, B2 and B3, are produced by the comparison between I1 and P4. To achieve this the first four frames are stored, and by motion adaptive comparison between I1 and P4 the speed and direction of the ball can be calculated. Then the positions in the two B frames can be predicted and placed. In this way only the motion data for B2 and B3 is required. In the transmission system B2 and B3 do not exist as pictures, only the motion data is sent providing a very high data reduction rate. Also the data volume for the P frame is reduced by sending only the motion

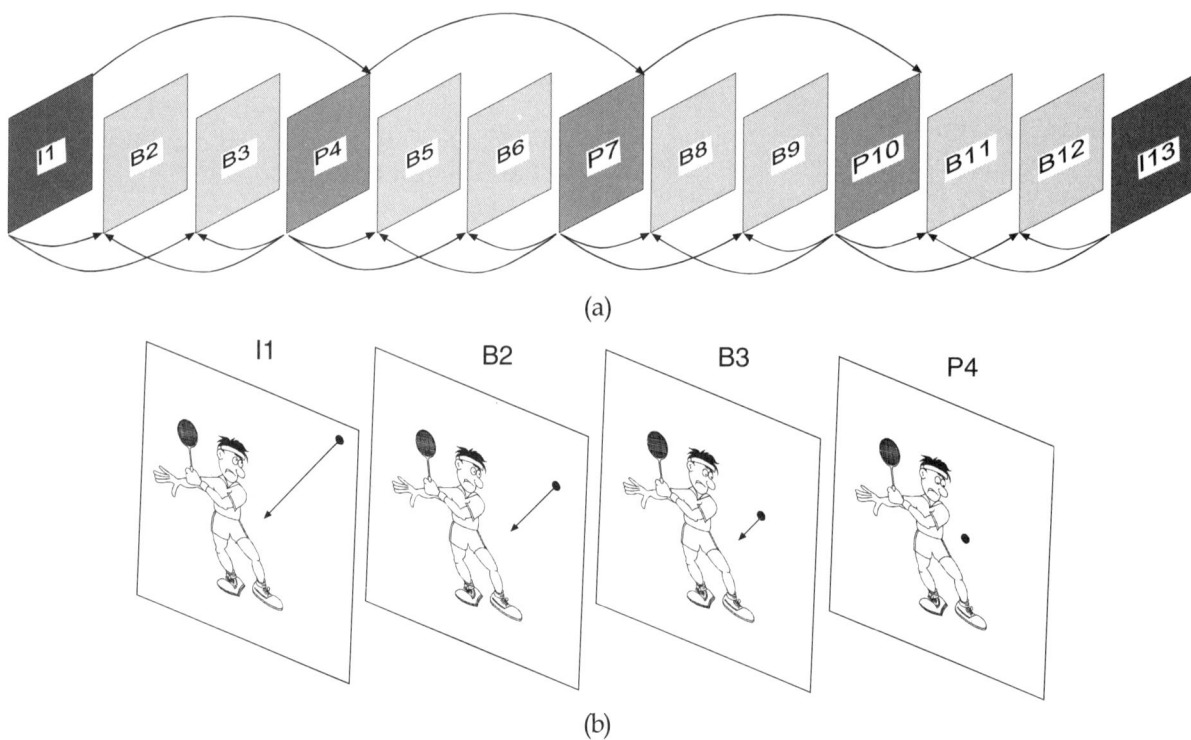

Figure 12.1 *Main profile MPEG frame construction and motion prediction*

prediction data calculated from the I frame. This carries through P7 and P10 until refreshed by I13. Each group is 12 frames consisting of one I frame, three P frames and eight B frames. Note that the largest numbers of frames (B) are those with the lowest data volume.

D-VHS recorder

A set-top box is generally used to receive satellite or terrestrial digital TV transmissions that use the MPEG transport stream and it may be available as an output signal from the set top-box. D-VHS has been developed to record MPEG encoded digital television programmes that have been selected by the tuning system of the set top box. This output is not the full transport stream nor is it strictly a programme stream. It is referred to as a partial stream as the transport packets are spaced apart; see Figure 12.3.

A multi-channel broadcast from a single transponder carries a number of programmes as a transport multiplex and the programme to be viewed or recorded is selected by the set top-box from this multiplex.

The D-VHS recorder is fitted with a coding/decoding interface called a CODEC and is capable of converting DVC digital signals to MPEG format as well as digitising input analogue signals. In order to make the purchase desirable a D-VHS recorder can also record and playback standard VHS, S-VHS and hi-fi audio to maintain backward compatibility.

Figure 12.2 shows the basic system diagram for such a video recorder. There is a partial stream connection between the D-VHS recorder and the set-top box via an IEEE1394 connector system using the MPEG TS protocol. A camcorder or digital video recorder, using DVC can also be connected by an IEEE1394 port using the DVC-SD protocol. Analogue signals can be routed via an MPEG encoder for digital recording or the signals can be recorded directly as VHS or S-VHS.

A D-VHS recorder can accommodate digital or

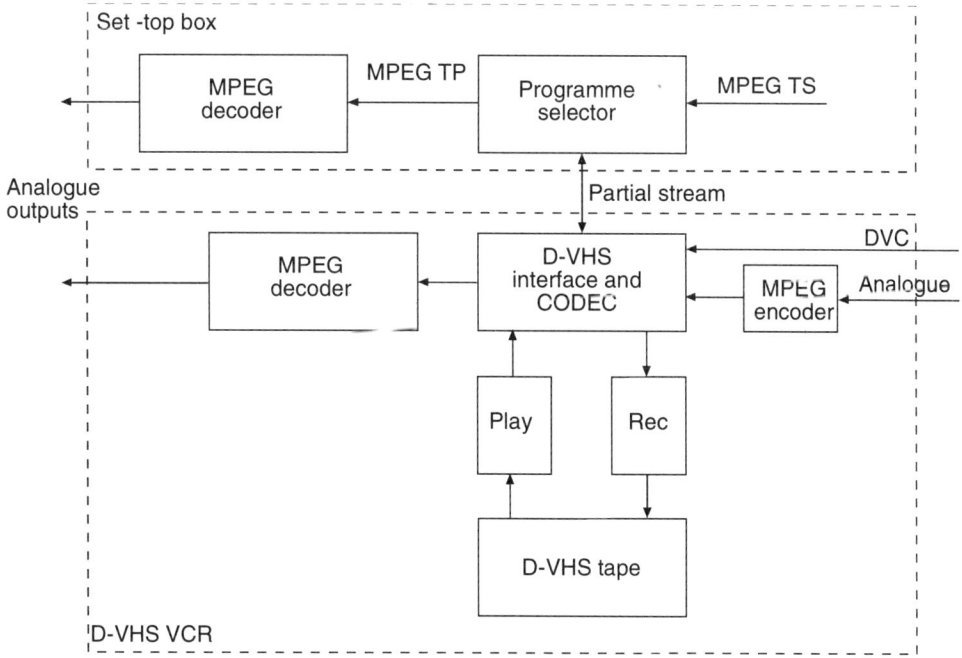

Figure 12.2 *Basic D-VHS system*

Video and Camcorder Servicing and Technology

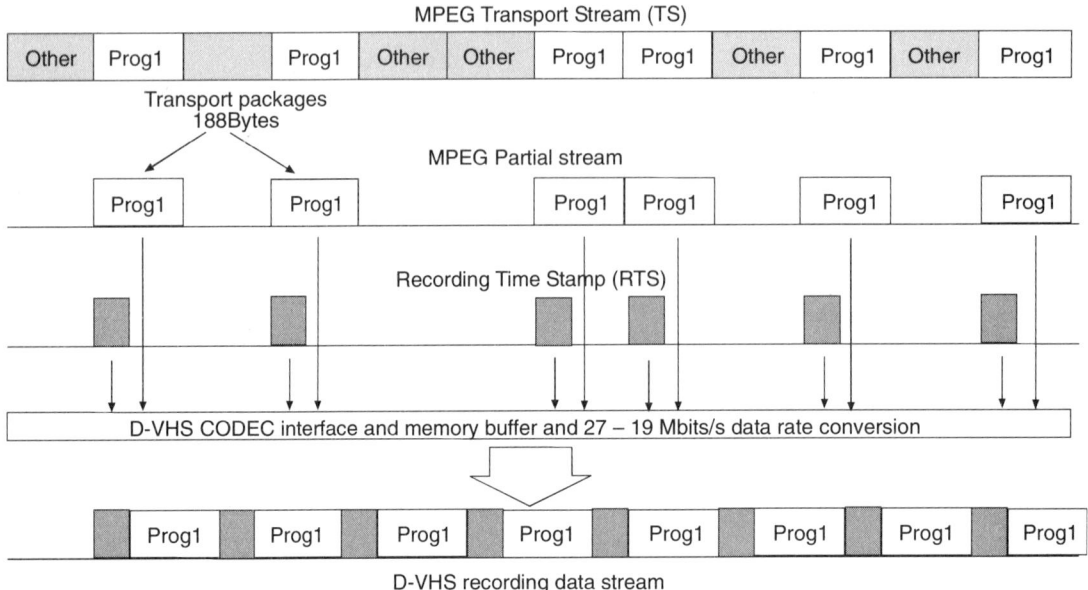

Figure 12.3 *Derivation of the recording data from an MPEG transport stream*

standard analogue recordings due to a multiple head drum assembly and by the method of D-VHS recording being similar to LP VHS in its basic drum and tape speeds.

MPEG programme stream selection

Figure 12.3 shows the MPEG transport stream on the top row. A selected programme that is fed between the set-top box and the D-VHS video recorder as a partial stream is shown in row 2. Each transport packet of data has its place in the partial stream and to ensure that its timing is maintained in replay a recording time stamp (RTS) is added to each packet prior to recording. It is not efficient on tape usage to record each packet with blank sections between them and so a variable memory buffer, part of the CODEC, places the transport packets and recording time stamps in a contiguous serial format. Each transport packet in the partial stream has a clock of 27 MHz although due to the spacing the average data rate is around 8 Mbits/s. For efficiency the data recorded onto the time has to be at the optimum volume. Too much data will cause massive errors, too little is a waste of tape. In the CODEC section the intermittent TPs are clocked at 27 MHz into a memory with a variable capacity, as the data volume varies between I, P and B frames, and then clocked out at 19.1 Mbits/s for recording.

D-VHS tape format

D-VHS tape tracks are similar in angle and length to those of VHS LP; see Figure 12.4 There are two tracks per head rotation and each track has a

MPEG and D-VHS

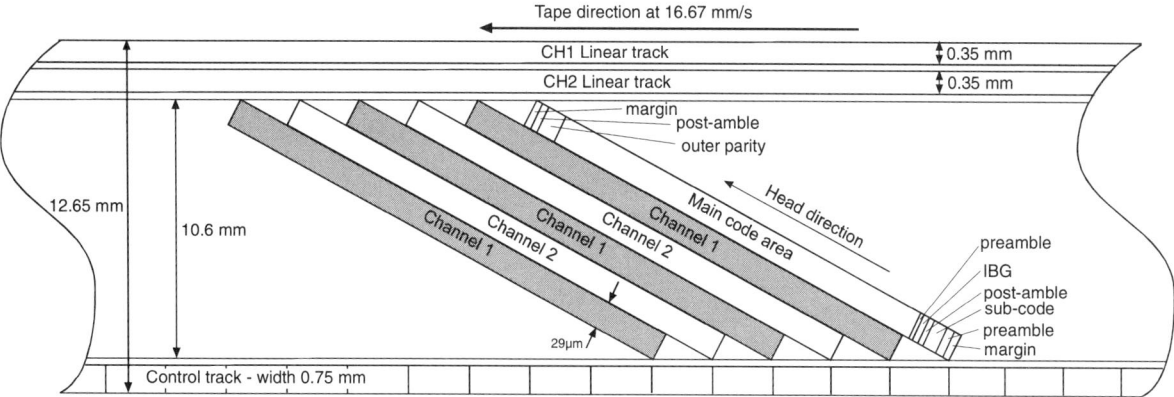

Figure 12.4 *D-VHS tape format*

different azimuth: these are +30° and −30° for each head. The record/replay rate is 60 tracks/s and the tracks are grouped into sets of six for servo purposes. Each track has a width of 29 μm and the pitch is also 29 μm, so there are no gaps between the tracks. Note that for PAL VHS LP there are 50 tracks/s and the track pitch is 25 μm (see p.150).

Where the DVC tracks in Chapter 11 have gaps for the data section separation, in D-VHS this is in the form of preamble and post-amble sections. These are sections with data present to keep the replay PLL clock running in sync by using the format 111000111000, etc. The margin sections at the start and end of the track are for head switching and also have non-valid data in the same format as the pre- and post-amble sections.

Sync blocks are used to quantify and organise the data. Each sync block (SB) contains 112 bytes of 8-bit data, or 896 bits. There are 356.356 sync blocks in a track of 112 bytes so a track is 39.9 kbytes. There are 60 tracks/s so the record/replay data rate is 2.39 Mbytes/s or 19.1 Mbits/s

As shown in Figure 12.5 each track is formatted into 356 SB, although the actual timing per track is 356.356 sync blocks as the last margin has an extra bit of a sync block due to the head switching timing.

There are two main data areas, the sub-code

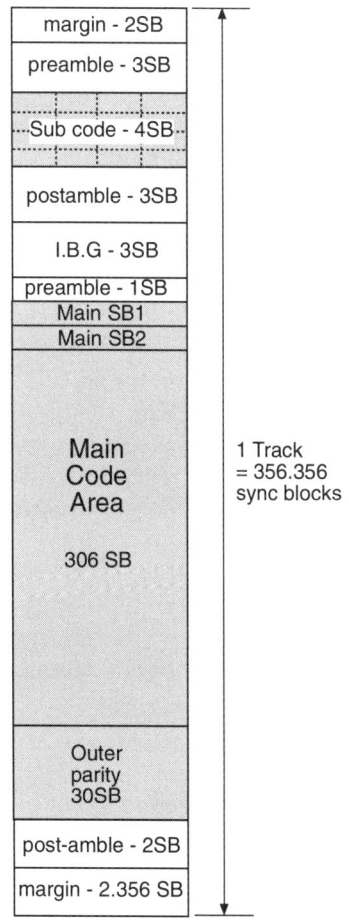

Figure 12.5 *Track data structure*

Video and Camcorder Servicing and Technology

consisting of 4 SB and the main code area of 306 SB with an additional 30 SB for outer parity for error correction.

IBG is an inter-block gap to separate two post-amble and preamble sections. There is no valid data, just the 111000111 PLL sync data pattern.

D-VHS sub-code

In the sub-code section each of the the four sync blocks are subdivided into four sub-code sync blocks (SCSB) of 28 bytes each, making 16 in total. The capacity of each SCSB is that of three pack data segments of 6 bytes each (18 bytes) the remaining 10 bytes of the 28 bytes are used for pack data identification and error correction parity. This gives 16 sub-code sync blocks each with three pack data segments, a total of 48 pack data segments, each of 6-byte capacity. Within this data the following typical information is contained.

Channel number, which is a recorded broadcast channel number.
Text header, which defines the type and amount of text data
Text data, the programme title.
Recording date and time
Programme time code, which is the elapsed time at any point in the recording
Programme time, the total recording time of the programme

It is possible for the type of data contained in the sub-code area to be changed to provide information other than that above.

Main code area

The first two sync blocks, SB1 and SB2, in the main code area are shown in detail in Figure 12.6. The reason for this is that an MPEG transport packet of 188 bytes is recorded over two tracks. Each sync block within the main code area is the standard 112 bytes capacity and 96 bytes of this are allocated to data. Two tracks have a capacity of 96 + 96 = 192 bytes, less 4 bytes for the packet header containing the RTS data, which leaves 188 bytes for the transport packet. This also means in terms of data streaming that it takes one head rotation to record or playback an MPEG transport packet and that the MPEG data is not compressed.

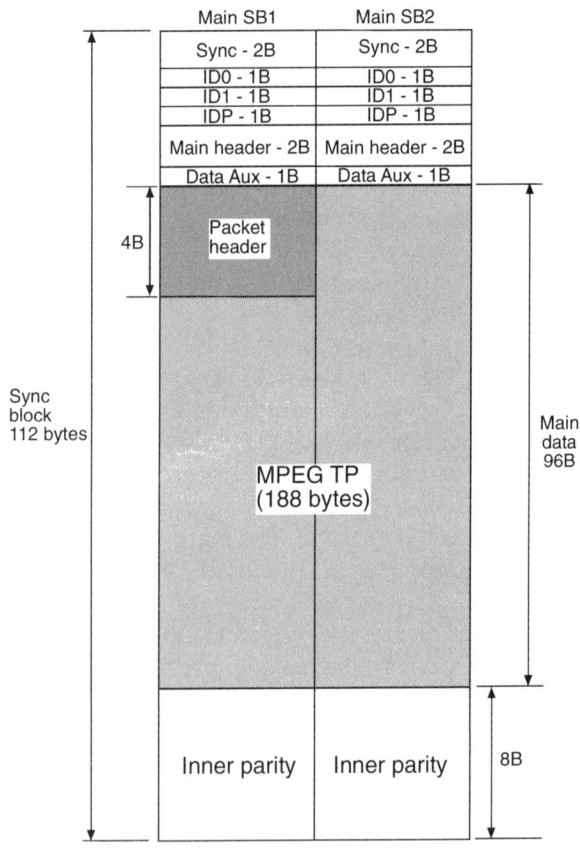

Figure 12.6 *Contents of the first two sync blocks in the main code area*

Recording time stamp

Within the packet header is the recording time stamp. It is clocked by a 27-MHz timing clock and is stored in the packet as counter data.

Figure 12.7 shows the packet format. It is 4 bytes or 32 bits in volume; 10 bits are reserved leaving a 4-bit counter and an 18-bit counter.

The first part of the count of a 27-MHz transport packet clock is the 18-bit counter RTS1. This is configured to count from 0 to 225224, a total count of 225225. Each time RTS1 reaches 225224 then RTS2 clocks on 1. RTS2 is a 4-bit counter and is configured to count from 0 to 11, a total count of 12, so the full count of RTS1 and RTS2 is 225225 x 12 or 2702700, and this is equivalent to six tracks. At the end of a six-track sequence both counters are set to 0.

RTS1 and RTS2 are not the only counters; ID0 and ID1 in each sync block count from 0 to 335 and identify each SB in a track. They also identify track and track pair to maintain transport package integrity in replay.

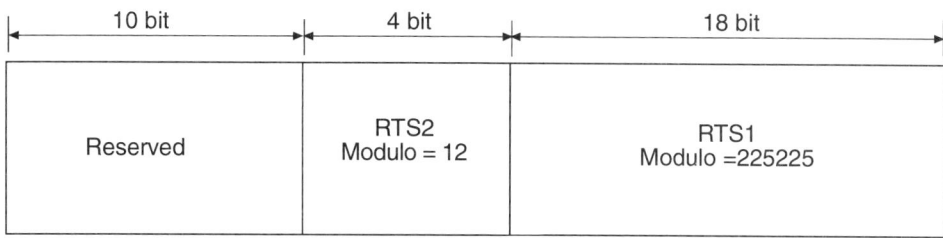

Figure 12.7 *Packet header (RTS) data structure*

13
Editing

The principle of editing in video recorders and camcorders was introduced in Chapter 4 (the section on servos). This is the operation known as 'backspace' editing. During the recording process the stop/start switch or record pause is selected. The system control takes over from the record/pause mode selected by the user and rewinds the tape, still threaded up around the mechanics, by 32 frames. This is counted by the system control using the control track pulses picked up through the control track head, which at this point is switched to a playback function. The video system, including the video heads, are in a 'pseudo-playback' mode. This reference to pseudo-playback is because the user sees that the video recorder or camcorder is in record mode, displaying E/E or camera pictures. The servo is playing back control track pulses and locking them to input or camera syncs ready to switch to full record. This sets up the control track so that there is a smooth transition between the last CTL pulse that is replayed and the next new one that is recorded at the edit point.

It is a useful function to have 32 frames of playback in reverse. Camcorders in the field are able to provide the user with a short review function as confirmation of the previous recording by reverse playing, sometimes for a longer period, before switching to forward wind up to the edit standby point of -32 frames or 1.3 seconds (according to the type and model).

At the end of the 32-frame review/rewind the system control enters record/pause awaiting the start signal. Once the start control signal is received a short playback period is again entered whilst the servos lock up and then the full recording function is entered with all of the signal circuits in record and the erase circuits active.

In earlier assemble edit video recorders the sequence from record/pause to full record was the pseudo-playback of 20 frames with five frames of overlap of the previous recording. There were then changes in design to reduce the time of the overlap period for a very good reason – overlap recordings rely heavily on the ability of the video heads to erase the previous recording by over-recording. It is reasonably effective on the luminance FM carrier but less so on the chroma signal. This causes visual problems at the assembly edit points and can ruin any use of edits. If a bright colourful scene is overlapped or over-recorded in the edit mode by one with less chroma or contrast then colour flickering can be clearly seen in the replayed picture over the edit.

In order to reduce the colour flicker effects at assembly edit points, the overlap period is reduced by a series of design improvements until flying erase heads and zero frame editing are finally fitted to eliminate the adverse effects completely.

Zero frame editing

Zero frame editing was introduced in order to reduce the visual disturbance assembly at edit points and reduce the overlap period from a few frames to zero. Camcorders were the first main beneficiary of zero frame editing as it virtually eliminated any disturbances during filming when regular use of the stop/start trigger is encountered. In the mains machines the overlap had been reduced to one to three frames by using the capstan FG signal as a counter source. Unfortunately, the low frequency response of the capstan FG amplifiers make counting erratic at the low tape speeds encountered in backspace editing, and it was found to be inaccurate as a counter reference for zero frame editing (ZEF). Use was therefore made of the control track pulse as an indicator of the precise edit point. Care had to be taken so as not to interfere with the VHS tape indexing system, replay capstan tape servo, or any other system using the control track.

When the recording sequence is ended by 'pause' the video recorder or camcorder runs on for four pictures or 80 ms. During this short time the recorded control track is slightly modified in its timing to identify these four pictures before the tape is then stopped and re-wound for 1.3 s worth of tape using the control track as a counting source.

When the recording is started again the video runs in the 'playback' mode for about 1.3 seconds looking for the specially modified control track pulses and the extra four pictures. During this time it synchronises replayed control track pulses and incoming video signal pulses to avoid disturbance of the control track at the edit point. Once the first of the four pictures is located there is a count of 3 and then the new recording commences exactly where the previous one left off and without errors.

On some earlier models of camcorder there was found to be a button marked 'edit'. This enabled the camcorder to control another VHS video recorder that conforms to two conditions: (1) that there is an external remote pause socket; (2) that it has backspace editing or zero frame editing.

The camcorder is connected to the mains machine with the special AV connector lead that has the extra remote 3.5 mm jack plug. Search is used on the mains machine, and the starting point or next recording to be added to the original is identified by visual search or play and then paused with a picture. Next, the pause button is held and record selected. This now puts the main video recorder into record/pause whereupon it will backspace 1.3 seconds' worth of tape.

The camcorder visual search and play is now used to find the beginning of the next sequence to be copied and picture pause selected. At this point a poor quality still picture will be evident unless the camcorder has a still picture function.

Now the 'edit' button is pressed. The camcorder will rewind 5 seconds' worth of tape from the paused point and then enter 'play'. After 3.7 s it will start the mains machine from record/pause to pseudo play. A further 1.3 s later full record is entered by the mains machine coincident with the precise edit point selected on the camcorder over a total 'run in' period of 5 s.

A small timing error will occur on mains VCRs that do not fully backspace by 1.3 s, and as this is classified as domestic equipment then the error margin is acceptable.

If virgin tape is used then there is no visual disturbance at the assembly edit point. However, if a previously recorded tape is used then a small portion of tape between the erase head and the video head drum will be over-recorded and colour flicker will occur.

For many years the technique for deriving the control track pulses in VHS video recorders was achieved by dividing the incoming vertical signal pulses by 2 (Figure 13.1). The division ratio was not designed to produce a symmetrical waveform of 20 ms/20 ms from the total period of 40 ms, as a non-symmetrical mark/space ratio was required. Discrete monostable trigger circuits were first used and, when triggered by a vertical signal pulse, suitably separated and cleaned, the 'on' period was set to 28 ms. From a total period of 40 ms this left a reset period of 12 ms – a ratio of 70%/30%. Eventually the VHS specification was rewritten to accommodate the proposed VHS program video indexes search System (VISS).

In this new specification the logic 0 is a duty cycle of 60% ±5% and logic 1 is 27.5% ±2.5%. For VCRs without any index systems the general control track duty cycle is also 60% ±5%. However, this does not seem to include the camcorders, which have their own duty cycle of 55% for normal control track ratio and a modified 65% for zero frame edit timing.

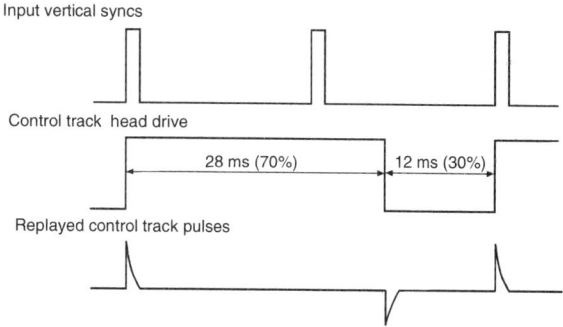

Figure 13.1 *Standard control track ratio*

The large and more sophisticated table VCRs with editing systems make more use of the index coding for zero frame editing. Normal control track duty cycle is 60%, or logic 0 and logic 1 is utilised for edit timing in these cases.

Example of a camcorder zero frame editing system

Refer to the timing chart in Figure 13.2. The camcorder is recording a scene when the stop/start button is depressed by the user, 'user pause', and the system control sets from record to record/pause. As the transition from record to pause is random, a point of operation is selected by the system control 1 ms after the falling edge of the next recording control track signal after the pause function is selected. At this point a control signal designated 'record control duty cycle' switches high. The effect of this is to change the ratio of the duty cycle of the recorded control track from 55% to 65%.

After a count of four control track pulse periods, the video recording signal is terminated and is timed by the falling edge of the drum flip-flop signal. This ensures that the video heads are switched off in vertical blanking so that a full field or video track is recorded and that the video signal is not disrupted in mid-scan. Control track recording continues for another few ms to provide for an overlap. This is to prevent any disruption to control track recording. The control track must be continuous over the edit point during replay or the servo will 'kick' and make the picture jump about. After the capstan motor is braked the tape is backspaced for 1.3 s and stops in record/pause awaiting the next 'start' instruction.

Once the trigger button is depressed the tape

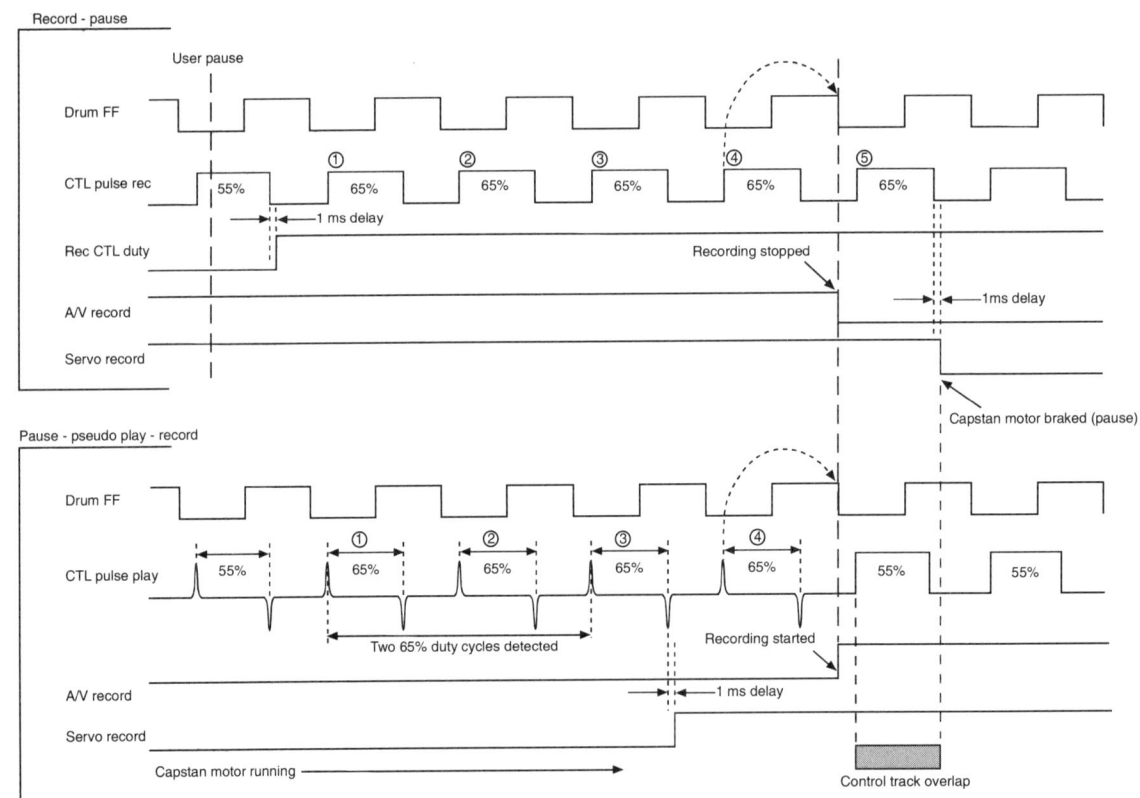

Figure 13.2 *Basic zero frame edit timing*

Editing

starts in pseudo-playback and the system control monitors the replayed control track pulses at 55% duration looking for a change. During the search period of approximately 0.9 s the vertical syncs from the camera are compared to replayed control track in the capstan servo in order to lock-up and stabilise the servo. When the first 65% duty cycle is detected a count of 2 starts and the servo system is primed for 'servo record' which commences 1 ms after the third 65% duty cycle period. After the fourth 65% period has commenced the video (AV) recording signal is switched ready and timed 'on' with the next fall of the drum flip-flop. The result of the accurate timing is that the first field of the new recording starts on the next video track after the original recording ceased. On previously recorded tape there is no overlap of video recording signals and hence no chroma interference particularly if a flying erase head is fitted. Audio recording is paralleled to the video switching logic. Control track timing is returned to 55% duty cycle and the remaining period of the control track is easily over-recorded with no problems, the overall backspace editing timing flow diagram is shown in Figure 13.3.

It may be noted that although this zero frame edit ensures a noise- (or glitch-) free edit a small price has to be paid. After user pause, the camcorder will record a further four control tracks or, to put it another way, eight video tracks. Whilst this has little effect in a domestic situation, it can cause minor problems when two zero frame editing units are connected together in a master edit control system where accurate scene timing is important.

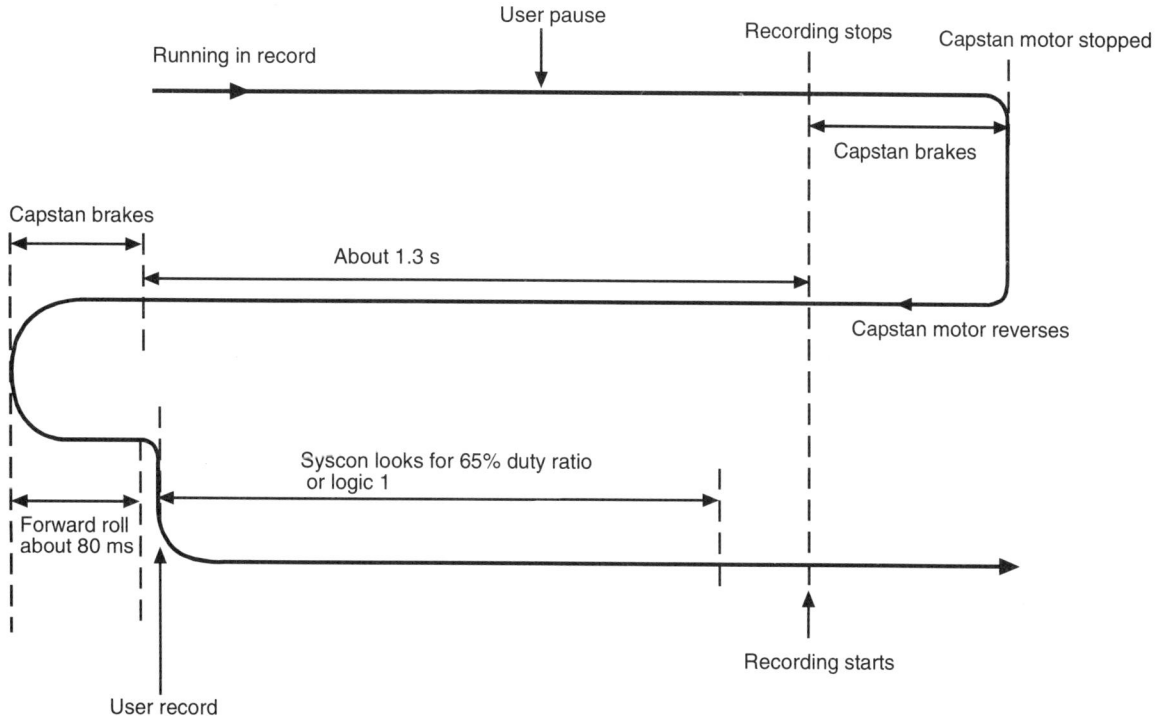

Figure 13.3 *Editing timing and control sequence*

Master edit control system

Some camcorders are fitted with a master edit control system and these can be identified by having an extra button marked 'edit'. Connection between the camcorder and the VCR is by a three-way cable carrying audio and video signals and an additional remote pause control on a 3.5-mm jack plug, as shown in Figure 13.4. The output signal is buffered from the camcorder by an open collector transistor, and a zener diode is fitted as a safety device to protect from excessive voltages. Fault finding note: it can short circuit and prevent operation of the main VCR by the camcorder.

When using the camcorder to copy tapes onto the main VCR, such as a number of 30-min cassettes onto a single E180, the edit control system makes the job very easy. Control of recording VCR is done by the camcorder system control in time with the camcorder backspace edit timing. The camcorder will output a signal to the dubbing VCR which will switch it from record/pause to record 1.3 s before the playback camcorder reaches the selected edit point. A description of the method operation is as follows (refer to Figure 13.5 for the precise timing).

(a) The main recording deck is used in play, cue and review modes to select the 'edit in' point where play pause is selected. It is then put into record/pause by holding 'pause' and selecting

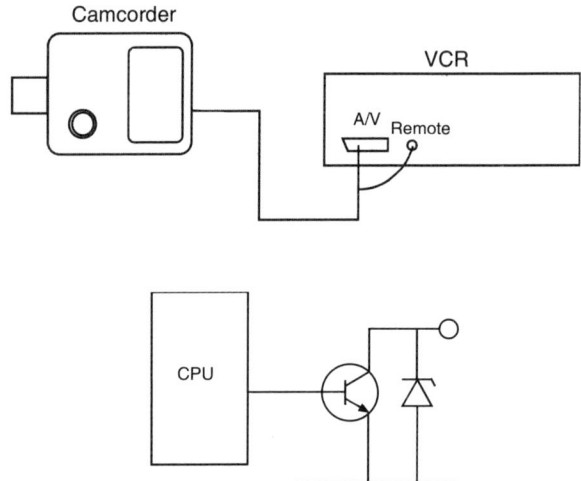

Figure 13.4 *Master edit control connections*

'record', whereupon it backspaces by 1.3 s and stops.

(b) By using play, cue and review on the camcorder the 'edit out' point is selected first and is marked by zeroing the tape counter. The 'edit play in' point can then be found whereupon play/pause is selected.

(c) Now by selecting the 'edit' button on the camcorder the user can sit back and watch whilst the camcorder does the work.

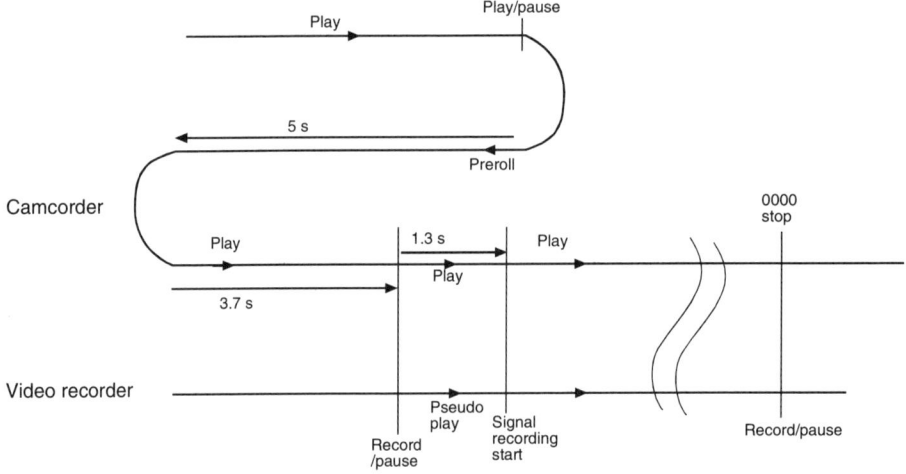

Figure 13.5 *Master edit control sequence between camcorder and VCR*

Operation sequence

The camcorder now backspaces in review mode for 5 s and then enters 'play'. After 3.7 s of playback has passed an output signal is sent to start the recording VCR, which now has 1.3 s to stabilise the servo, reach the edit point, and switch to record at the same time as the camcorder reaches its selected 'edit play' point. The result is a smooth, well-timed edit. It may also be possible to end the copying by setting the camcorder counter to 0000 and selecting 'M' for memory to identify a cut-out point on the camcorder. When the camcorder reaches the counter 0000, M point it stops in play/pause and sends out a signal to put the main VCR back into record/pause.

Operation error note

If the main recording VCR is not in itself a ZFE machine then the operation works well. However, if the main machine is a ZFE version then an anomaly can occur. Referring back to the ZFE timing shows that after the pause signal is received the ZFE machine runs on for eight fields to record the modified control track. However, the camcorder can stop almost instantly and it is possible for the mains machine to carry on recording one or more tracks of poor quality still picture produced by the camcorder in play/pause. This shows up as a 'flash' at the point when replaying. A solution to the problem is to manually pause the main VCR at its edit-out point before stopping the camcorder.

Random assembly editing; automatic edit function

This is a subsystem of zero frame editing and is activated in timer recordings to join two programme segments together. As some time may elapse between two or more sequential timer recordings the pinch roller pressure must be released to prevent tape damage. This precludes the use of zero frame editing as the precise timing of the edit points by the control track will be lost.

Use is made of the capstan FG counter by the system control microcomputer. At the record/pause point the tape is backspaced by approximately 1.3 s and the pinch roller is released. When the next recording commences the capstan FC is forward counted up to two tracks prior to the pause point. Recording then commences with an overlap error of up to five frames due to tape shift when the pinch roller is released. Automatic edit function (AEF) will also be engaged if the tape protect time out of record/pause is reached, which will also release the pinch roller.

Once zero frame editing was established as a system, a further refinement was added to maintain compatibility with the VHS index search system (VISS). In the VISS system a '0' pulse is indicated by a control track pulse mark/space ratio of 60/40. A '1' pulse is indicated by a control track mark/space ratio of 27.5/72.5. For VCRs without VISS the standard control track mark/space ratio is given as 60/40 ±5% or '0' (Figure 13.6). A 10-bit code can be put onto the control track using this logic without affecting control track operation.

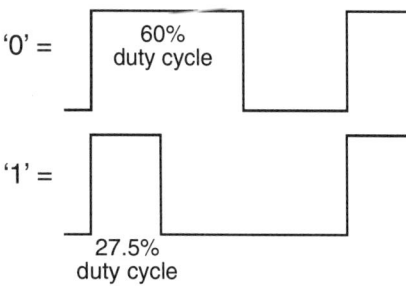

Figure 13.6 *Control track logic ratios*

Operation of zero frame edit using VISS data

Refer to Figure 13.7. When the VCR receives a 'user pause' control signal (G) whilst in the recording mode, the system control switches the servo 'S data' line from 'index detect mode' to 'duty detect mode', and so the data line is sent 'high' at point (B).

Next, timed by the second successive drum flip-flop falling edge (C), the recorded control track ratio is altered in logic from '0' to '1'. After another count of two flip-flop pulses on the fourth edge (E), recording effectively ceases. However, the capstan motor drive is maintained for a further two counts up to the sixth edge (F) and two more control track '1' pulses are recorded, making four control track '1' pulses in total.

In hi-fi video recorders the hi-fi audio heads lead the video heads, if both audio hi-fi FM signals and video FM signals were cut off at the same time it would leave a audio track partially recorded. To avoid this, the FM audio record drive is cut off one count later on the third drum FF, allowing the audio head to complete a full track. The recording VCR then backspaces from (F) by 1.3 s by counting capstan FG pulses and stops in record/pause.

When a recording 'start' signal is received, possibly from a camcorder, then a count of the capstan FG pulses is commenced. This is to time the servo up to a point where the original 'pause' (G) command was given; this is a starting run-up period. Once point (G) is reached, an index 'H' search period of 210 ms is initiated and the system control then looks for a logic 1 from the replayed control track. The normal period for the search is usually no longer than four flip-flop periods (160 ms). As soon as the second logic 1 is detected the servo index at (H) goes 'high' to indicate a detected index pulse. The duty detect mode at point (I) is then switched off as it is no longer needed. Using the index (H) as an indicator, recording commences on the next fall of the drum flip-flop at point (E) thus commencing the

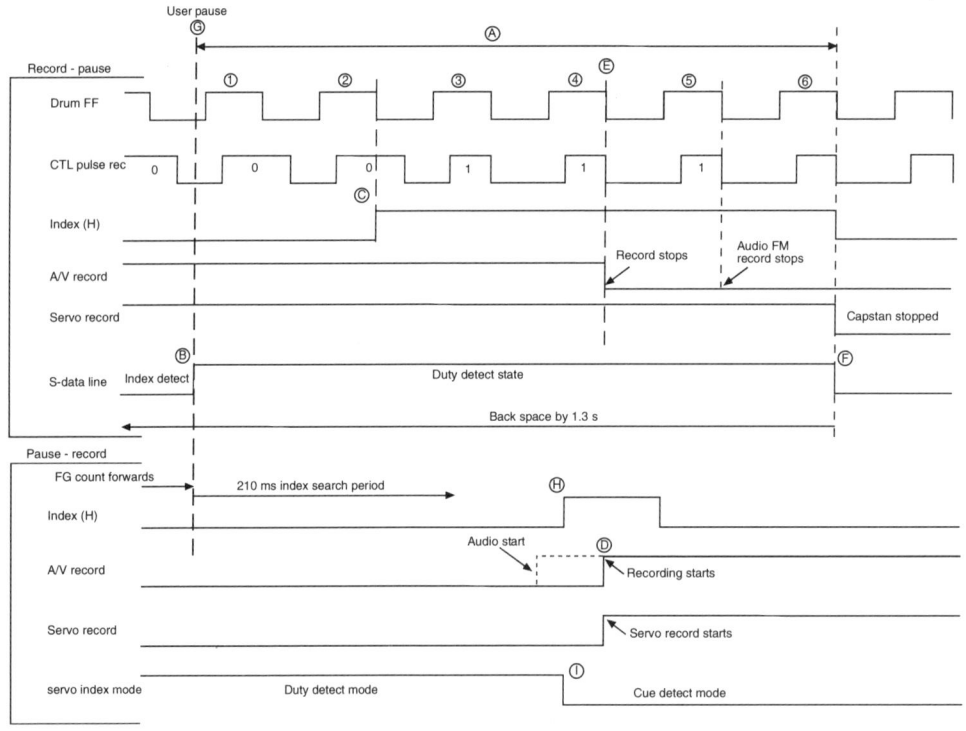

Figure 13.7 *VISS editing timing diagram*

recording on the next video track after recording ceased at point (D). Hi-fi recording starts earlier than point (D) (both start and stop timings are controlled from the audio flip-flop, which leads the phase of the drum by 138°).

When a complex head drum is used as shown in the previous chapter on Hi-fi recording then the timing of the head signal switching becomes more complex, as shown in Figure 13.9. Note the drum flip-flop and audio flip-flop have been used for both SP and LP where a Ch.1 head in SP is a Ch.2 head in LP (LP references are in brackets). The switching edges are more important than knowing which head is active.

Head timing

As discussed in the audio Hi-fi chapter the advance angle and corresponding height of the flying erase head and audio heads put them in advance of the video heads and so the signal on/off timing for noise-free edits becomes different for each head.

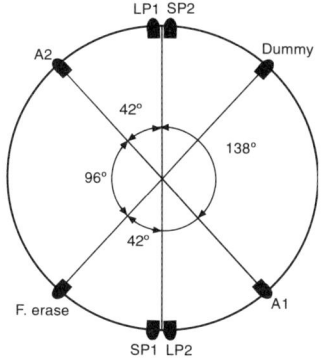

Figure 13.8 *Head mounting positions*

After a record start signal to switch the video recorder from pause to record, the next available flying erase head switching transient from low to high activates the erase head. The full erase head starts by erasing two tracks in advance; it is timed to be 42° after a Ch.2/Ch.1 flip-flop edge in SP (a Ch.1/Ch.2 edge in LP). Following it by 96° is the first audio head A2 aligned upon a Ch.1 track, and on a Ch.1/Ch.2 edge the audio recording starts. Then the first video head, delayed by 318°,

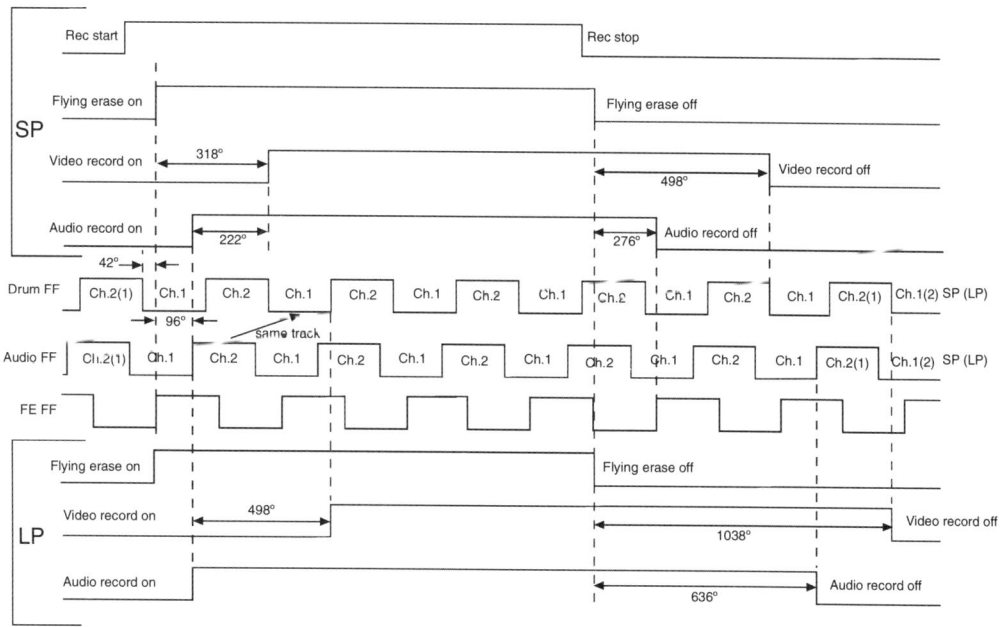

Figure 13.9 *Head switching timing at start and stop of editing points*

Video and Camcorder Servicing and Technology

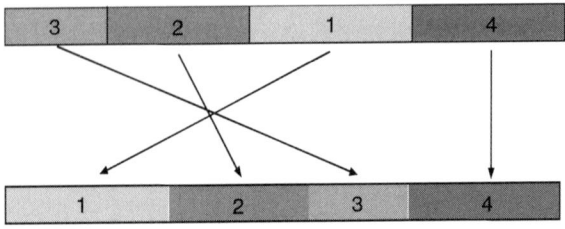

Figure 13.10 *Assembly editing*

Random assembly editing

Some analogue camcorders have a random assembly (RA) function where up to eight segments on a tape can be marked for cut in and cut out. The sequence in which the segments are selected becomes the sequence of the assembly edits even though they are in different parts of the camcorder recording.

When the camcorder is connected to a video recorder with the remote start/stop function, it can search for each segment in turn and copy it onto the video recorder by a sequence of stopping, searching and starting, as previously outlined. The quality of the edits will be dependent upon the editing system of the video recorder and are restricted to assembly editing, as shown in Figure 13.10.

As part of this function there may be a small timing adjustment of ± 1 s to the remote output connection to the video recorder. This allows the mechanical backspace edit timing period of 1.3 s to be optimised to the camcorder.

Insert editing is more specialised as this involves placing a scene within an existing recording, it can be achieved with analogue systems but the original segment is erased by the inserted segment... until digital came along.

records the Ch.1 video track and partially erases the deep layer audio. In LP mode the start of video recording has a longer delay from the start of audio, as the heads are more than a track advanced, 496°.

At the end of a recording the flying erase switches off almost immediately, timed with an edge on the flying erase switching sequence. Here is a delay of 276° before the audio switches off, timed by an audio flip-flop edge and 498° before the video heads switch off. In LP the timings are longer: 636° for audio and 1038° for video after the flying erase has been switched off.

All of the head timings are coincident with their respective flip-flop signals as this ensures that the start and stop of recordings are at the beginning or end of tracks to minimise edit disturbances.

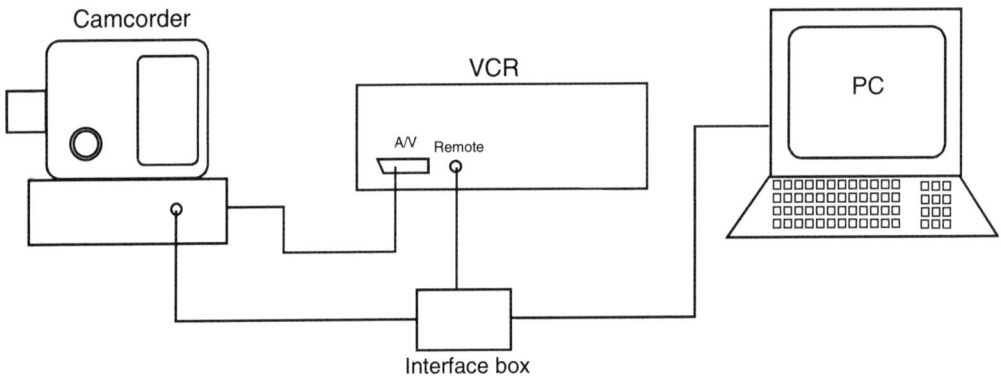

Figure 13.11 *Digital camcorder assembly editing with PC software control*

Digital camcorder assembly editing

One useful function of digital recording is that of the time code. This is the real time count of each frame buried within the time code section of each track. It does not change with each track, only each frame – every 12 tracks.

A docking station provided with the digital camcorder has all of the necessary connections to a PC and video recorder, as shown in Figure 13.11. Either standard or S-VHS signal paths can be used according to the functions of the video recorder. A remote control connection is for the stop/start of the video recorder between record/pause and record.

The user can then build up an editing list by selecting video segments on the tape by their time code for cut in and cut out, and this information is displayed on the PC, as shown in Figure 13.12. There is control over the camcorder operations so that searching for a selection can be done via the

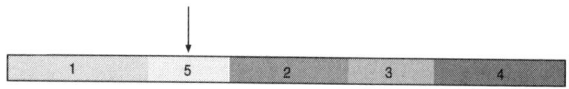

Figure 13.13 *Insert editing where other segments move along the timeline*

PC and the list can be fine tuned by editing of the time codes in the list. Once completed the list can be saved as a file for future copies, all identical to the first one. Within the menu there is the facility to optimise the video recorder remote control stop/start timing that is outputted 1.3 s before the cut in time code. It can be varied to match the video recorder pre-roll timing.

Editing with IEEE1394

The next step is to download real time movie segments onto the PC hard drive. A large hard drive is required as 4 GB can store around 20 min, so for 1 hour a 13 GB hard drive is the minimum. 20 GB plus is recommended, and it must also have a fast read/write time, commensurate with a high rotation greater than 5200 rpm. The data stream is the same as off tape, around 40 Mbits/s.

With suitable software the video segments can be edited into a timeline. They can be changed around with new segments being placed in between earlier ones without any loss, as they move up to make room; see Figure 13.13.

Once the editing is completed the digital edits can be up-loaded back to the digital camcorder and recorded onto a new tape. VHS copies can then be made from this digital master.

From a service point of view, non-recognition of the camcorder being connected can be a problem. This could be a software problem or a failure of the camcorder IEEE1394 interface.

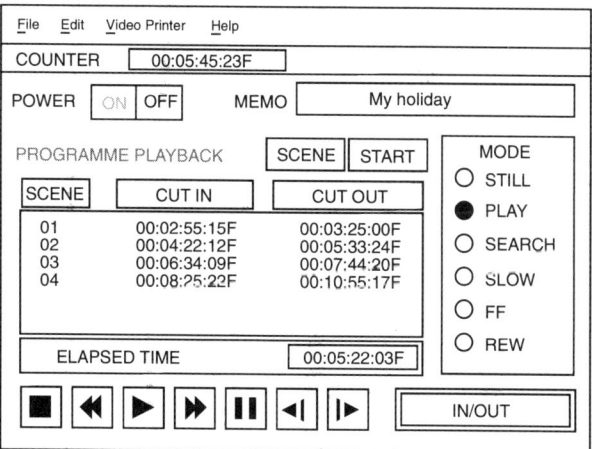

Figure 13.12 *Software editing list menu with camcorder/VCR control*

14
PDC and IEEE1394

Programme delivery control (PDC)

Programme delivery control is an extension to Teletext and carries control codes to identify the TV station, date and time, and to start the video recorder to record the programme even if the programme schedules are changed.

If the user has set the video recorder timer to record a given programme at a set time, say a football match on ITV at 3pm, and then the match is delayed, PDC will delay the start time of the recorder until the football starts. If the game goes into extra injury time then the recording will continue until the end of the transmission. Timer recordings can overlap: if the football is set to finish at 5pm and snooker is on another channel at the same time, then the football recording will continue until the final whistle. The first part of the snooker will be missed! All timer recordings have to be entered manually or via Videoplus and it is important to enter a 'stop' time, even with PDC, in case of a change or failure of the PDC stream.

Teletext is formatted into 800 pages numbered 100–899, each set of 100 pages are combined into a magazine, and there are eight magazines 100–199, 200–299... 800–899. Each Teletext page has a Header and 23 rows of 40 characters; row 0 is the header and row 23 is the last line of character text. Each character is formed from an 8-bit byte resulting in 40 bytes of data for a row of 40 characters. Each row is preceded by 5 bytes of synchronisation data; comprising 2 bytes of clock run in, 1 byte frame code, data start and 2 bytes for magazine and row information, totalling 45 bytes of data. In transmission terms the 45 bytes of data that make up a row is called a packet.

There are in addition some 'invisible' rows of data which are not displayed these are termed packets and are equivalent to non-displayed rows 24–30.

Packet 24 carries the text for the prompts at the bottom of the screen in 'Fastext'.

Packet 25, is not used.

Packet 26 is used on the TV listings pages of a broadcaster who is transmitting PDC to provide the programme identification labels for VCRs that use Teletext to set timer programming; also for character replacement in Spanish and display control in Level 2 decoders (defined before PDC).

Packet 27 contains links for the Fastext buttons.

Packet 28, if used, contains character set information (Level 2).

Packet 30 is the one that we are interested in for PDC. It is transmitted in magazine 8 and is termed packet 8/30 and contains the PDC information.

There are two possible formats for packet 8/30. Format 1 is the television service data packet (TSDP) to carry the television 16-bit network identification codes, with date and time, and short programme labels. It is used by TVs and video recorders that have self-tuning and on-screen display channel identification. Format 2 is for programme delivery control (there can be up to four format 2 packets each second). Each packet is transmitted once a second and the two formats are interleaved every 200 ms.

Formats

The first 12 bytes of each format are similar. Bytes 1–3 contain the standard clock run-in and framing code to synchronise the decoder clock for accurate data acquisition and keep the decoder phase-locked loop timing stable. Bytes 4–6 carry the magazine and packet numbers; in this case they are 8 and 30, respectively. Bytes 7–12 carry the

8/30 FORMAT 1
Teletext 16-bit Network Identification Codes

Byte	Data Description		1 LSB	2	3	4	5	6	7	8 MSB
1	Clock Run In		1	0	1	0	1	0	1	0
2			1	0	1	0	1	0	1	0
3	Framing Code		1	1	1	0	0	1	0	0
4	Packet address		Magazine Number			Packet Number				
5			Packet Number							
6	Designation Code		Designation Code							
7	Initial Teletext Page		Page Number (Units)							
8			Page Number (Tens)							
9			Sub Code A (Minute, Units)							
10			Sub Code B (Minute, Tens)			Relative Magazine No.(0)				
11			Sub Code C (Hour, Units)							
12			Sub Code D (Hours, tens)		Relative Magazine No.(1)	Relative Magazine No.(2)				
13	NI (Network Identification) (Station Name)		NI b_5-b_8				NI b_1-b_4			
14			NI b_{13}-b_{16}				NI b_9-b_{12}			
15	LTO (Local Time Offset)		1	Offset in ½ hours					Negative flag	1
16	UDT (Unified Date & Time)	MJD (Modified Date & Time)	MJD b_1-b_4				1	1	1	1
17			MJD b_9-b_{12}				MJD b_5-b_8			
18			MJD b_{17}-b_{20}				MJD b_{13}-b_{16}			
19		UTC (Co-ordinated Universal Time)	Hours (Units)				Hours (Tens)			
20			Minutes (Units)				Minutes (Tens)			
21			Seconds (Units)				Seconds (Tens)			
22	SPL1 (Short Programme Labels 1)		SPL1 b_5-b_8				SPL1 b_1-b_4			
23			SPL1 b_{13}-b_{16}				SPL1 b_9-b_{12}			
24	SPL2 (Short Programme Labels 2)		SPL2 b_5-b_8				SPL2 b_1-b_4			
25			SPL2 b_{13}-b_{16}				SPL2 b_9-b_{12}			

Figure 14.1 *Format 1 configuration*

initial page for automatic display on text selection, e.g. 400 on Ch.4.

All of the bytes 4–12 are 8/4 Hamming error protection codecs. Hamming error protection 8/4 means that 8 bits of data are sent but only 4 bits are data; the other 4-bits are parity bits. For clarity only the data bits are shown in the diagrams.

Format 1

Format 1 carries information for a TV or VCR to identify the broadcasters unique network identification code (NI) and a comprehensive date and time; see Figure 14.1.

Bytes 13 and 14 are a 16-bit network identification (NI) code that identifies the country and station name. Byte 15 is the local time offset and is used to add or subtract half-hour segments from the UTC to allow for local or regional variations. such as British Summer time.

Bytes 16–18 is the modified julian day and is the number of days counted from 16th November 1858. The current day and time is calculated from this value. Bytes 19–21 are the co-ordinated universal time in the standard format HH MM SS. Where local times may given by the variation of LTO to UTC it is possible for the broadcaster to set LTO to 0 and transmit UTC as the normal local time.

Bytes 22–25 are for use by the broadcaster to label programmes with a wide range of options such as 'test page' or 'CH5 Teletext'.

Network identification data is unique to each broadcaster and can be BBC1, BBC2 CH5 or Carlton TV, each prefixed by the country code; for the UK this is 44. Format 1 provides all of the information required the automatic tuning installation of 'plug and play' TVs and video recorders.

Bytes 13–25 are not Hamming protected so all 8 bits are data bits. This is because if the data was lost for one or two packets then it would not be a problem, as a high refresh rate is not important and the data can be picked up from the next clean packet.

Bytes 26–40 are for status display and usually carry the station name in parity protected plain text: BBC1, BBC2, Channel Four, Channel 5, CNN International.

Format 2

PDC works by the video recorder matching the transmitted programme identification label with the date and time set in the video recorder's memory. When a match is found then the video recorder will enter record mode. It will stay in record as long as the PIL is sent during transmission. Once the PIL is discontinued, or if the recording inhibit is sent, the recording stops.

8/30 Format 2
PDC Network Identification Codes

Byte	Data Description	1 LSB	2	3	4	5	6	7	8 MSB
1	Clock Run In	1	0	1	0	1	0	1	0
2		1	0	1	0	1	0	1	0
3	Framing Code	1	1	1	0	0	1	0	0
4	Packet address	Magazine Number			Packet Number			
5		Packet Number							
6	Designation Code	Designation Code							
7	Initial Teletext Page	Page Number (Units)							
8		Page Number (Tens)							
9		Sub Code A (Minute, Units)							
10		Sub Code B (Minute, Tens)				Relative Magazine No.(0)			
11		Sub Code C (Hour, Units)							
12		Sub Code D (Hours, Tens)		Relative Magazine No.(1)		Relative Magazine No.(2)			
13	Label Channel Identifier / Label Update Flag / Prepare to Record Flag	LCI b_1	LCI b_2	LUF	PRF				
14	Status of analogue sound / Mode Identifier / Not yet defined	PCS b_1	PCS b_2	MI					
15	CNI (Country)	CNI b_1	CNI b_2	CNI b_3	CNI b_4				
16	CNI (Country) Day	CNI b_9	CNI b_{10}	PIL b_1	PIL b_2				
17	Day Month	PIL b_3	PIL b_4	PIL b_5	PIL b_6				
18	Month Hour	PIL b_7	PIL b_8	PIL b_9	PIL b_{10}				
19	Hour	PIL b_{11}	PIL b_{12}	PIL b_{13}	PIL b_{14}				
20	Minute	PIL b_{15}	PIL b_{16}	PIL b_{17}	PIL b_{18}				
21	Minute Country	PIL b_{19}	PIL b_{20}	CNI b_5	CNI b_6				
22	Country Network or Programme Provider	CNI b_7	CNI b_8	CNI b_{11}	CNI b_{12}				
23	Network or Programme Provider	CNI b_{13}	CNI b_{14}	CNI b_{15}	CNI b_{16}				
24	Programme Type	PTY b_1	PTY b_2	PTY b_3	PTY b_4				
25	Programme Type	PTY b_5	PTY b_6	PTY b_7	PTY b_8				

Figure 14.2 *Format 2 configuration*

Dependent on the MI flag, 0=PIL starts 30 s before the programme; '1' indicates that it starts at the same time as the programme. In order to obtain a match the video will, in standby, continuously scan all channels. If a transmission is delayed then the programme label is not transmitted until the programme starts and then recording commences. As there can be up to four labels it makes it possible to overlap recordings by sending a new programme label before the current one has ceased

Format 2 is for PDC. Bytes 1–12 are the same as for format 1; see Figure 14.2.

Byte 13 (LCI, bit 1 and LCI, bit 2) is the label channel identifier, the two bits allow for four programme labels (bytes15–25) so that up to four sets of data can be transmitted simultaneously.

Within PDC there are other commands that can be carried, directly or indirectly. 'Prepare to record flag' (PRF) is a direct command: when set to 1 the video recorder is put into record/pause and then when PRF drops to 0 recording starts. MI (mode identifier) is used to indicate the label starts at the same time as the programme. The LCF (label update flag) is used to replace and update the video recorder's timer memory if a programme that has been set is postponed to another day or switched to another channel. LCI, LUF, PRF and MI are delivered by bytes 13 and 14. A further data set within these bytes is PCS (programme control status) that sets the status of the sound channel, mono, stereo or dual language.

CNI (country network identifier) is a 20-bit data stream sent as Hamming protected 8/4 and is a 4-bit data 'nibble' spread over several bytes, 15, 16, 12, 22 and 23. Examples of this are 2C 7F (44 BBC1), 2C 1D (44 Carlton TV).

It is possible to view this data by selecting a clock cracker page and pressing the red fastext button; Channel 4, page 399 plus red Fastext jumps to page 3AA, PDC data display.

PIL (programme identification label) data bits are sent over bytes 16–21 and carry the date and start time of the broadcast.

Bytes 23 and 24 are the programme type (PTY) data. PTY codes can be used as a two-digit number to identify a programme type. For example, 81 may be used to label a daily serial and 82 to label the weekend omnibus edition. Video recorders with PTY capabilities are able to carry out multiple daily recordings.

During a recording the PDC data may change, the broadcaster may wish to switch the video from PDC to timer mode, or interrupt the recording. Some indirect codes are generated by modifying the date and time to values outside the normal months and years, month 15 is popular.

00/15	28:63	Continuation – tells the video recorder to start recording after a pause.
00/15	29:63	Interrupt (Int) – Pause.
00/15	30:63	Recording inhibit/Terminate (RI/T) – Stop.

00/15   31:63   Timer Control (TC) – Revert to timer control.
31/15   31:63   No PIL; use PTY

These codes allow the broadcaster to pause a recording and then continue. This is valid only if the pause duration is shorter than the video recorder's pause-timeout, i.e. approximately 6 min. For a longer period, for example a film being segmented by a News broadcast, then the video recorder is stopped and restarted again.

Auto set–up (plug and play)

Many modern video recorders have an auto set-up function that activates when power is first applied. When new or when the back-up memory supply is exhausted, the system control microcomputer (syscon CPU) resets. This enables the video recorder to scan the tuning band, find broadcast programmes and tune them into its channels. Tuning data, frequency and programme name, or identification, are stored in the video recorders permanent memory.

Packet 8/30 Format 1 is used for this purpose and for clock updates as it carries all of the information required to do this. Broadcast programme names are decoded from bytes 13 and 14, time and date from bytes 15, 19, 20 and 21.

Figure 14.3 is the system diagram. Tuning starts at the low end of the band and increases slowly. Two inputs are monitored and these are the AGC and AFT values from the tuner/IF circuits as an indication of reception of a broadcast programme. When a programme is found and confirmed by the syscon CPU it instructs the character generator to look for NI codes for the programme name. Once this has been decoded it is compared to a list set in the syscon CPU ROM for that location. After confirmation, the name and tuning data are stored in the system RAM and tuning recommences.

Once tuning is complete and all receivable broadcasts are stored, attention is turned to the programme stored in channel 1 to set the VCR clock. Data bytes MJD and UTC are decoded and written into the time RAM and then the clock runs off a crystal reference as normal.

Auto clock adjustment takes place only when the video recorder is in standby mode and the time is on the hour (i.e, minutes are 00) except the two times 23.00 and 0.00. When conditions are met the video recorder powers up, tunes to channel 1, decodes the data from 8/30 format 1, and writes the time into RAM again and powers off.

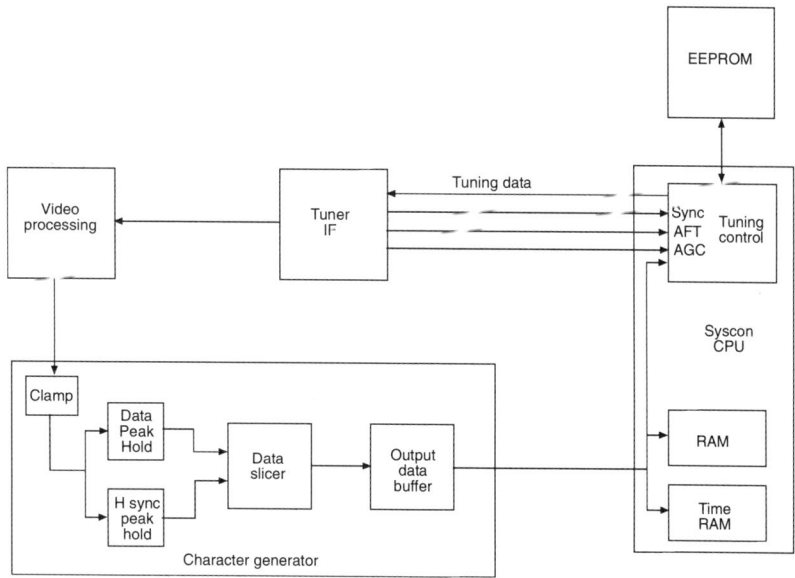

Figure 14.3 *Auto set up tuning, time and date data aquistion*

Video and Camcorder Servicing and Technology

Digital communications IEEE1394

Digital data transfer between digital products is possible via a communication transfer protocol designated IEEE1394, FireWire or i-link, according to the manufacturer.

There are three data transfer rates: 93.304 Mb/s (S100), 196.608 Mb/s (S200) and 393.216 Mb/s (S400). Any equipment specified for the higher rates can also communicate at the lower speeds.

Cable lengths are confined by the communication speeds and cable construction to 4.5 m although repeaters can increase this; S100 may work up to 10 m with a suitable high quality cable. There are two connector/cable types: two pairs with a 4-pin connector, most common for DVC, and a 6-pin connector for two wire pairs with two power wires. Where power for repeaters and PC peripherals can be provided, a typical cross section is shown in Figure 14.4.

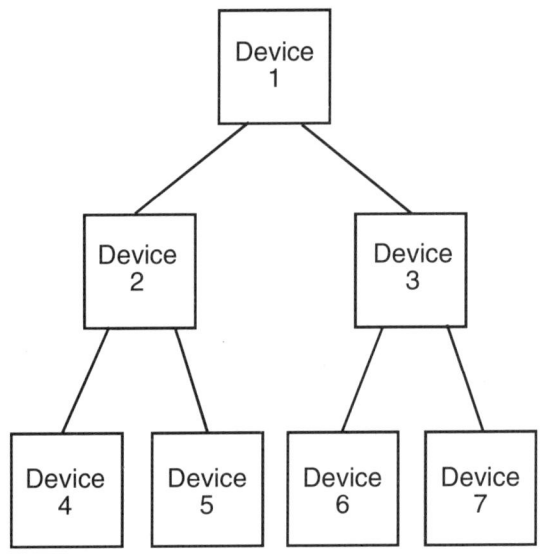

Figure 14.5 *Device connection tree arrangement*

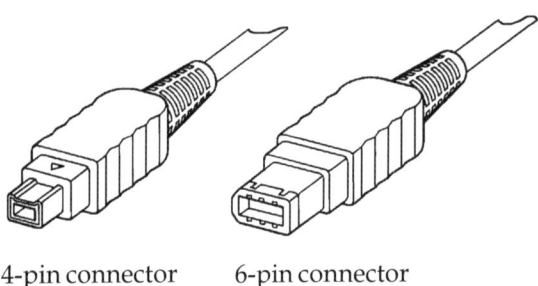

4-pin connector   6-pin connector

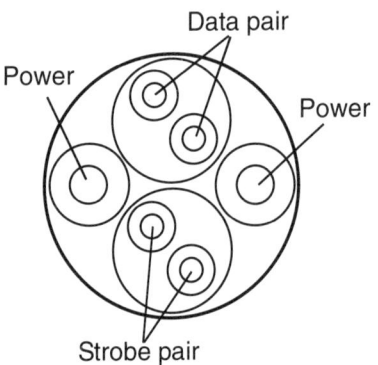

Figure 14.4 *IEEE1394 connectors and cable cross-section*

Up to 63 devices can be connected in a tree-type connection network (Figure 14.5). If there are 64 devices then device recognition fails and communications will fail as there will be too many devices for the number of available identifications. There are limitations in that there must not be more than 17 devices in a serial chain. If there are 18 then the data transfer time is extended beyond the device capabilities and communications will fail. This also means that there cannot be more than 17 devices in a path where one device communicates with another via two or more tree networks.

A unique advantage with IEEE1394 is that the connections can be made with the power on, known as 'hot plug-in', as each product will search for the specified recipient before data transfer takes place.

Data transfer must be of the same protocol in order to succeed. For instance, MPEG2 transport streams cannot be sent to a DVC product or vice versa as the data is not recognised by the recipient.

IEEE1394 protocols

There are two main protocols for DV and PC communications. One is the AV protocol for transferring real time video and audio signals and basic system operation controls: play, record, fast forward etc. The other is the SBP-2 protocol for personal computers and peripherals.

Isochronous (synchronised) data transfer is used for the AV protocol for real time digital video data packets or an MPEG2 transport programme stream (TS). Asynchronous (non-synchronised) data transfer is used for computer peripherals and file transfer.

AV protocol

The AV protocol has rules for DVC-SD (standard definition) digital video communications, MPEG2, and audio or music transfer between devices.

DVC-SD: Audio and video data signals in the DV digital video format.

MPEG2 -TS for broadcast, audio and video data signals in the MPEG2 transport stream as used by digital broadcasting services, satellite and terrestrial.

Audio and music, audio and data signals from CD and mini disc.

Operational control commands for camera, VCR, DBS receiver, CD and MD, and TV and monitor display controls.

Serial bus protocol 2 (SBP-2)

SBP-2 is a protocol for linking computer peripherals such as CD & DVD ROMs, scanners, printers, cameras and hard drives, It is envisaged that IEEE1394 will replace SCSI due to the higher transfer speeds available.

Where an AV product is connected to a PC then the PC must have suitable IEEE1394 software running with AV protocols to have successful communications.

A PC can be connected to any number of

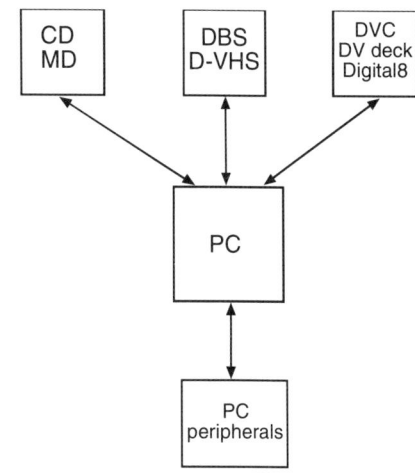

Figure 14.6 *PC peripheral options*

peripheral devices by IEEE1394 communications up to 64. There is control data and audio data linking to mini disc and CD players, DVC camcorders, Digital8 camcorders and DV main decks, for non-linear editing and assembly of video and audio segments into completed productions. External high capacity hard drives can be linked via IEEE1394 along with other peripherals such as printers, scanners and still cameras.

A whole range of audiovisual products can be interconnected by IEEE1394 as a single cable. Audio and video data can be routed as required either as DV-SD data or MPEG2 data, and be controlled by the PC or a single remote control.

IEEE1394 data communications

The AV protocol DVC-SD is shown in Figure 14.7.

In the digital camcorder section, each of the 12 video tracks that make up a frame has 135 macro blocks of data, an audio data section, sub-code and auxiliary video data sections. These are kept together in a data transfer block called a sequence. As there are 12 tracks then there are 12 sequence blocks: sequence 0 – sequence 11.

Each sequence block consists of 25 data packets labelled DP0 to DP24 (up to DP275 – DP299). Each data packet has six digital interface (DIF)

Video and Camcorder Servicing and Technology

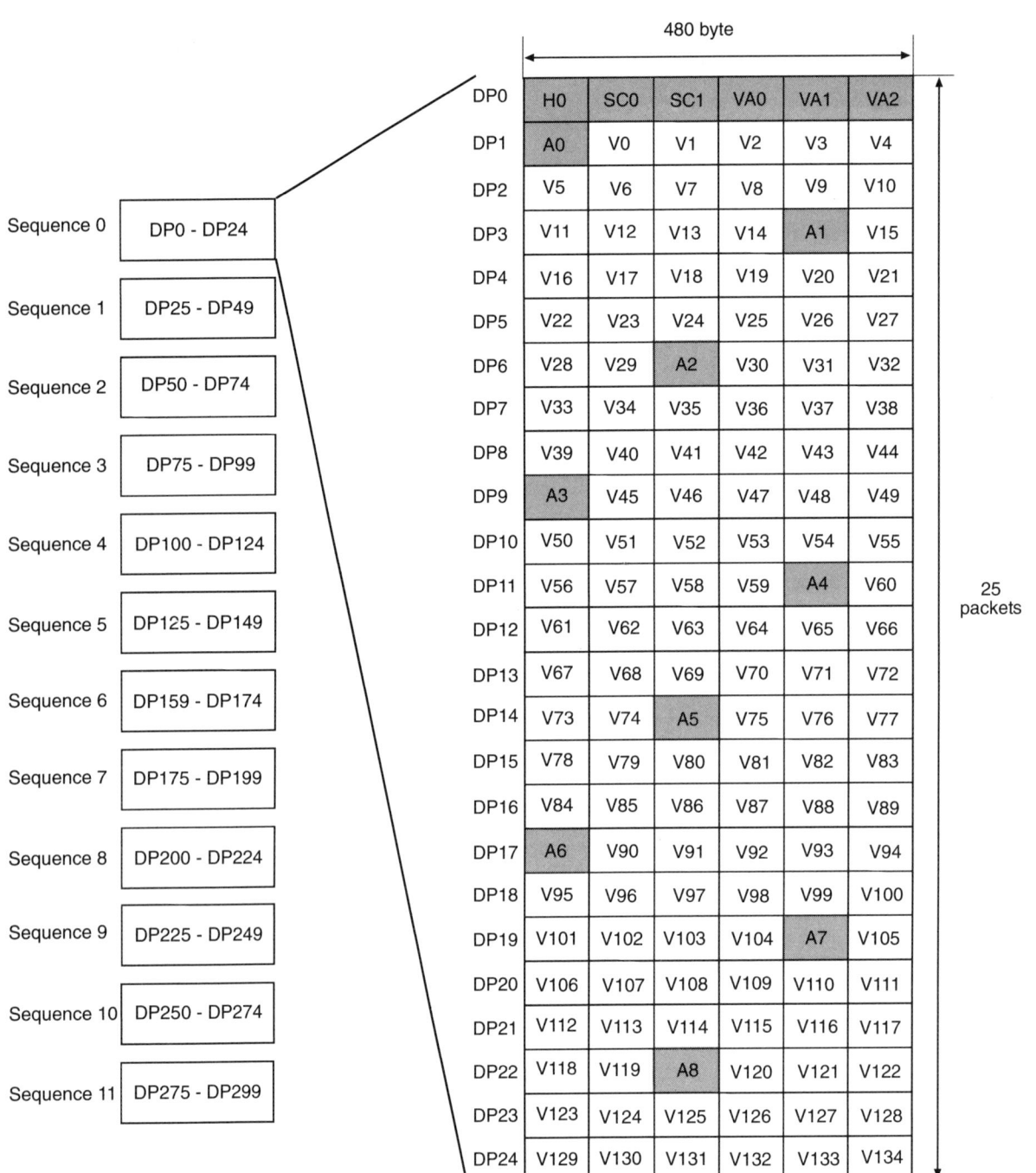

Figure 14.7 *IEEE1394 AV protocol DVC-SD*

blocks. Each DIF block has 80 bytes of data making a packet of 480 bytes of data.

There are 150 DIF blocks (6 x 25) in a sequence consisting of a sequence header, 135 video data blocks, 9 audio data blocks, three video aux blocks and two sub-code blocks.

Each sequence commences with a packet containing header H0 block to identify the sequence. This is followed by two blocks of sub-code and three blocks of auxiliary video data in the first packet. The following packets contain video locks V0 – V134 interspersed with audio in blocks A0 – A8.

This first sequence is followed by 11 more to complete the frame.

Figure 14.8 shows the communication interface consisting of a link IC, PHY IC and control microcomputer. These may be contained within a single IC package.

Control of the AV protocol is carried out by the link IC and the micro. The DVC-SD signal from the camcorder internal data bus is fed to the link IC where it is converted to the AV protocol sequence blocks and the header is added. Also within the link IC are the data transmit and receive sections to the cable via the PHY IC.

Within the PHY IC are the IEEE1394 matching circuits: a transmit encoder and receive decoder for the twisted pair cables. It also performs the function of transceiver control to other devices providing bus initialisation, and arbitration to ensure that only one device is sending data at any time.

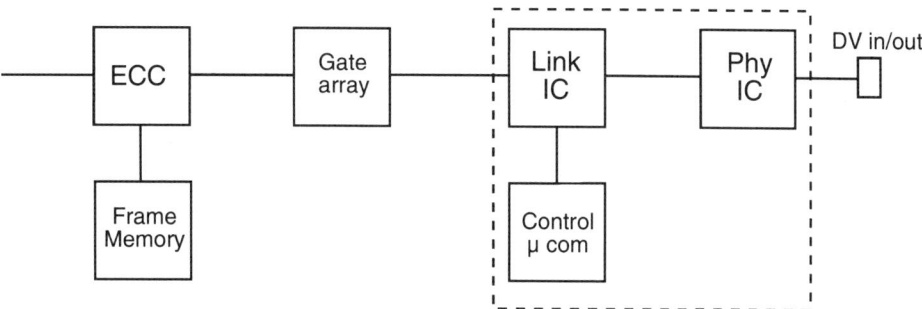

Figure 14.8 *IEEE1394 communication interface ICs*

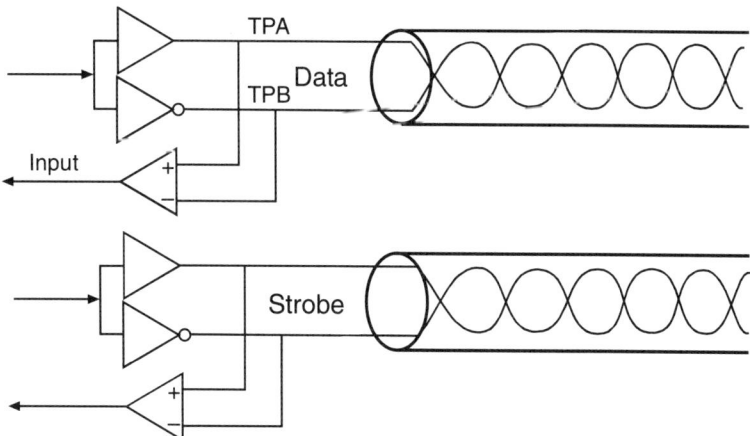

Figure 14.9 *Twisted pair transmit/recieve matching and drive circuit*

Video and Camcorder Servicing and Technology

IEEE1394 twisted pairs

Data is transmitted along the twisted pair cables by sending the data in normal and inverted form along each cable of the pair. In Figures 14.9 and 14.10, TPA is normal and TPB is inverted. Any noise picked up along the cable route has the same polarity on both TPA and TPB. At the receiving end TPB data is inverted and added to TPA data. The noise on TPB cable is also inverted and cancels with the noise on TPA while the data is supported by addition. This applies to both data and strobe twisted pairs.

DS coding

IEEE1394 signal transmission uses two twisted pairs for data transmission: one pair carries the data and the other a strobe signal. By use of an exclusive OR gate the data and strobe signals can be combined to form the clock signal as shown in Figure 14.11.

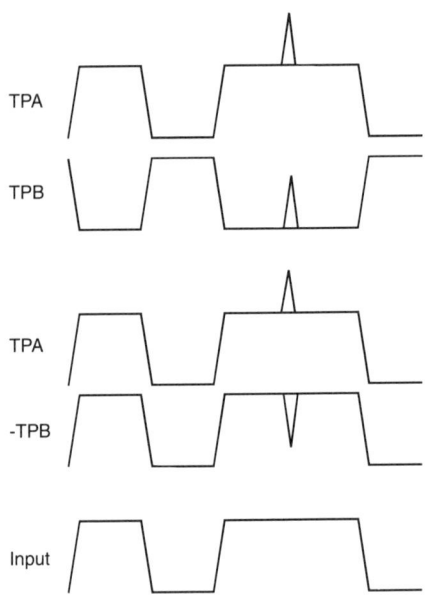

Figure 14.10 *Cable noise pick up cancellation*

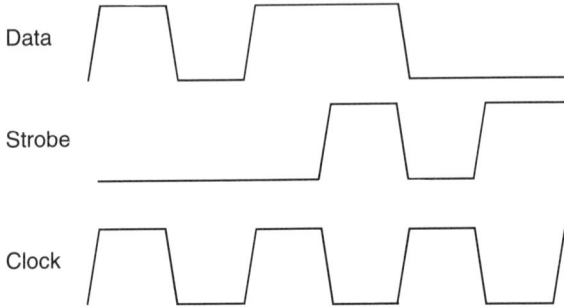

Figure 14.11 *Data/strobe coding to reproduce the data clock*

15
Soldering and Desoldering

Soldering and desoldering is a specialised skill and the equipment is such that its use requires training. Consideration has to be given to the workload and the type of work.

For service work the operator must be trained and skilled enough to control the soldering and desoldering process and utilise soldering equipment that can be flexible enough to be used in a variety of applications.

In this chapter we will consider the types of desoldering and soldering equipment that can be used in the service department by skilled operators. The choice of equipment will be decided by the tasks performed in the service department and the level of skill available, although in some cases training will be required in order to get the best service from the equipment.

A choice of suitable equipment can be made by considering the usage options and the type of work.

Repetitive usage includes:

- Re-soldering of a quantity of dry joints on a single PCB;
- Fitting numbers of components to a PCB;
- Changing a number of components on a PCB;
- Replacing a large through hole IC;
- Replacing a large flat surface mount IC;
- Replacing a chip sized package.

Occasional usage includes:
- Re-soldering a broken connection;
- Re-soldering a connector;
- Removing a single component and replacing it;
- Removing a component with a large surface area or contacts, i.e. a tuner.

For an example, consider a single soldering iron. Do you leave it switched on all of the time because it is frequently used? Do you suffer from deterioration of the tips because of this? Would a quick heat soldering iron that is switched off for long periods be a better choice?

Surface mounted components

In order to remove a surface mount passive component, such as a resistor or capacitor, heat must be applied to both ends simultaneously. It is not good practice to heat one end and try to lift it off as component or print damage may result.

A basic surface mount component removal method is to apply two soldering irons either side of the component, as shown in Figure 15.1. This is cumbersome and not very practical, although reheating of both ends simultaneously is achieved. However, lifting the component is difficult, needing a third hand!

A better method is to employ a special soldering iron tip, as shown in Figure 15.2, such as a specialised soldering iron with a number of special tip attachments.

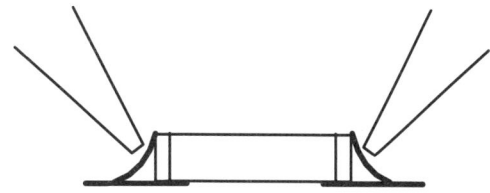

Figure 15.1 *Removing a surface mounted component with two soldering iron tips*

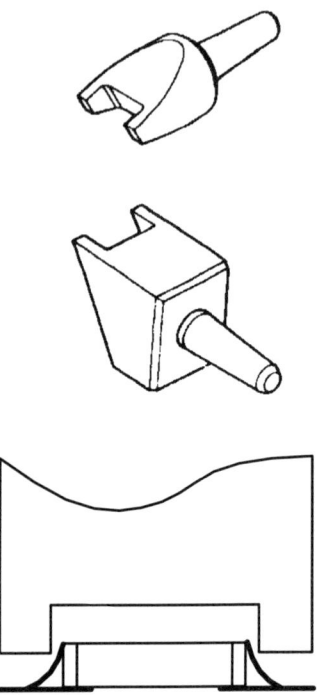

Figure 15.2 *Removing s surface mounted component with special tips*

With such a device both ends of the component are heated at the same time and it can be lifted off. A small component may come off due the capillary action of molten solder, others may require tweezers. Slightly larger attachments can therefore be used to remove surface mounted diodes and transistors that have three legs in triangulation, by heating all three legs at once. There are some transistors, particularly power switching transistors, which have the collector terminal soldered to the print. These transistors are constructed with a larger tab as a heat sink for the collector and then the print land is utilised for additional heat sinking. Greater care is required to remove such components to avoid print damage by either insufficient heat or too much heat. Insufficient heat will cause damage by dragging up the print when the component is pulled off because the solder has not re-flowed. Too much heat will cause the print to lift off the PCB by expansion of the copper or the PCB blistering!

For small components solder is a problem. The smallest commercially available is 28 standard wire gauge although some equipment manufacturers have a smaller metric size available, typically 0.3mm, from their service department supplies. A small tip size can cause a problem by overheating and baking the rosin in the solder onto the tip if it not temperature controlled. The tip then becomes tarnished and dirty and will not 'wet' properly. Solder joints then become messy. A useful hint is to use a damp sponge or a tin of tip cleaner and then dip the iron tip into it frequently to clean it.

Removal of a flat surface mounted IC is much more of a problem and it can be approached in two ways. The most basic is that of removing the IC by mechanical force, the other, a more sophisticated way, is done by de-soldering and removal. Brute force is dangerous, some technicians use a sharp modelling knife, this is not good because it puts too much pressure upon the IC legs and the print. and if the knife slips then the print is ruined.

A slightly better method is to use some very small cutters and gently clip each leg right where it meets the IC body. Small cutters can be ground down to fit between the legs of the IC. It is then a compromise between size and strength, where cutters costing more than £40 still have to be reduced in size. Care must be taken when grinding, and it is important not to overheat the metal. When the IC's legs are cut the cutters must be stable so that when the cut is made it is central to the leg and the cutters do not jump away from the cut position. If they do jump then strain is put upon the remaining part of the leg and the print may be damaged.

Precision cutters

A pair of cutters are available as shown in Figure 15.3. They are Swiss made by EREM (Type No. 670E) as precision cutters for surface mount components, in particular the removal of DIL or quad pack ICs. With cutting edges on each side the tips are ground to form a symmetrical delta-

Soldering and Desoldering

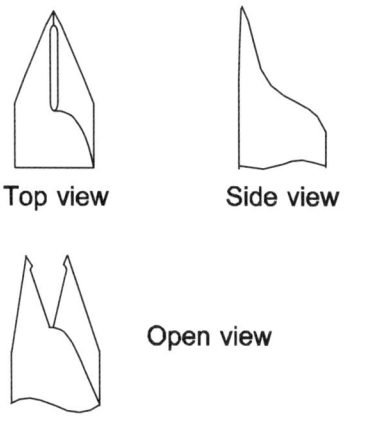

Figure 15.3 *EREM precision cutters with a specially contoured tip*

shaped cutting tip. With these cutters the IC leg is penetrated from either side and is cut to its centre where the two edges meet. By grinding to a delta shape the cutting tip is made very small and it is strengthened by the contoured head.

The cutting edge is only about 2 mm long with most of the cutter blade ground back to form a non-cutting clearance area. This cutting tip is ideal for surface mounted ICs; each leg can be quickly and carefully snipped in turn and then the IC body removed. Only a small cleaning up operation is required with solder wick or even just a wetted iron tip to remove the remaining leg debris.

IC removal

It is not good practice to remove and then replace the same integrated circuit. Once it is removed it should be considered as unusable. Often the IC must be sacrificed to preserve the print and PCB. It does not matter how much a new IC may cost, it is low compared to the cost of a new PCB and subsequent alignment, or the risk of writing off someone's equipment.

Re-flow

Service engineers have become used to multicore solder that supplies flux to the joints as it heated. When it comes to surface mounted devices (SMD) they have to re-train in the techniques of reflow technology. This means learning more about the use of flux as a separate application, where and when it has to be used and the working temperatures involved.

Rosin cored solder is generally too big for an IC with lead spacing less than 0.25 mm pitch, except for the very small 0.3-mm cored solder mentioned previously. The main problem when using solder is that of bridging the legs and subsequent cleaning up. Using flux makes the job a whole lot easier, but care must be taken and a learning curve followed for best results. Do not expect to achieve perfect results the first time or every time.

If all soldered surfaces are coated with liquid flux then solder will run onto those surfaces as it is attracted to them due to the wetting effect. In this way a small reservoir of solder can be held on the solder iron tip and applied to the joint in small quantities.

Method for passive component replacement

First, remove the component carefully, avoiding print damage. Second, clean up the excess solder from the solder pads, using an absorbent solder wick.

Always store the wick inside its reel away from fresh air at all times until required, this prevents it from tarnishing or drying up. Use a hot iron at about 390°C to heat the solder wick and avoid overheating the PCB print.

Then use a flux remover cleaning solvent, do not spray it everywhere, squirt it onto a cotton bud so that it is well soaked and then start to clean up the print, after that task there should be a crisp clean section of the PCB.

Fitting passive components

1.) Using a very thin cored solder apply some of it to one pad to form a small mound, and then brush with liquid flux. Hold the component with tweezers and sit it on the PCB, at the same time re-flow the solder pad – this secures the component while the other end is soldered. The results should be as illustrated in Figure 15.4. Excess solder can be removed with the de-solder wick.

2.) Coat the solder pads with liquid flux, spray or from a bottle with a brush in it; do not liberally spray it everywhere. As it dries it becomes sticky and this property can be used to hold the component. When the component is sitting upon the PCB apply a little more flux to its ends. Then using a very small, clean soldering iron bit apply a small amount of solder to the iron tip, but not too much. Practice may be required to judge the amount. Apply the iron tip with its small reservoir of solder to one end of the component, it my require stabilising with tweezers. Solder will flow off the iron tip, and due to the 'wetting' effect around and beneath the component, the resulting joint will look just like new.

3.) Using solder paste. Their are two types of solder paste or cream, some versions that contain rosin flux and others that do not, or not enough. For paste without rosin flux, apply flux in a thin coat by brush, or apply sticky flux as a small blob.

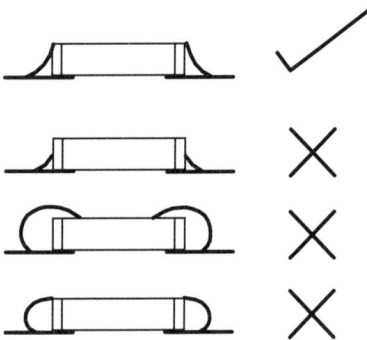

Figure 15.4 *The correct way to solder surface mounted components*

Add a small ball of solder paste to each print pad, place the component into the paste, hold the component and then heat gently with a hot air pencil.

Large scale ICs

There are three possible methods of fitting integrated circuits. The one to use is the one that suits the operator best, and it may also be determined by the level of soldering equipment that is available.

1.) As with the passive components using a small soldering iron bit such as 0.4 mm and fine solder with a diameter of 0.3 mm. This obviously depends upon the IC and its leg spacing. Double-sided tape or blue tack could be used to allow the IC to be temporarily stuck down and positioned before anchoring the corner legs with the soldering iron. However, it is not good practice to do so. According to the steadiness and eyesight of the operator each leg may then be soldered. Any excess solder on joined legs can be removed later with the de-soldering braid. A hot air pencil is useful to finally flow the solder to give a decent finish; otherwise the job may look a bit rough.

2.) Use liquid flux to paint the PCB which will allow the IC to be temporarily stuck down while the corner legs are soldered. Then apply the flux to all of the legs to coat them, not soak them. As with the passive components a reservoir of solder upon the soldering iron bit can be re-flowed by wiping it across the legs and print. It is possible to experiment with larger bits to hold more solder, and surprisingly enough, the solder will re-flow without running across the legs and joining them. There are special tips for this job that have a small curved bowl in the tip that acts as a solder reservoir.

3.) The third method is to use solder paste. It is still a good idea to apply flux to both the PCB and the IC to help with the tinning process, even though the solder paste may contain flux. If too much solder paste is applied then there may be

Soldering and Desoldering

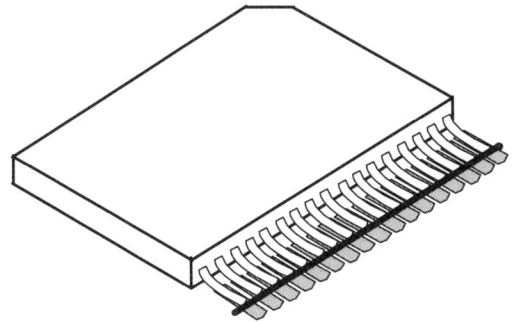

Figure 15.5 *Applying solder paste to the legs of an IC*

lots of solder balls rolling around the PCB and a surfeit of flux. Use flux cleaner to clean up the PCB.

It is another learning process with practice to decide exactly how much paste to use. Apply the solder paste in a thin line across the edge where the leg meets the print so that it bridges leg and PCB, as shown in Figure 15.5. Then gently heat it with a hot air pencil or blower at about 300–350°C. The paste will first change colour and start to 'set' just before it melts and flows. Care has to be taken to ensure that the paste, as it forms tiny balls of solder, does not run beneath the IC. I have had problems where this has happened and it has shorted out bits of print around through plated holes. Then the IC has to be removed and another fitted to correct the error.

Solder paste does not lend itself to intermittent use over long periods. It does not store well, although extended life is possible through refrigeration. Fresh solder paste does not tend to form small balls as its flux content is still active, it will flow onto the IC pins. A good idea is to dispense small amounts of the solder paste into syringes and use it as required leaving the main supply stock in cold storage.

Using solder paste and a hot air pencil can be very messy with waves of flux and solder balls running everywhere.

Whatever method is used it is down to personal preference, skill and available tooling. The following sections include deatils of some soldering and de-soldering stations and tools that are available, together with their advantages and disadvantages.

Pace system

The basis of the system is hand pieces for the different types of work, temperature control power supply, and a pump. Temperature control is done by a SensaTemp™ temperature management system. SensaTemp is a closed-loop system that senses the thermal load, or heat drain, by monitoring the tip temperature, in order to develop thermal energy to compensate for the load requirement, increased energy is then supplied to the heating element from the power unit.

Light loads, such as surface mounted components, require only a light energy requirement and so power delivery is reduced to prevent over heating. Heavy loads such as large components and large areas of print, where heat is conducted away from the component can pose a problem. Rapid power delivery is required to increase the tip temperature to ensure good solder reflow. The precise control of the SensaTemp system allows the work to be completed at much lower temperatures than with a basic soldering iron due to precision temperature control and high power delivery. There are obvious advantages here for precision work, and the fine print on a PCB will not suffer damage through overheating and is less likely to part from the board.

Soldering iron

A variety of tasks can be accomplished with a lightweight soldering iron, shown in Figure 15.6, ranging from light soldering to 'through hole' work. Tip sizes are from 0.4 mm conical shape through to 4.8 mm chisel shape, with up to 50 other types of options. One of these options is that of wide blade tips to heat SM IC's along one edge for insertion soldering. Double-sided U shaped tips are available for 'dual in line' (DIL) types of IC's. Tips are available for passive SM components with sections cut out of them so that the tip can fit either side of the component.

The soldering iron provides for component insertion and removal at lower controlled temperatures and a wide variety of quick change tips for through hole and SMD components.

Video and Camcorder Servicing and Technology

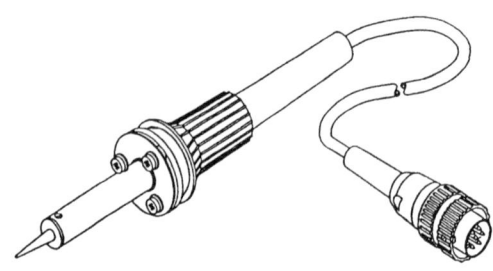

Figure 15.6 *Lightweight soldering iron*

Solder extractor SX-70

This hand piece, shown in Figure 15.7, operates off the vacuum provided by the pump that is within the power unit. Control over the suction is via a small push-button mounted upon the handle so that the solder can be re-flowed by the heated tip before suction is applied. A large glass tube is fitted into the handle of the hand piece. It has a large capacity and can easily be removed for cleaning by unhooking the rear grip. A pipe is fitted to the output end of the glass tube; a 'snaplock' connector attaches this pipe to one on the power unit. The tips for the desoldering hand piece are heated by the SensaTemp thermal system and can be set to any temperature within the range allowed. A small bore hole runs through the tip. Melted solder is maintained in a liquid state through the tip and into the glass cylinder where it rapidly cools on a heat exchanger. Airflow is switched on/off by a push-button switch, with further control upon the power unit for flow rate.

The solder extractor provides for continuous surface, through-hole desoldering and printed circuit land cleaning. Temperature control is much better for the print than a very hot solder mop technique.

Thermojet hot air pencil

Both airflow and temperature can be accurately controlled on the Thermojet hand piece shown in Figure 15.8. A small curved tip is supplied as standard, but a dual version (dual-jet) tip is available for re-flowing DIL ICs. Safe installation of SMDs is achieved by a slim line focussed airflow, actuated by a finger switch mounted upon the handle similar to the solder extractor. It allows the user to achieve accurate and rapid solder paste re-flow without affecting adjacent components.

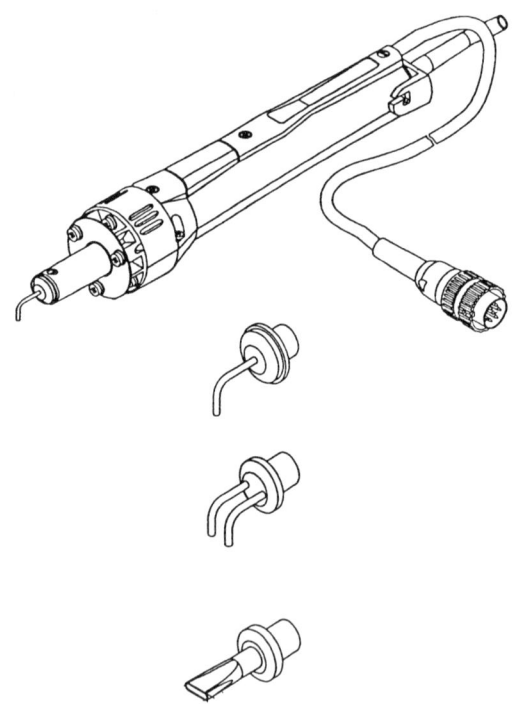

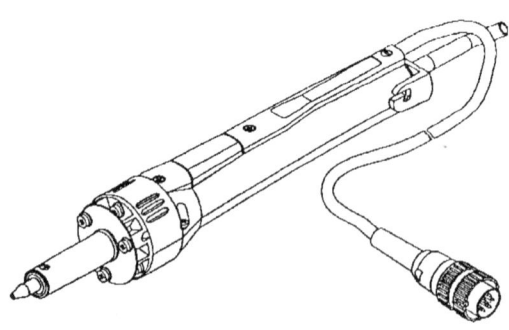

Figure 15.7 *Solder extractor*

Figure 15.8 *Hot air pencil and a selection of tips*

Thermotweeze dual tip iron

The Thermotweeze is shown in Figure 15.9. It is a dual element soldering iron in the construction of a pair of tweezers, and each element is temperature controlled. Tips come in pairs for removing ICs and SM components. Leadless ICs that have 'J' legs curled beneath them are particularly easy for the Thermotweeze. Thermotweeze tips for ICs are triangular shaped and the IC is approached from each corner, clamped and heated and then simply lifted off the PCB.

Small tweezer shaped tips can be used to heat and lift most flat SM resistors, transistors and capacitors. Flat parallel blades in the tweezer tips can be used to heat and lift DIL ICs also. Single-handed component removal in tight situations is easily achieved without the danger of harming adjacent components.

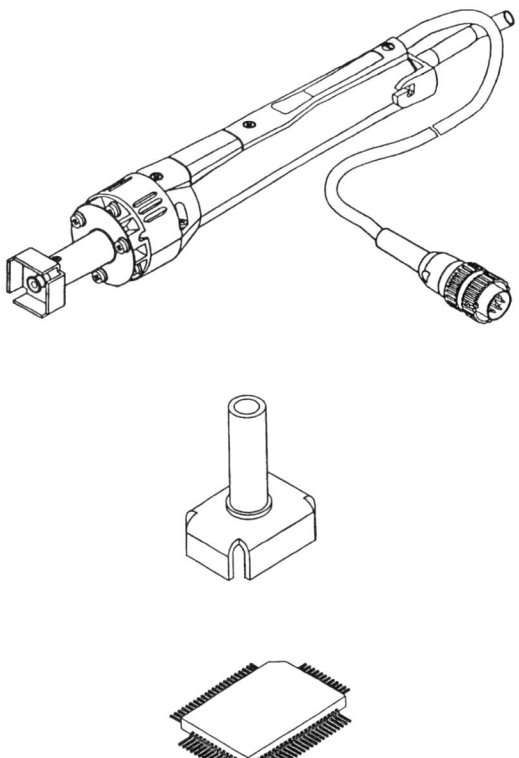

Figure 15.10 *Heated rectangular tips with central suction IC extractor*

Thermopik hand piece

Square or rectangular shaped tips similar to those on the soldering iron are used on the Thermopik hand piece, the difference being the ability to lift the component. It is one thing being able to re-flow the legs around a flat pack IC, but it is another problem to get underneath the IC simultaneously to be able to lift it. The Thermopik hand piece solves the problem by utilising a self-adjusting centre mounted suction pad, see Figure 15.10. Suction is provided by the power unit pump. Once the IC has been re-flowed the suction pad is operated by the handle-mounted push-button and the IC gently lifted off.

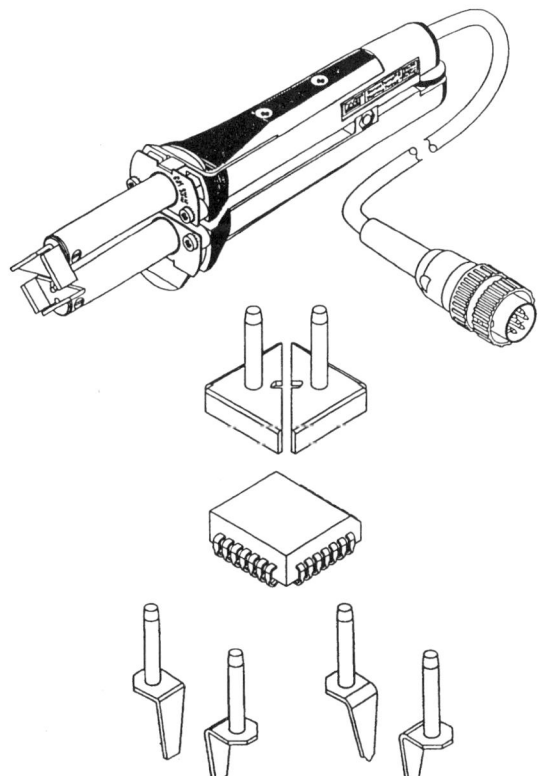

Figure 15.9 *Twin element hot tweezer and tips*

Hot air

Up to this point the removal of SM components has been by using the least expensive method of cutters and the more expensive method of specialised tips and tools.

There are two problems when using shaped tips for the removal of QFP (quad flat pack) IC's. The problems are that of contact and accessibility. In order to remove the IC without damaging the print, the user has to ensure that the solder on ALL pins is re-flowed before there is any physical attempt at removal. The larger the IC and the higher the number of legs, then the greater the danger of damage. With an increase in the number of legs, such as a 128 QFP, then the smaller and more delicate the PCB pads are. Then there is the question of access, once the tool is set around the IC it is not possible to see if all legs are re-flowed; 99 may have but one may not have. If the tool is then twisted to break the surface tension before re-flow is achieved the print will suffer. By using the PACE thermopik or other tips that have small suction pads the danger of damage is reduced, but not eliminated. ICs in brown goods, as opposed to computers, may be adhered to the PCB and this makes the judgement between it not being fully re-flowed, or just stuck, more difficult to make.

The common method previously outlined

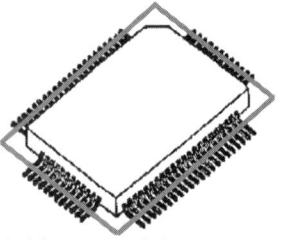

Figure 15.11 *Solder around the perimeter of an IC to aid reflow*

requires the user to apply flux to the ICs' legs and a generous layer of solder tinning on the tip. Another method is to place a length of solder around the perimeter of the IC upon its legs to aid re-flow when the shaped tip is applied, as shown in Figure 15.11.

Leister Hot Jet

One has to be very cautious of hot air being blown onto high density PCBs as it is always possible that adjacent micro components may go astray. A Hot Jet hot air blower can be used as the engine of the workstation for removal and replacement of surface mount components. A first impression is of size and weight, but first impressions are not always the best - it is deceptively light in use. The hand piece contains the heater, blower motor, controls and mains lead, and it is self-contained. Controls at the rear include a small on/off switch,

Figure 15.12 *Leister 'Hot Jet' hot air blower*

a temperature controller, an airflow regulator, mains cable and air intake; this is the cool end. The hot end is about 20 mm in diameter and 60 mm long with an output grill. Onto this fits the largest range of nozzles ever devised, far too many to list. I stopped counting at 90, but the manufacturers boast some 400 in all; see Figure 15.12.

Each nozzle, or tip, is shaped to the IC, for a QFP package the nozzle has four side vents that match the IC size. In practice it was found that that the flexibility is enough to remove ICs that were a couple of millimetres larger or smaller than the given nozzle size. Nozzles are made of a high-grade stainless steel and very durable. Unless you run over a nozzle with a steamroller or subject it to similar abuse, the nozzle with probably last for life. On the down side, the high quality is synonymous with cost as additional nozzles are expensive.

In use there is no direct contact with the IC, and the method is simple. Hold the tool above the IC with one hand and a pair of tweezers or a vacuum pick-up pencil in the other hand. Watch the solder re-flow and then lift the IC off the PCB.

The temperature calibration is not designed to be accurate; the user soon learns that a reading of 3/4 on the temperature knob is about right for most average video/audio/camcorder QFPs. Airflow control is excellent as it is a smooth flow rate down to the lowest setting.

Hot air, adjacent component masking

To help with the problem of overheating adjacent components, hot air blowers can use a variety of shaped tips that are available including rectangular ones with four jets. For J-leg ICs there are nozzles with tilted vents to flow the hot air beneath the IC. Gentle lifting pressure is applied to ensure that all legs are re-flowed before lifting; a lightweight vacuum pen is useful for extraction.

The disadvantage with hot air occurs on high-density circuit boards where the peripheral components the size of a pin head (0.3 x 06 mm) are mounted around the IC and adjacent to its legs. These components, unless protected will go walkabouts and if you can't see them you can't find them. One technique is to mask off the area around the IC to protect peripheral components.

The first product to make headway into the service industry protecting peripheral components is a desoldering station model JT6040 made by JBC, a Spanish company. Generally this consists of a hot air blower with variable temperature and flow, and a vacuum sucker to lift the IC and end pieces called extractors and protectors. Coming in a number of large sizes, the extractors consist of a rectangular-shaped bowl with a spring-loaded vacuum cup mounted over the centre. The bowl is placed around the IC to be removed and the vacuum cup is lowered onto the IC where it sticks when the vacuum pump is activated. The extractor is lightly spring loaded so that the IC is gently lifted when its legs have been re-flowed with hot air, as shown in Figure 15.13.

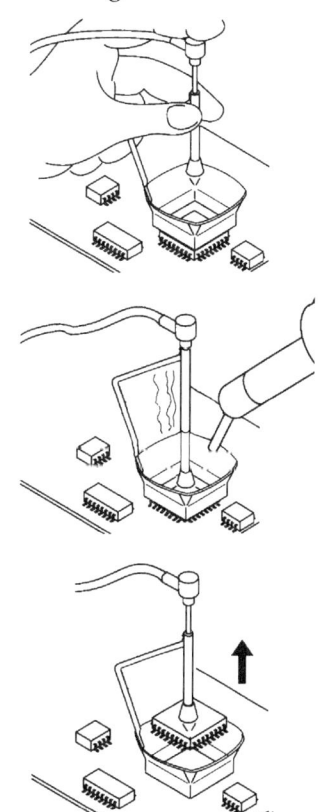

Figure 15.13 *Protecting adjacent components and applying lightweight lifting pressure in a single assembly*

Video and Camcorder Servicing and Technology

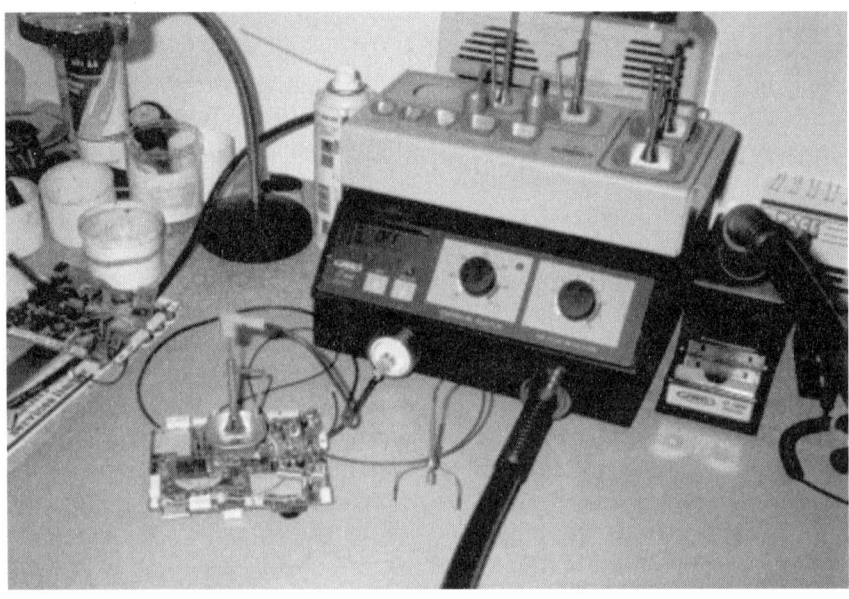

Figure 15.14 *JBC desoldering station and an asortment of surface mounted IC removal assemblies*

What about the problem of the peripheral components? Well there does not seem to be one. Those components outside the bowl are protected from the hot air while those inside are subject to eddy currents rather than a sideways blast and do not easily blow away, that is provided that you do not go mad with the air flow. It is such a magical sight to see the IC quietly and gently lift off the PCB that it makes you want to take more off just to watch it again!

Another accessory is called the 'tripod', as shown in Figure 15.13. This has a vacuum extractor in the centre of the three legs. It can be placed over a small IC on its own if there are no local peripheral components, or with one of the smaller single protector bowls around the IC to protect any adjacent components.

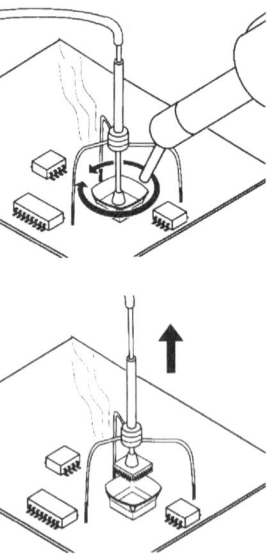

Figure 15.15 *Removing a surface mounted IC with a separate tripod and protection bowl*

Soldering and Desoldering

(a)

(b)

Figure 15.16 *(a) An IC surrounded by Chemask ready for hot air removal. (b) The same PCB without the IC and the Chemask peeled off showing that adjacent components are unaffected by the heating process*

Chemask

'Chemask by Chemtronics is used mainly in production flow soldering to protect certain parts from being soldered, such as edge connectors and screw holes etc., However, it can perform a useful function in IC extraction.

Chemask is a bright pink fluid that sets over a short period of time, around 30 minutes, less if warmed. Chemask sets to a rubbery texture that can be peeled off the PCB without leaving any residue on any of the components. The Chemask is spread around the IC and over a large number of surrounding components, some of which are adjacent to the IC legs. Once set Chemask protects the adjacent components from the hot air flow while the IC is removed, see Figure 15.16.

The quoted maximum temperature for Chemask is 268°C, but work is normally carried out at higher temperatures in order to heat up the IC quickly before conduction and damage to the PCB print occurs. Chemask may change colour and shrivel up or burn and the situation may look very gloomy. Once cooled the Chemask can be peeled off to reveal unscathed components, and no re-flow has occured. The Chemask may suffer badly, but all of the adjacent components will be as good as new, even the ones right next to the IC legs.

A word of warning - do not allow the Chemask to flow between and beneath the legs of any adjacent ICs because it will leave the compound underneath the legs when it is peeled off. It is a good idea to mask any local ICs with tape before applying the Chemask because as a liquid it does spread out after application before it sets.

Tip heaters

With the advent of digital technology and very high density PCBs there can be a problem of accessibility to remove and replace very small components. The requirement here is for a very small tip in the region of 0.2–0.4 mm, which is conical or chisel shaped. If a large tapered tip is used that is heated from one end then there is a problem of heat loss through the length of the tip, resulting in inadequate control at the sharp end. A tip that relies on thermal conduction through its mass requires a reasonably large mass to achieve this efficiently. A long taper, ending in a very small tip, causes major problems with heat conduction down to the tip.

If the soldering task has a component with a connection on a large print land or it is a very small connector with metal tags, it is very difficult to maintain re-flow temperature at the tip. At best there is a 'cold' joint of semi-crystallised solder and at worst the iron itself may adhere to the PCB as the tip rapidly cools.

Tip life is very much reduced through burning if the iron is left powered during idle periods. as the heat has plenty of time to travel down to the tip. Tip costs are high. Obviously the iron can be switched off when it is not in use. However, switching off can be very frustrating for service staff wanting to complete a quick resolder when the reheat time is nearly 2 minutes.

A solution is soldering equipment that utilises the tip as the heater element.

Figure 15.18 illustrates the difference between a conventional rear heated tip and a directly heated tip. Part (a) shows the heat up time as 2 s as opposed to 90 s for a standard iron. Part (b) illustrates the temperature variant for soldering of a consecutive number of joints. In practice when resoldering very small digital camera connectors with a 0.4-mm tip, the temperature variant can be much less than this, approximately 20°C.

Heat can be delivered exactly where it is required, by manufacturing the heater and the temperature sensor into the soldering iron tip. Each tip cartridge is a heating element and the soldering pencil simply the holder. JBC have

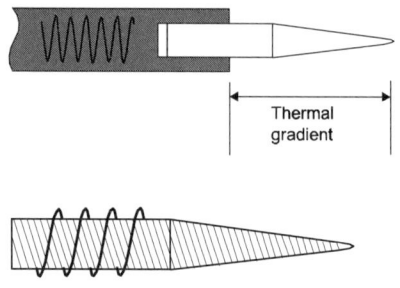

Figure 15.17 *The difference between a heated tip holder with a thermal gradient (top) and a heated tip.*

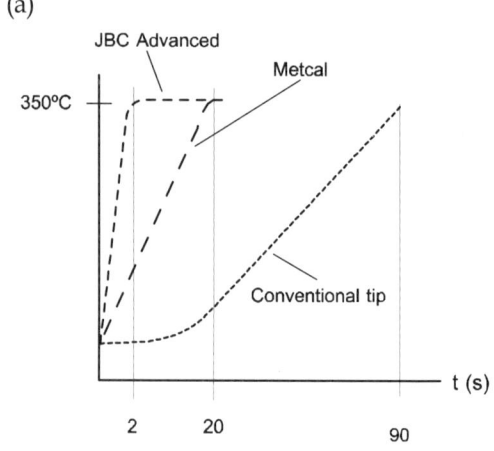

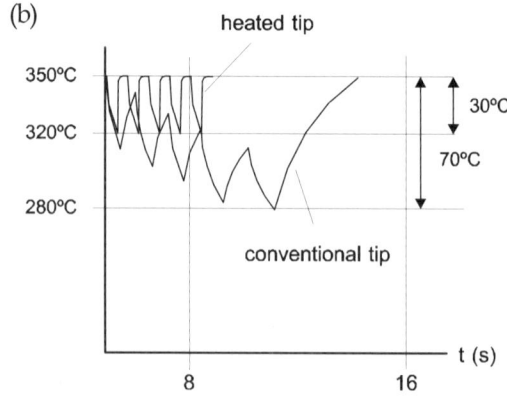

Figure 15.18 *Typical graphs for heated and conventional tips showing approximate heat and re-heat timings.*

Soldering and Desoldering

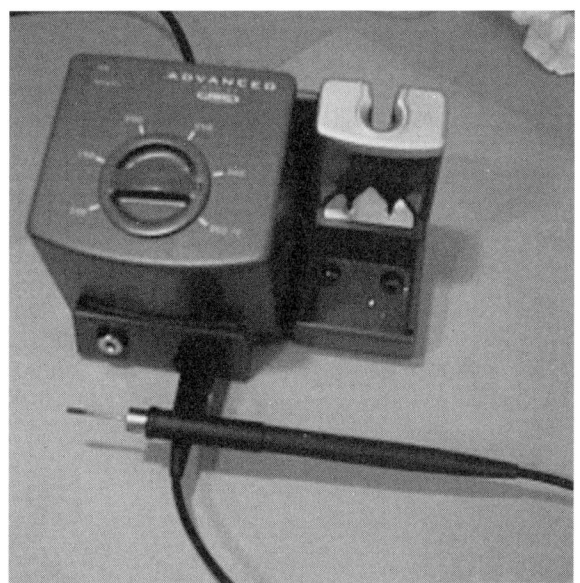

Figure 15.19 *JBC 'Advance' soldering iron*

Figure 15.20 *JBC tip resoldering a connector, the graduations on the ruler are 0.5 mm*

improved the tip heating time from around 90 s to 2 s with the consequential improvement in tip temperature control while in use. By delivering the heat directly to the tip the working temperature can be reduced. Whereas a rear heated tip power unit would be set to 350/375 °C the iron can be set to just over 300°C for the same task using a 0.4- mm tip.

An additional advantage for tip life is the intelligent soldering iron stand. When the solder pencil is docked in the stand its presence is sensed and the tip temperature reduced to around 120°C. This prevents the tarnishing and burning of microfine tips and extends tip life as a result. Also part of the docking stand is utilised to change the tip cartridges. The tip is placed on a 'V'-shaped groove and the soldering pencil is retracted, leaving the tip. At this time the intelligent stand senses the presence of the soldering pencil and removes the power. The pencil can then be moved to an adjacent tip and pushed onto the tip cartridge shaft thus inserting the next tip. When the pencil is moved away from this position power is restored and the new tip quickly heats up. The soldering pencil can be docked in this position as a power-off function.

Metcal RF heating system

Each cartridge tip is of a bimetallic construction. There is an inner core of a material with a high thermal conductivity and low electrical resistance, and an outer layer of a ferromagnetic alloy with a high electrical resistance.

Skin effect

An electrical conductor is a basic cylinder, and when a voltage is applied a current flows that is proportional to the electrical resistance. Current flow is through the whole cross-section of the cylinder, evenly distributed through the volume. If the voltage applied is of a very high frequency then the current flows through the outer layer, the skin of the conducting cylinder. This is called the skin effect; see Figure 15.21.

SmartHeat™

Metcal have designed their system as a constant current power supply at a frequency of 13.5 MHz with the connecting cable of the soldering pencil being a co-axial wave-guide. Tip cartridges are manufactured with a low resistance conductive

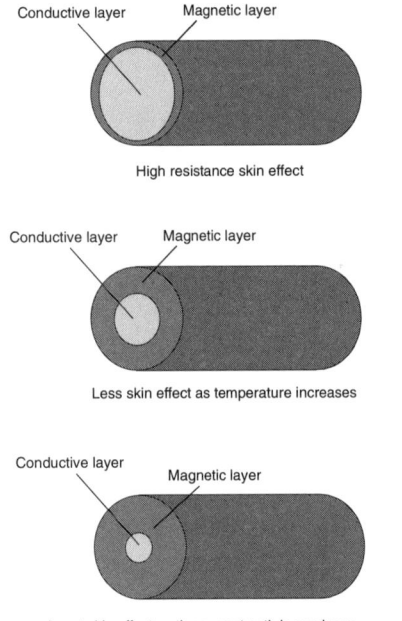

Figure 15.21 *The 'skin effect' heating principle used by Metcal*

core and a high resistance outer layer of a magnetic material. For the heating process the current flows through the outer layer of magnetic resistive material and the heat for the tip flows through the central core. As the outer layer heats up, the material reaches its Curie point; this is the physical temperature at which the material ceases to be magnetic. When the Curie point is reached and the magnetic properties cease, then so does resistance to the RF constant current. As the resistance of the skin falls close to that of the core current flows through the core, and the heating effect ceases. When the tip cools and falls below the Curie point then heating is restored. Tip temperature is therefore set by Curie point and the materials that the tip is made from. As the Curie point of a material is well defined then the temperature is controlled closely, within 1°C. Metcal have called this SmartHeat™ technology. Advantages include:

- Constant tip temperature
- No need for calibration or temperature setting
- High output at lower temperatures
- Easy to use
- Dual power supply.

It takes about 15–20 seconds to heat up the tips from cold, depending upon the tip size, after which time the temperature remains stable. The system is suitable for continuous and occasional use. Although the heating time is longer than JBC, it is still much quicker than a conventional soldering iron and is not a problem during use.

Temperature is set by the tip cartridges: 500 series 270°C (20W); 600 series 330°C (20W); 700 series 395°C (30W).

There are surface mount removal tips with different sized slots for different component sizes: twin sided tunnel-shaped tips for dual in line ICs and four-sided quad tips for PLCC, SQFP and PQFP surface mount ICs, and around 70–80 different tip sizes in each temperature range.

Standard soldering pencil tips range from 0.2 to 1.6 mm conical and from 0.4 to 5.2 mm chisel, along with other variations of bevel and bent-shaped tips in each temperature range.

In addition there is a blade tip. This can be used on DIL ICs or more readily for heating desoldering braid when cleaning up a PCB.

Care has to be taken with the microfine tips at 330°C to prevent burning and because of the quick heating the unit can be switched off between tasks. The MX 500 can be set to switch off after an idle period of around 30 minutes to save power and the tips.

Metcal Talon tweezers

Another benefit with the MX500 is the additional Talon hand piece. This is effectively heated tweezers with each tip being controlled by SmartHeat™ and the Curie point. This regulates the tip temperature to within 1°C in idle mode ready for use. The designated temperature is reached within a few seconds so the power does not have to be left on. One pair of tips is 0.4 mm and another 15.8 mm, which has small chamfers on each corner for small components.

Soldering and Desoldering

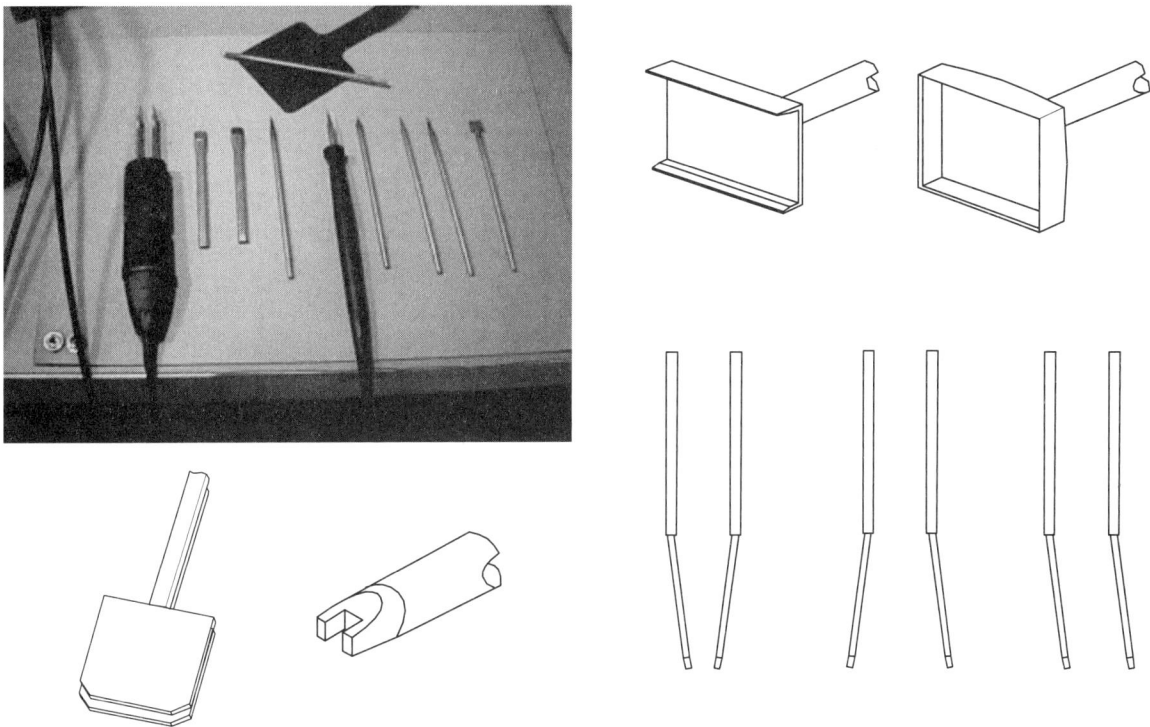

Figure 15.22 *Metcal soldering pencil and Talon tweezer with sample tips*

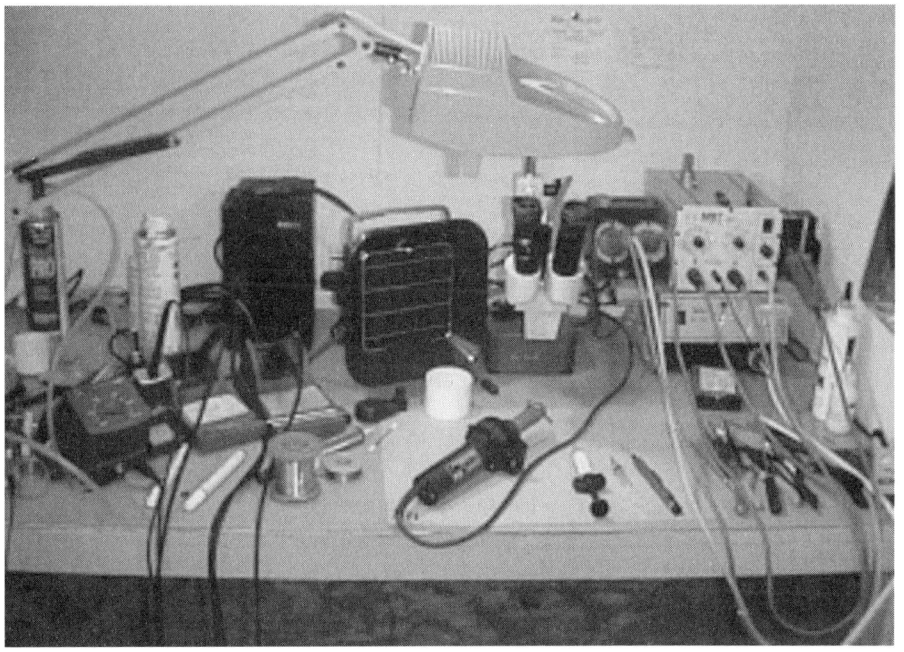

Figure 15.23 *Comprehensive soldering and desoldering station (courtesy of Newark Video Services)*

311

Chip sized integrated circuit packages

A newer and smaller IC chip package is gradually being introduced into products that are digital technology. They are hand in hand with six-layer laminated printed circuit boards.

Where an original QFP IC with 168 legs has overall dimensions of 20 x 20 mm including the legs then the area of PCB that it covers is 400 mm^2. This approximates to a pin area of 3.13 mm^2. A chip size package (CSP) IC sometimes referred to as a ball gate array (BGA) has dimensions of 11 x 11 mm giving an area of 121 mm^2 and a pin area of 0.72 mm^2.

The space that this IC takes up is about 25% of the former QFP/SSOP package and is easier to solder to the PCB during production. On the down side the connections are beneath the IC, making fault finding and measurement difficult or impossible. If one of these ICs is replaced then the quality of the soldering cannot be verified by inspection.

Each IC comes with small 0.5-mm balls of Eutectic solder already applied, as shown in Figure 15.25. Cleaning of the PCB and an application of flux is all that is required for service work. The IC is aligned to the PCB, which has the frame of the IC printed on it. The IC is placed accurately within the printed frame and then heated to melt the solder. Capillary action of the solder running onto the PCB pads will align the IC by pulling it into place.

To remove a faulty CSP IC and replace it with a new one takes time and needs care even though the operation is much more simple than for a QFP, which requires all of its legs soldering.

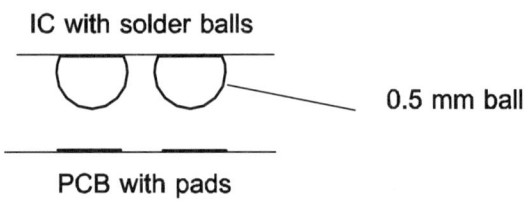

Figure 15.25 *CSP lower surface with Eutectic solder balls*

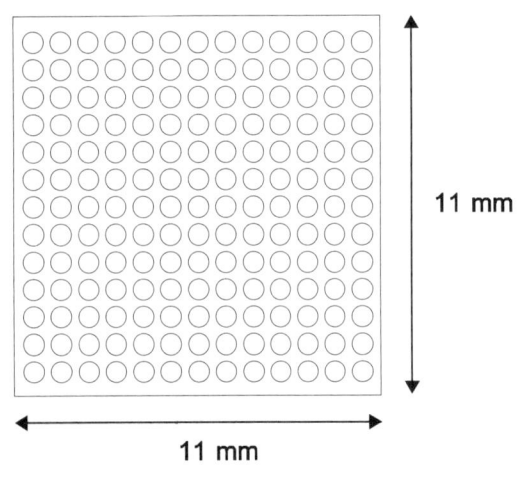

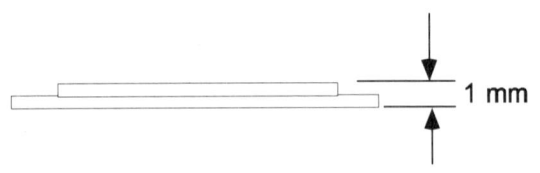

Figure 15.24 *CSP IC connection layout and profile*

Technique for replacing a CSP IC

The CSP ICs can stand much more heat than their QFP counterparts and in order to prevent thermal damage to its laminated layers the PCB must be heated also. Obviously there are many other components upon the PCB that must stay put and not fall off or blow away.

Soldering and Desoldering

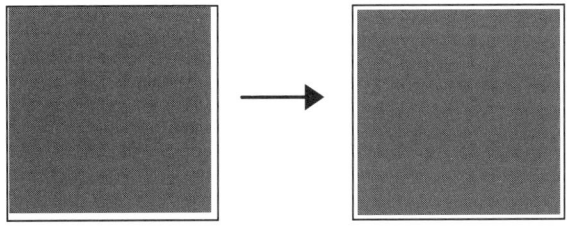

Figure 15.26 *CSP IC slides into alignment when heated*

Two hot air sources are required, one to heat the PCB and the other to heat the CSP. There are two methods to heat the PCB; one is to heat only the area beneath the IC and the other is to heat the whole PCB. The first task is to heat the PCB up to a temperature where the solder does not re-flow; about 50 to 80°C, which takes about 90 s. Then the hot air from a second source, which is above the IC that is to be replaced, is turned on to heat the IC. This takes about 45–90 s according to the size of the component.

Method

- Mask off the adjacent components to the IC with Chemask and allow it to set.
- Warm up the PCB with hot air from below, the time period for this is about 90 seconds.
- Then heat up the BGA IC for the same period, 90 seconds and remove with a suction pad.
- Remove all heat and allow the PCB to cool.
- Clean up the PCB pads of the IC with solder braid. If necessary, remove the Chemask and replace with a fresh application over the adjacent components.
- Apply flux to the PCB and place the IC.
- Reheat the PCB again for 90 seconds.
- Then heat the IC for the same period to ensure reflow of the Eutectic solder. Finally see if the IC functions. If not - you have either not fully soldered it or you may have changed the wrong one!

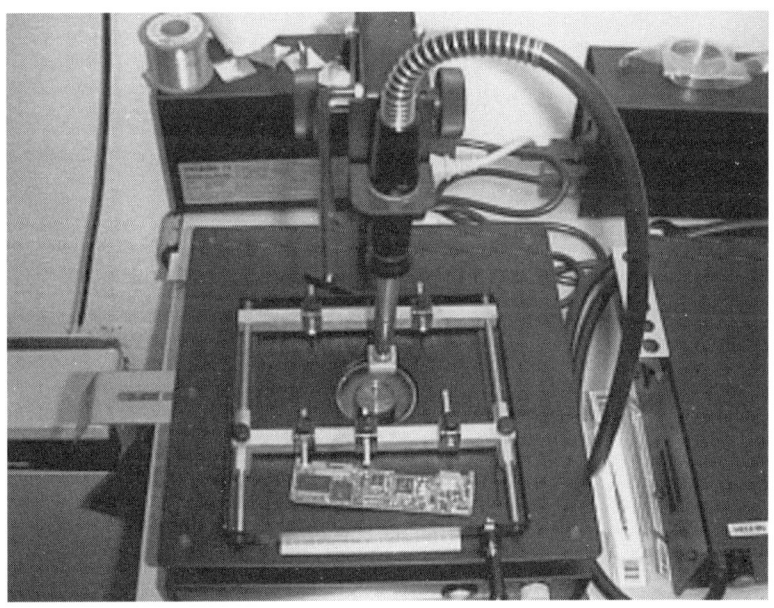

Figure 15.27 *CSP/BGA IC removal jig showing lower board heater and upper IC heater*

Index

Symbols

2D/3D filter 54
3-dimensional, 4-line delay comb filter 51
3D frame memory filter 54
8 mm format 185
8mm colour recording system 124
8mm colour system 124
8mm format
   Automatic tracking 187
   Recording frequency spectrum 186
8mm phase control, colour 125
8mm video recording spectrum 31

A

AC bias 6
Adaptive quantisation 249
Advanced crosstalk Y comb filter 43
Analogue drum servo 72
Analogue/digital camera processing 208
Aperture correction 39
Audio
   Compander 160
   Drop out compensation 164
   DVC audio 253
   Head drum construction 155
   RMS detector 162
Audio head switching 162
Auto iris control 228
Auto set-up (Plug and Play) 291
Auto white balance 210, 229
Auto white operation 229
Autofocus detection 226
Azimuth tilt technique 18

B

B/H loop 3
Betamax chroma record system 120
Betamax chroma replay system 121
Betamax delay line 123
Bode diagram and phase correction 64

C

Camcorders 178
Camera tube 189
   Bias light 191
   Colour separation 192
   Colour signal 192
   Colour spectrum 194
   Crystal filter 191
   Gamma correction and tracking 195
Camera tube construction 190
Capstan motor 105
Capstan motor cicuit 103
Capstan motor servo 82
Capstan servo 85
Cassette adapter 179
CCD (charge coupled device) 44
CCD and memory image stabiliser 211
CCD delay line 44
CCD imager 197
   Chrominance derivation 203
   Delay line sample and hold 205
   High grade image stabiliser 211
   Horizontal resolution 206
   Image stabiliser 210
   Line transfer CCD 197
   Partition area CCD. 211
   Transfer gate 198
   Triple CCD imager block 206
   Vertical shift register 200
CCD imager colour matrixing 201
CCD imager structure 198
Characteristics of feedback servo control 62
Chemask 307
Chip sized integrated circuit packages 312
Chip sized integrated circuit packages
   Technique for replacing a CSP 312
Chroma correlation 108
Colour
   8mm APC loop 128
   8mm comb filter 129
   90°/line replay correction 119
   Balanced modulator 118
   Betamax system 120
   VHS phase correction 117
   VHS record system 113
   VHS replay phase correction process 115
   VHS replay system 114
Colour crosstalk 19
Colour recording and playback 106
Colour separation 192
Colour under system 111
Comb filter crosstalk cancellation 40
Comb Filter, TV 109
Comparison of FM carrier frequency spectra 29
Correlated double sampling 202
Critical damping 65
Cross-talk of FM carrier 40
CTL duty cycle 79
Cyclic drop out compensator 36

D

D-VHS
   D-VHS sub code 276
   Main code area 276
   Recording time stamp 277
D-VHS recorder 273
D-VHS tape format 274
Damping, servo 64
DC motor 60

DCT frequency distribution 247
DCT processing 249
Demagnetising 4
Demodulation 38
Depth multiplex recording 158
Detail enhancer 50
Digital 8mm 262
    Audio 264
Digital binary counter 86
Digital camcorder assembly editing 287
Digital camcorder block diagram 233
Digital camcorders 230
    ATF pilot signals 238
    Audio error correction 254
    Audio tape track layout 255
    Camera processing 241
    Cassette construction 230
    Compression 245
    DV data recording and playback 256
    DVC audio 253
    DVC LP system 237
    ITI (Integrated Tracking Information) 235
    Start sync block area 235
    Tracking error bit 240
    Tracking signal 239
    Video heads 236
    Video tape data map 245
Digital chrominance circuits 132
Digital Communications IEEE1394 292
Digital control track 91
Digital drum servo operation 87
Digital effects
    Digital strobe 214
    Mixes and wipes 214
    Negative function 214
    Solarisation 214
Digital phase correction 89
Digital processing 51
Digital ramp 87
Digital record phase. 92
Digital servo 86
Digital servo faultfinding 99
Digital speed control 89
Digital still camera 268
Digital still imager 267
Digital still picture camera 265
Digital YNR 55
Digital zoom 212
    Interpolation and inter-compensation. 212
Digitisation of a video signal 231
Discrete cosine transformation 246
Discrete record processing circuit 25
Double limiter 36
Drop out 31
Drop out compensator 35
Dropouts 35
Drum motor 104
Drum motor circuit 102
Drum motor servo 81

Drum servo system 84
Dual loop servo 80
DVC tape format 235
Dynamic aperture correction 45
Dynamic drum system 176
Dynamic range 12

E

Eddy current loss 7
Editing 278
    Head timing 285
Editing with IEEE1394 287
Extinction frequency 11

F

Fault finding in the colour circuits 130
Fault finding on servos 95
Feedback speed control 61
FG/PG control 80
FM carrier crosstalk 19
FM luminance carrier 19
FM modulation
    FM magnetic track 18
FM modulator 16
FM recording amplifier 27
FM recording and playback 20
FM replay and demodulation 31
Focus
    Automatic focusing 218
    Frequency detection 219
    Infrared focusing 218
    Lens construction. 223
    MR sensor 225
    Through camera lens 219
    Ultrasonic focussing 218
    Zoom/focus tracking curve 226
Frequency detection auto focus system 223
Frequency modulated recording 20
Frequency modulation 13
Frequency modulator 16, 24
Frequency spectrums 20
Frequency to voltage converter 65

G

Glass delay line 44

H

Hall effect device 100
Hi-Fi audio 154
Hi-Fi recording and playback 158

I

IEEE1394
    AV protocol 293
    DS coding 296
    Serial bus protocol 2 (SBP-2) 293
IEEE1394 data communications 293

IEEE1394 protocols 293
Indexing 79
Interleaved NRZI (non-return to zero inversion) 257

J

Jump circuits 151

K

Keyed AGC 28

L

Leister Hot Jet 304
Limiter 77
Line correlation 108
Long play 150
LP video track 150
Luminance line interpolation 47
Luminous intensity 209

M

Macro blocks 242
Magnetism 1
   Head gap 15
   Magnetic field 2
   Magnetic force 1
   Magnetisation loss 7
   Magnetising force 2
   Magneto motive force 1
   Penetration loss 7
   Transfer characteristic. 4
Master edit control system 282
Mechanical image stabiliser 214
Metcal RF heating system 309
   Skin effect 309
   SmartHeat 309
Metcal Talon tweezers 310
Mode cam switch 146
Modulation index. 14
Motor drive amplifier 77
Moving lens stabiliser 216
MPEG and D-VHS 271
MPEG compression 271
   B frames 272
   I frames 272
   Main profiles and levels 271
   Motion prediction 272
   P frames 272
MPEG programme stream 274
Multimedia memory card 269
Multipole motors 99

N

Noise correlation 47
Non-Linear network 77
NRZI operation for a long string. 260

P

PAL 1½ line offset 106
PAL colour system 108
Parity error correction 254
PDC
   Format 1 289
   Format 2 289
Phase control circuit 68
Pre-amplifier 32
Programme delivery control (PDC) 288
Progressive scan CCD imager 265

Q

'Q' network 32

R

Random assembly editing 283, 286
Re-flow 299
Record amplifier 22
Recording equalisation 8
Recording frequency response 8
Recording head impedance loss 7
Recording tracking pulses 79
Replay characteristic 10
Replay signal level 9

S

S-VHS 30
Sample and hold 75
Servo
   Damping 64
   Error voltage limiter 76
   Frequency to voltage converter 65
   PG circuit 74
   PG pulse shaping 72
   Phase control 67
   Record phase 73
   Sampling 73
Servo control 74
Servo IC, LSI 93
Servo operation frequencies 83
Slow motion 171
Solder extractor. 302
Soldering and desoldering 297
   Chemask 307
   Fitting passive components 300
   Hot air 304
   Hot air, adjacent component masking 305
   IC Removal 299
   Large scale ICs 300
   Method for passive component replacement 299
   Pace system 301
   Precision cutters 298
   Surface mounted components 297
   Tip heaters 308
Soldering iron 301
Still frame 172

Still picture 168
    FM noise 167
    Four head drum 173
    Video head width 169
Still pictures
    Video head path 166
Still pictures and slow motion 166
Subtractive mixer 42
Super blocks 243
Synthesised vertical sync pulse 169
System control 138
    Input expander 142
    Operation controls 143
    Output expander 145
    Pause 139
    Safety function 139
    Serial-to-parallel converter 141
Systems control 135

T

Tape bias (audio) 5
Tape protection 135
Temporal redundancy 271
The DVC system 241
Thermojet hot air pencil 302
Thermopik hand piece 303
Thermotweeze dual tip iron 303
Tracking 78
Transfer gate 198

V

Variable length coding 251

Variangle lens stabiliser 215
VHS colour system 111
VHS demodulator 38
VHS HQ technology 45
VHS system 29
VHS-C
    Drum speed 181
    Overlap 181
    Timing 183
    Video head switching 181
    Videomovie 188
VHS-C format video head drum 179
VHS-C system cassette 178
Video data compression 243
Video frequency modulation 15
Video head switching 32
Video recorder head drum servomechanism 69
Video signal bandwidth 29
VISS data 284
Visual search compensation. 84
Viterbi error correction 261

Y

Y comb filter 40, 47
Y detail enhancer 50

Z

Zero frame editing 278
Zero frame editing, example 280